FLUID METERS

FLUID METERS

THEIR THEORY
and APPLICATION

Report of ASME
Research Committee on
Fluid Meters

SIXTH
EDITION
1971

edited by
Howard S. Bean

THE AMERICAN SOCIETY OF MECHANICAL ENGINEERS

United Engineering Center   345 East 47th Street   New York, N. Y. 10017

Library of Congress Catalog Card No. 45-44685

CONTENTS

Foreword   *v*
Committee Personnel   *vi*
Preface   *vii*
Introduction   *xi*

PART ONE   THEORY AND MODE OF OPERATION   1

Chapter I-1,   Classification of Fluid Meters   3
    Quantity Meters   3
    Rate Meters   5

Chapter I-2,   Units, Reference Conditions and Letter Symbols   9
    Letter Symbols   12

Chapter I-3,   Special Terms   19

Chapter I-4,   Elements of Quantity Meters   37
    Weighing Meters   37
    Positive Displacement Meters: Volumetric Types   38

Chapter I-5,   Differential Pressure Meters: Theory of Fluid Flow in
    Terms of Differential Pressures, and Equations for
    Differential Pressure Meters   47
    Principal Primary Elements   47
    Theory of the Flow of Fluids in Terms of Pressure
    Differences   47
    Definite Relationships Between C, R_d, β, and D:
    Venturi Tubes   65
    Sonic Flow Primary Elements   65
    Modifications of the Three Basic Primary Elements:
    The Nozzle-Venturi   72
    Eccentric and Segmental Orifices   73
    Centrifugal Meters: The Elbow Meter   75
    Linear-Resistance Meters   77

Chapter I-6,   Area Meters   81
    Tapered Tube and Float   81
    Cylinder and Piston   82
    Orifice and Plug   83
    Flow Coefficients and Equations for Actual Rates of Flow   87
    Operating Adjustment Factors   88

Chapter I-7,   Fluid Velocity Measuring Instruments and Meters   91
    Anemometers   91
    Current Meters   93

Concepts Applicable to Velocity Meters with
Rotating Primary Elements, Particularly Anemometers
and Current Meters   94
Pipe Line Flowmeters   97
Instruments for Determining Point-Velocity and
Mean-Velocity   101
Pitot and Pitot-Static Tubes   101
The Hot-Wire Anemometer   105

Chapter I-8,   Head-Area Meters   113
Weirs   113
Flumes   122

Chapter I-9,   Other Meters and Methods of Determining Rates of Flow   125
Magnetic Flowmeters   125
Force or Vane Meters   128
Transverse-Momentum or Mass Flowmeters   130
Thermal Meters   131
Tracers   132
Method of Mixtures   136
Pressure-Time or "Gibson" Method   137
Sound and Light Velocity Methods   144
Floats and Moving Screens   145
Miscellaneous Meters and Methods   146

PART TWO   APPLICATION OF FLUID METERS—ESPECIALLY
            DIFFERENTIAL PRESSURE TYPE   149

Introduction   151

Chapter II-I,   Conversion Factors, Constants and Data on Fluids and
Materials   153

Chapter II-II,   General Requirements for Fluid Metering: Installation   179

Chapter II-III, Primary Elements and Equations for Computing Rates of Flow   197
Thin-Plate Square-Edged Orifice   198
Eccentric and Segmental Orifices   210
Small Precision Bore Orifice Meters   211
Flow Nozzles   216
Venturi Tubes   230
Nozzle-Venturi   232
Sonic Flow Primary Elements   234
Elbows   255
Electromagnetic Flowmeters   255

Chapter II-IV,   Examples   257

Chapter II-V,   Tolerances   265

Index   269

FOREWORD

WHEN the Research Committee on Fluid Meters was organized in 1916, one of its stated objectives was "the preparation of a textbook on the theory and use of fluid meters sufficient as a standard reference." In carrying out this objective the first edition of Part 1 of this report was published in 1924 and received immediate approval and wide usage by the users of fluid meters and by educators. As originally planned by the committee, the report was to be issued in three parts, and Part 1, "Theory and Application," was the first one published. It was to be followed by Part 2, "Description of Meters," and Part 3, "Installation." After its publication, Part 1 was so well received that the number printed sold so rapidly that the second and third editions of this part were needed before time could be found to prepare the other two parts of the report. The second edition of Part 1 was considerably different from the first; however, it followed about the same format and arrangement while the third edition was very little different from the second. These were published in 1927 and 1930, respectively.

Part 2 of the report was published in 1931 and contained a complete description of the physical characteristics of the meters then being manufactured. However, it was found that the material in this part became obsolete so rapidly that it was decided not to try to keep it up but to tell anyone interested in these descriptions that they should be secured from the manufacturers, since their literature must necessarily be up to date.

Part 3, published in 1933, gave instructions for correct installation of meters and discussed the effect of incorrect installations. However, Part 3 was abandoned also because the committee decided the material in it should be an integral part of the complete report of the committee.

The fourth edition of Part 1 was prepared in 1937 and was a completely new draft of this part of the report. It was altered because there had been considerable criticism of the fact that the material presented was difficult to put to practical use. The changed format and additional material presented apparently corrected this condition, since this edition went through many printings.

The fifth edition, issued in 1959, followed the same general format as the fourth, and included material gained in the long interval between the two editions.

Another publication by the committee is a manual "Flowmeter Computation Handbook," which was issued in 1961. The procedures in it can be adapted to computer programming.

The format of the sixth edition differs slightly from that of the fourth and fifth editions. Each chapter is complete in itself, so that altering one chapter will not affect preceding or following chapters. Also, somewhat like the third edition and Part 3, the material on installation and application will be both a part of the complete report and a separate publication.

PREFACE

FIRST Edition, 1924. After six years of effort on the part of the Research Committee on Fluid Meters of The American Society of Mechanical Engineers, it now presents its first progress report entitled "Fluid Meters, Part 1."

This report takes the form of a reference book on fluid meters of all kinds. It contains not only such practical instruction and information, including formulas, constants, and the like, as may be needed by the actual or prospective user but also more general information—the physical principles of design and operation—which may be useful to students, designing engineers, and inventors.

Part 1 treats the general types of fluid meter as well as the principles and methods involved and gives information which may, in many cases, be applicable to various commercial meters. In this part, instruments of individual makers are not discussed in detail but are referred to only incidentally or for illustrative purposes. The general physical principles are in the body of the text, while the derivation of formulas and the refinements of the theory involved has been placed in the appendixes of the report.

Fluid meters are of great and rapidly increasing importance, but, hitherto, the information available on this group of instruments has been incomplete. The material forming Chapters 1, 2, 3, 4, 5, 6, and Appendix C, were recently rewritten. This material contains a new presentation of the subject based on a mathematical analysis which is more specifically applicable to fluid flow than Bernoulli's theorem. This analysis did not include certain additional experimental data now covered in the report, but these data required no serious modification of the text.

The most important modern advance in experimental aerodynamics and hydraulics is the application to them of dimensional analysis. This is absolutely indispensable to an understanding of the behavior of moving fluids, for the phenomena of fluid motion are so complicated as to defy analysis by any other known method. The use of this method is especially valuable, in that it makes possible the reconciliation of data obtained from experiments with the venturi tube and the thin disk orifice which were formerly thought to be irreconcilable. These data are now shown to be mutually confirmatory.

The personnel of the committee, which prepared this report, was as follows: Messrs. R. J. S. Pigott, *Chairman*, J. M. Spitzglass, *Secretary*, H. Bacharach, E. G. Bailey, M. M. Borden, G. S. Coffin, C. A. Dawley, L. M. Goldsmith, F. G. Hechler, Horace Judd, Leo Loeb, P. S. Lyon, H. H. Mapelsden, H. N. Packard, C. G. Richardson, and T. R. Weymouth. Dr. Edgar Buckingham served as a member of the committee until 1922 and made valuable contributions to the early drafts of the report.

SECOND and Third Editions, 1927 and 1930. Continued demand for the report necessitated the publication of a second edition in 1927, and a third in 1930. Before publishing the second edition, the reorganized Fluid Meters Committee carefully reviewed and revised practically all of the chapters of the report, exclusive of the Pitot Tube and Flow Nozzle sections. Subsequent experimental work has provided data for the revision of these two sections which are incorporated in the third edition together with certain minor corrections throughout the report.

Work on Parts 2 and 3 of the report is progressing. The former is designed to present a brief description with illustrations of the various commercial meters on the market. Part 3 will discuss the proper methods of installation and care which meters must receive to function satisfactorily.

FOURTH Edition, 1937. A consistent demand for this report exhausted the third edition of 1600 copies in approximately one year and a half and made the printing of a fourth edition necessary. Thirteen years have elapsed since the first edition was published, and, while the three different editions have had a flattering reception, the committee feels that the fourth edition should be an improvement over the others as regards convenience in practical use. Part 2, "Description of Meters," was published in 1931 and Part 3, "Selection and Installation," was issued in 1933.

The majority of the criticisms of the several editions of Part 1 have had to do with the difficulty of putting the theoretical equations into practical form for every day use. In preparing this fourth edition, therefore, a radical change has been made in the arrangement and in certain parts of the text. This part has been subdivided into three sections—A, B, and C—of which the following is a brief description. In Section A, the classification and nomenclature of fluid meters, as used throughout the reports of this committee, are given, together with the definitions of special terms and other general information. In Section B, the theory of fluid measurements is presented. In most cases, the assumptions made and the steps taken to develop practical working equations from the theoretical relations are set forth. Section C contains figures and tables for use in solving practical fluid measurement problems, and their proper use is illustrated with examples.

The personnel of the subcommittee that prepared this edition was as follows: Messrs. H. S. Bean, *Chairman*, W. W. Frymoyer, Louis Gess, W. S. Pardoe, Ed S. Smith, Jr., R. E. Sprenkle, E. C. M. Stahl, and T. R. Weymouth. This subcommittee acknowledges with appreciation the assistance rendered in the preparation of this edition by Mr. M. A. Goetz and Mr. T. H. Smith of Mr. Stahl's staff.

FIFTH Edition, 1959. In the twenty-one years which have elapsed since this report was last revised there has been a great deal of growth in the science of fluid mechanics and a manyfold growth in the interest in fluid metering. This increase in interest and knowledge has made necessary many changes in the material covered by this report and made available many more accurate determinations of the constants and other material not appearing in the fourth edition. Realizing this and wishing to incorporate the material which had been published in Part 3 of the committee's report, it was decided that instead of trying to revise the fourth edition that the fifth edition should be a completely new publication, and so the material incorporated in it has been completely rewritten.

In the writing of the material all of the knowledge of modern fluid mechanics which was available to the committee has been used in an attempt to explain the phenomena which are used in the science of fluid metering. While no attempt has been made to go into the fundamentals of fluid mechanics, the results of the study of these fundamentals are available because of the use of more rational formulas and explanations and the use of less empiricism. In addition, the experience of many years in the solution of fluid metering problems by the members of the committee has made the material of the greatest practical use. It is believed that this edition will carry on the great tradition established by the preceding editions.

The arrangement of this edition is similar to that of the fourth. Also, the method of presenting the material is the same, with two exceptions: First, to conform with the universal teaching practice of engineering schools in this country, the gravitational system of units has been used in the development of equations and the presentation of physical data. Second, the coefficients of differential head meters are presented on the basis of the pipe Reynolds number in the belief this will be of greater convenience to the user, and will aid in promoting international standardization.

The personnel of the subcommittee which prepared this edition is as follows: Messrs. H. S. Bean, *Chairman*, L. K. Spink,† *Vice Chairman*, E. E. Ambrosius, R. C. Binder,* L. Gess, C. S. Hazard, V. P. Head, A. L. Jorissen,† E. J. Lindahl, I. O. Miner,* and R. E. Sprenkle, while many other members of the main committee cooperated by offering material and constructive criticism.

The success of the work is primarily the responsibility of Mr. Howard S. Bean, the Chairman of the subcommittee; and he has worked hard and long as a labor of love in preparing this material. His service has been of incalculable value in preparing the fourth and fifth editions of this report, and the committee wishes to take this opportunity to state its appreciation for these services.

S IXTH Edition, 1971. In preparing this edition the Committee has endeavored to include some mention, if only by illustration, of the meters which have come to the attention of the members during the ten years since the preparation of the fifth edition. At the same time most of the types of fluid meters and metering procedures included in preceding editions are still in use, hence it is necessary to continue to include them. Furthermore, there has been a notable increase in the interest and application of fluid meters in the fields of aeronautics and cryogenics, and this has influenced the presentation of parts of the text on differential pressure meters.

During the planning stage for this edition an agreement was made between this Committee and the ASME Standing Committee on Performance Test Codes to include in this edition all of the material in Instruments and Apparatus Supplement Part 5 on Measurement of Quantity of Materials, except Chapter 1 on Weighing Scales. To be sure, much of the material in the other four chapters of Part 5, displacement meters, velocity meters, flow measurement (i.e., differential pressure type meters) and other meters and methods had been included in varying degrees in previous editions. However, in assembling this edition much of these chapters has been included without change, supplemented with additional material as needed to make the treatment as complete as possible.

The arrangement between the two Committees provided that all of the material on application previously included in I and A Part 5-4, Flow Measurement, would be made a Part II of this edition, with some supplemental data. Thus, this Sixth Edition is composed of Parts I and II as a complete volume, and also Part II is available separately under the title "Flow Measurement Application: Part II of Fluid Meters Sixth Edition; (Replaces I and A Part 5-4)." In this way the plant engineer concerned with application does not need to be burdened with the complete volume.

Throughout the text, illustrations are used to show the application of various principles and methods of fluid metering. Some of the illustrations show proprietary equipment. The inclusion of such figures does not constitute nor imply an endorsement of the equipment by this Committee nor the Society. Furthermore, no operating data are given for such proprietary equipment.

The members of the subcommittee which prepared this edition are: Messrs. H. S. Bean, *Chairman*; H. V. Beck, *Vice Chairman*; E. E. Ambrosius, B. T. Arnberg,

---

*Retired from the Committee before completion of report.
†Deceased.

R. B. Dowdell, L. P. Emerson, H. J. Evans, Louis Gess,* E. J. Lindahl, J. V. Moore, J. W. Murdock, R. M. Reimer, R. E. Sprenkle, E. F. Wehmann; and K. C. Cotton, L. A. Dodge, J. R. Jordan, A. S. McDaniel and E. L. Upp (for J. V. Moore). The last five, under Mr. Cotton's leadership had the primary responsibility for Part II. The subcommittee was aided by material and comments from other members of the main Committee and other subcommittees.

The committee wishes to acknowledge the assistance of Mr. D. R. Keyser of the Naval Engineering Center in the preparation of parts of several chapters.

INTRODUCTION

Fluid meters can be divided into two functional groups. One measures primarily quantity; the other measures primarily rate of flow. All fluid meters, however, consist of two distinct parts, each of which has a different function to perform. The first is the primary element, which is in contact with the fluid, resulting in some form of interaction. This interaction may be that of imparting motion to the primary element; the fluid may be accelerated; or there may be an exchange of heat. The second or secondary element translates the interaction between fluid and primary element into volumes, weights or rates of flow and indicates or records the result.

For example, a weigher will have weighing tanks as its primary element and a counter for recording the number of fillings and dumpings as its secondary element. In an orifice meter, the orifice, together with the adjacent part of the pipe and the pressure connections, constitute the primary element, while the secondary element consists of a differential pressure gage together with some sort of mechanism for translating a pressure difference into a rate of flow and indicating the result, in some cases also recording it graphically and integrating with respect to time. The same sort of combination will be observed in other types of meters.

The secondary devices may obviously be varied almost without limit, but the primary elements depend for their operation on a few simple physical principles. Therefore, fluid meters may best be classified with regard solely to the nature of the primary element or to the physical principle involved, and this plan has been adopted here.

The conditions for proper installation and operation, the errors, and other characteristics of the primary element are usually altogether distinct from, and independent of, those of the secondary element, so that it is convenient to treat the two separately so far as is possible. Therefore, throughout the two parts of this edition of the report the discussions are directed particularly toward the primary elements, and secondary elements are included only to the extent necessary for an adequate description of the meter or the principle of the metering procedure.

The material on differential pressure meters occupies a considerable portion of the report. This is due, in part, to the nature of the information to be presented and, in part, to an endeavor to present the development of the theoretical equations required and the correlation of experimentally determined factors in as complete a manner as possible. Moreover, since the compiling of the preceding editions the use of various forms of flow nozzles under sonic flow conditions has become an important tool in aeronautical and aerospace programs. This has led to using a development of equations of flow from the critical or "choked" condition, and thence to the subcritical condition. This procedure of developing equations for flow nozzles (including Venturi tubes), as well as the classical procedure that starts with the incompressible and subcritical states, are presented in order that the reader may have the use of either.

In order that a user may be able to use the final working equations and factors with justifiable confidence, recommendations on construction, installation, and operation are given in considerable detail in Part II.

Numbers in brackets [ ] refer to like numbered references listed at the end of each chapter.

Part One

THEORY AND MODE OF OPERATION

Chapter I-1

Classification of Fluid Meters

I-1-1 A study of the literature shows that different writers on fluid meters have used different principles of classification and that the terminology adopted has sometimes been used very loosely. For example, the terms "positive," "inferential," "total," "partial flow," and "proportional" have been used ambiguously or inconsistently. An attempt has, therefore, been made here to adopt a classification which shall be as clear and logical as practicable, and particular attention has been paid to making the definitions precise and complete and to using them consistently.

The classification adopted for the purpose of this report is given in Table I-1-1 and is supplemented by the following brief descriptions.

Quantity Meters

I-1-2 The term "quantity" is here used to designate those meters through the primary element of which the fluid passes in successive and more or less completely isolated quantities, either weights or volumes, by alternately filling and emptying containers of known or fixed capacities.

I-1-3 The secondary element of a quantity meter consists of a counter with suitably graduated dials for registering the total quantity that has passed through the meter.

Note: An attachment with a clock mechanism may be added to the register of a quantity meter. Thus, the time-rate of registration is obtained, that is, the quantity meter becomes a rate meter also.

I-1-4 Weighing Meters. The term "weighers" is applied to those quantity meters in which the equilibrium of a container is upset by a rise of the center of gravity as the container is filled or to those which employ a container suspended from a counterbalanced scale beam. The former are affected slightly by the temperature of the liquid, because a change in temperature produces a change in the density of the liquid and thereby alters the height of the center of gravity for a given weight. Weighers are not used for gases and vapors.

I-1-5 Volumetric Meters. This class of quantity meters measures volumes instead of weights. These meters may be subdivided into two groups, those in one group being designed for metering liquids, while those in the other are for gases. In both groups, the measurement is effected by separating the fluid into individual portions by mechanical means and counting these segments. These meters are called Positive Displacement Meters or simply Displacement Meters.

I-1-6 Tanks. This is a very elementary form of meter of limited commercial importance. As the name implies, these meters consist of one or more tanks which are alternately filled and emptied. The height to which they are filled can be regulated manually or automatically. In some cases, the rise of the liquid operates a float which controls the inflow and outflow; in others, it may start a syphon. Occasionally, some tank meters have been erroneously classified as weighers.

I-1-7 Reciprocating Piston. These meters use one or more members having a reciprocating motion, which operate in one or more fixed chambers. Adjustment of the quantity per cycle can be effected either by varying the magnitude of movement of one or more

Table I-1-1 Classification of Fluid Meters and Methods of Fluid Measurement

Division	Class		Type
1. Quantity Meters	Weighing		Weighers
			Tilting Traps
	Volumetric		Tank
		Liquids	Reciprocating Piston
			Rotating Piston or Ring Piston
			Nutating Disk
			Sliding and Rotating Vanes
			Gear and Lobed Impeller (Rotary)
		Gases	Bellows
			Liquid Sealed Drum
2. Rate Meters	Differential Pressure		Venturi
			Flow Nozzle
			Nozzle-Venturi
			Critical Flow Nozzle
			Thick-Plate Rounded-Edge Orifice
			Thin-Plate Square-Edged Orifice
			Concentric
			Eccentric
			Segmental
			Gate or Variable Area
			Centrifugal
			Elbow or Long-Radius Bend
			Turbine Scroll Case
			Guide Vane Speed Ring
			Pitot Tube (Impact Tube)
			Pitot-Static Tube
			Pitot-Venturi
			Linear Resistance or Frictional
			Pipe Section
			Capillary Tube
			Porous Plug
	Area (Geometric)		Gate
			Cone and Float
			Slotted Cylinder and Piston
	Velocity		
	Open or Closed Channel		Cup (Anemometer)
			Propeller
	Closed Channel		Turbine
	Head-Area		Weirs
			Flumes
	Force		
	Open or Closed Channel		Hydrometric Pendulum
			Vane
	Closed Channel		Transverse-Momentum
			Flow Meters
	Thermal		
	Open or Closed Channel		Hot Wire
			Total Heating
	Other Meters and Methods		
	Open or Closed Channel		Electromagnetic
			Tracers
			Mixtures
			Floats and Screens
	Closed Channel		Pressure-Time
			Sound Velocity
			Light Velocity

of the reciprocating members or by varying the relation between the primary and secondary elements.

I-1-8 Rotary (or Oscillating) Piston. Meters of this group have one or more vanes which serve as pistons or movable partitions for separating the fluid segments. These vanes may be either flat or cylindrical and rotate within a cylindrical metering chamber. The axis of rotation or annular movement of the vanes may or may not coincide with that of the chamber. The portion of the chamber in which the fluid is measured usually includes about 270 deg. In the remaining 90 deg, the vanes are returned to the starting position for closing off another segment of fluid. This may be accomplished by the use of an idle rotor or gear, a cam, or a radial partition. The vanes must make almost wiping contact with the walls of the measuring chamber. The rotation of the vanes operates the secondary element, or counter.

I-1-9 Nutating Disk. Meters of this type have the disk mounted in a circular chamber with a conical roof and either a flat or conical floor. When in operation, the motion of the disk is such that the shaft on which it is mounted generates a cone with the apex down. However, the disk does not rotate about its own axis; this is prevented by a radial slot which fits about a radial partition extending in from the chamber sidewall nearly to the center. The peculiar motion of the disk is called "nutating." The inlet and outlet openings are in the sidewall of the chamber on either side of the partition. Adjustment of these meters is usually effected by changing the relation between the primary and secondary elements.

I-1-10 Geared or Lobed Impeller (Rotary). Usually, these meters consist of two rotors, each with two or more lobes, mounted so as to rotate in a manner similar to gears with almost rolling contact. The inner surface of the case and the impeller tips are machined carefully to reduce clearance to the minimum.

I-1-11 Bellows. These meters have three or four measuring compartments separated by two or more movable partitions, called diaphragms, attached to the case by an impervious, flexible material, so that each partition may have a reciprocating motion. This motion of the diaphragms operates not only the secondary element but also the valves controlling the flow of gas in and out of the meter. Adjustment is effected by regulating the magnitude of the movement of the diaphragms.

I-1-12 Liquid Sealed Drum. In one group of these meters, the measuring compartments are formed in a drum by approximately helical partitions which do not extend to the center of the drum. The drum is mount-

ed on a horizontal shaft so that it can revolve within an exterior casing which is filled with a sealing fluid (usually water) to a sufficient height to seal completely the measuring compartments.

I-1-13 In another group, a bell, divided into several compartments by radial partitions, is mounted on an inclined shaft in an outer case that is filled with a sealing fluid (usually oil) to a depth always sufficient to seal the bottom edge of the bell and partitions. The bell has a nutating motion that causes the upper end of its shaft to follow a circular path and operate both the valve mechanism and the secondary element.

Rate Meters

I-1-14 The term "rate," or "rate of flow," is applied to all meters through which the fluid does not pass in isolated (separately counted) quantities but in a continuous stream. The movement of this fluid stream flowing through the primary element is directly or indirectly utilized to actuate the secondary element. The quantity of flow per unit of time is derived from the interactions of the stream and the primary element by known physical laws supplemented by empirical relations.

I-1-15 In rate meters, the functioning of the primary element depends upon some property of the fluid other than, or in addition to, volume or mass. Such a property may be kinetic energy (head meters), inertia (gate meters), specific heat (thermal meters), or the like. The secondary element is designed to utilize a change in the property, or properties, concerned for obtaining an indication of the rate of flow and usually embodies some device which draws the necessary inferences automatically, so that the observer can read the result from a dial or chart. In some cases, the secondary element indicates or records pressures, such as static and differential, from which the rate of flow and time-quantity flow must be obtained by computation. In others, the secondary element not only indicates the rate of flow but also integrates it with respect to time and records the total quantity that has passed through the meter. In some cases, the indications and recordings of the secondary element are transmitted to a point some distance from the primary element.

I-1-16 Differential Pressure Meters. With this group of meters the stream of fluid creates a pressure difference as it flows through the primary element. The magnitude of this pressure difference depends upon the speed and density of the fluid and features of the primary element.

I-1-17 The Venturi tube, the flow nozzle, and the orifice plate are examples of primary devices from which the static pressure is taken off at two points between which the cross section and, therefore, the linear velocity change. The Pitot or impact tube makes use of the difference between the static and total pressures at a single point. The centrifugal device utilizes the centrifugal change of velocity pressure across a curved stream—illustrations of this being the metering elbows and the turbine scroll case taps. A section of pipe, a capillary tube, or a porous plug might be used as a primary metering element. When so used, they would be termed "friction" meters since the pressure difference between inlet and exit is a measure of the external and internal (viscosity) friction overcome in moving the fluid through the pipe, tube, or plug.

I-1-18 In all these types, the secondary element of the meter may, first of all, include a differential-pressure gage or two separate gages from which the difference of pressure can be found. It may also contain a mechanism for translating the pressure difference into rate of flow, in accordance with the known equation for the primary device. A mechanism for integrating to give total flow can be added also. Alternately, the pressures may be recorded on a chart, and the total flow for the period of the chart computed manually or with any of various computing mechanisms.

I-1-19 Area Meters. Area meters are those in which a variation in the cross section of the stream under constant head is used as an indication of the rate of flow. For example, a float is suspended in a vertical tapered tube (with the larger diameter uppermost) and moves axially with the flow, with the flow being vertically upward. The weight of the float, less its buoyancy in the fluid, is balanced by the drag force of the fluid upon the float, which rides at such a height as will maintain this balance of forces. The secondary element of the meter translates the changes in area into changes in rate of flow, or the tapered tube may be transparent and marked with a scale to show the flow rate by the position of the float.

Another design utilizes a vertical cylinder with many holes or a longitudinal slot. As the float or piston rises, the total area of holes or slot through which the fluid may leave the cylinder increases. The movement of the piston may be retarded by means of a spring or added weight, resulting in an appreciable pressure drop across the meter, thereby increasing its capacity.

I-1-20 Velocity Meters. Velocity meters are those in which the primary element is some device that is kept in continual rotation by the linear motion of the stream in which it is immersed. The secondary element is, essentially, a revolution counter, and the instrument measures the total distance of travel of the fluid past the primary device. By suitable graduation of the dials, this distance can be multiplied by the sectional area of the stream so that the meter reads directly the total volume passed through. Obviously, for a fluid of known density, the dials can be made to read weights instead of volumes.

In this respect, the velocity meters resemble the quantity meters which give the total quantities directly and not rates of flow. They are radically distinct from the quantity meters, however, in that they do not pass the fluid through in separate (isolated) portions. The primary device need not take up the whole cross section of the stream, and the fluid can pass through with little obstruction, even though the mechanism be blocked.

I-1-21 Head-Area Meters. The weirs and flumes are used in open conduits and involve a simultaneous variation of head and area. The secondary device is operated by the change in height of the free surface of the liquid.

I-1-22 Force Meters. The force meter for an open channel consists of a vane or "obstruction," submerged in the stream and moved by it against the resistance of a spring or gravity. In the closed channel types, the weight rate is obtained from a measurement of the force required to change the angular momentum of the fluid.

I-1-23 Thermal Meters. The primary element of a thermal meter incorporates a device for dissipating heat to a part or all of a fluid stream and a device for measuring either the cooling effect of the stream upon the heating element or the change in the average temperature of the stream. In the hot-wire form (e.g., the hot-wire anemometer), the wire, which is held approximately normal to the stream, is heated electrically; and the flow of the fluid over the wire tends to cool it and change its resistance. This change of resistance can be interpreted in terms of the velocity of the fluid over the wire. An alternate procedure is to maintain the temperature and, therefore, the resistance of the wire constant and to determine the fluid velocity by measuring the electrical energy required.

In the closed channel form of thermal meter, the heat added changes the average temperature of the entire stream. The heat input may be kept constant and the temperature difference allowed to vary with the flow, or the temperature difference may be kept constant and the heat input varied. In the only known commercial example, the secondary element measures the electrical input for heating and determines the rate of flow from the known specific heat of the fluid and the known difference of temperature which is maintained constant.

I-1-24 Electromagnetic Flow Meters. When a conductive fluid flows through a magnetic field, an electromotive force will be induced which will be normal to both the magnetic field and the fluid channel. The strength of the induced emf increases as the fluid velocity increases, and a relation can be established between the strength of the emf and the average fluid velocity. Thus, for a given size of pipe, a recording voltmeter can be scaled to record the quantity of fluid flow.

I-1-25 Tracer Method. A substance is injected into the fluid stream, and the time interval required for the substance to be carried past one or more detection sections downstream is determined. From this time interval and the distance to or between the detection sections, the average velocity of the fluid stream may be computed.

I-1-26 Mixture Method. This consists of introducing a suitable substance, such as a soluble salt, another liquid, or a gas, at a known rate into the channel carrying the fluid to be metered. After the original fluid and the introduced substance have become thoroughly mixed, the percentage of the introduced substance in the mixture is determined. From these data, computing the rate of flow of the original fluid stream is possible.

I-1-27 Pressure-Time Method. The pressure-time method is based on the equation of impulse and momentum and the relation between change of pressure and change of velocity of a column of water expressed in terms of the velocity of the pressure waves which are propagated, during the change, from one end of the column to the other.

I-1-28 Sound-Velocity Method. The sound-velocity method depends upon determining, between two sections of the channel, or between three sections equally spaced, the differences in the time intervals required for sound to travel in the fluid, in the direction of flow and in the opposite direction, or in the fluid at rest. From this difference in the time intervals, the mean velocity of the fluid can be calculated.

The use of light in place of sound would be similar, to a limited degree, to the use of sound. It would be limited to use with gas flows.

Chapter I-2

Units, Reference Conditions and Letter Symbols

I-2-1 Fundamental Units. In the United States the units used extensively throughout industry are those of the gravitational system in which the primary units are the *foot* for length, the pound-weight or pound-force for force, and the second for time. However, in the general field of fluid mechanics, of which fluid metering is a part, as well as in some engineering schools, the "absolute" system of dimensions is used extensively. In any homogeneous system of dimensions in mechanics, only three primary dimensions may be used, and other dimensions are defined in terms of these three as secondary dimensions. The primary dimensions may be mass, length, and time or force, length, and time. In the absolute system, the primary dimensions are *mass* (not weight), length, and time.

On this subject, at its annual meeting in 1963, the Research Committee on Fluid Meters voted to use these absolute dimensions in the development of equations and the presentation of physical data in this sixth edition of its report. Furthermore, it was decided to use the name "pound" for the unit of mass and also for the unit of force. This requires the use of the terms "pound-mass" and "pound-force," or the respective abbreviations lb_m and lb_f, so that there may be no uncertainty as to which is intended. In connection with the pressure abbreviations, psi, psia, and psig, it is to be remembered that pressure is "force per unit area." Hence, the "p" in these abbreviations is *force* pounds, even though the subscript "*f*" is omitted.

I-2-2 The reasons for using *mass*, that is, the absolute system in preference to *force*, or the gravitational system, are, first, that the mass of a substance or body is the measure of its quantity and may be evaluated by its inertia. Neither mass (quantity) nor inertia is changed by simply changing location. On the other hand, force, as represented by weight, is a measure of the gravitational pull of the Earth upon the body and varies from place to place, depending upon the local value of *g*.

Secondly, in most scientific work both in this country and abroad, the metric absolute units are used extensively. The primary units of this system are: length, the meter; mass, the kilogram; and time, the second. The actual units very generally used are the centimeter, gram, and second, i.e., submultiples of the primary units. In this system, the unit of force, the "kilopound," is defined as the force required to accelerate a kilogram mass 9.80665 meters per sec². This system is now usually referred to as SI units (Systeme International d'Unites) [1].

I-2-3 The internationally adopted value of the acceleration due to gravitation and designated "standard gravity" is g_o = 980.665 cm/sec² = 32.1740 ft/sec² at mean sea level and 45 deg north latitude.

I-2-4 The relation between mass and force is derived from Newton's second law, namely, that force is proportional to the product of mass times acceleration, that is,

$$\text{force} \propto (\text{mass}) \times (\text{acceleration}) \quad \text{(I-2-1)}$$

or

$$\text{unit force} = (g_c)(\text{unit mass})(\text{unit acceleration}) \quad \text{(I-2-2)}$$

in which g_c is a proportionality factor. In the absolute

system, the value of the proportionality factor will be unity if a force unit is chosen that will give unit acceleration to a unit mass. Such a unit of force is a "poundal;" thus,

$$\text{(1 poundal)} = (1)\ (1\ \text{lb}_m)\ (1\ \text{ft/sec}^2) \qquad \text{(I-2-3)}$$

Alternatively, a mass unit may be so chosen that a one-pound force will cause it to accelerate at 1 ft/sec². Such a unit of mass is called a "slug," and

$$\text{(1 lb-force)} = (1)\ (1\ \text{slug})\ (1\ \text{ft/sec}^2) \qquad \text{(I-2-4)}$$

I-2-5 The use of neither the poundal nor the slug has gained popular acceptance, and, therefore, neither is used in this report. Instead, a combination of the absolute and gravitational systems is used. To do this a "standard pound-force" is defined as the force required to impart to a standard pound-mass an acceleration equal to g_o, or 32.1740 ft/sec². That is,

$$\begin{aligned} &\text{(1 std pound-force)} \\ &= \text{(1 std pound-mass)}\ (32.1740\ \text{ft/sec}^2) \qquad \text{(I-2-5)} \end{aligned}$$

Equating the products indicated by equations (I-2-2) and (I-2-5),

$$\begin{aligned} &(g_c)\ (1\ \text{std lb}_m)\ (1\ \text{ft/sec}^2) \\ &= (1\ \text{std lb}_m)\ (1\ \text{ft/sec}^2)\ (32.1740) \qquad \text{(I-2-6)} \end{aligned}$$

from which

$$g_c = 32.1740 \qquad \text{(I-2-7)}$$

thus assigning a definite value to the proportionality factor g_c. By using equation (I-2-7) with equation (I-2-3) it is seen that

$$1\ \text{poundal} = (1\ \text{std lb}_f)/32.1740 \qquad \text{(I-2-8)}$$

or combined with equation (I-2-4)

$$1\ \text{slug} = 32.1740(1\ \text{std lb}_m) \qquad \text{(I-2-9)}$$

Attention is again called to the statement that in this edition the equations and presentation of data are based on an absolute system of dimensions in which *mass* along with length and time are the primary dimensions. However, in line with common usage, the mass unit actually used is very much smaller than should be used for a homogeneous system of absolute dimensions. The difference in size is shown by equation (I-2-9). To take account of this difference in size, a numerical factor (multiplier) is required. For convenience, this factor is designated g_c; and, since it represents a number, its dimensions are unity. Alternatively, it could be said the size of the force unit in common use is much too large, as shown by equation (I-2-8).

It is important to note that the standard pound-force, as defined above, is independent of variations of local gravitational acceleration. Thus, analogous to the standard pound-force, a "local pound-force" or "local weight" may be defined as the force which will impart to a standard pound-mass an acceleration equal to the local value. Since weight is a force, this gives

$$\text{(local weight)} = g\text{(std pound-mass)} \qquad \text{(I-2-10)}$$

where g = the local acceleration due to gravity. Noting that the standard gravitational acceleration, g_o = 32.1740 ft/sec², it is seen by equations (I-2-5) and (I-2-10) that

$$\text{(standard weight)} = \text{(local weight)}\ (g_o/g) \qquad \text{(I-2-11)}$$

I-2-6 In the use of manometers, the difference between standard and local gravity values will affect the value of the actual pressure indicated by the manometer. At a local value of gravity, g, such that $g < g_o$, the pressure (i.e., force per unit area) represented by a given volume (i.e., a given mass) or height of a liquid in a manometer will be less than if the gravity were g_o. To express it another way, for a given pressure in terms of psi at standard gravity, the reading of a manometer will increase as the local value of gravity decreases. Therefore, the factors for converting manometer readings to psi at local gravities must be multiplied by g/g_o to give pressure in psi at standard gravity.

The curves of Fig. II-I-2, Part II, give the factors to convert inches of mercury and inches of water to psi at *local* values of gravity. To obtain pressures in psi at standard gravity, it will be necessary to multiply the factors of Fig. II-I-2 by g/g_o. Note that this is the reciprocal of the multiplier required to convert "local" weight to "standard" weight, equation (I-2-11).

I-2-7 Temperatures. With English units, temperatures are expressed in degrees Rankine, which is the scale of absolute temperatures in Fahrenheit degrees. The zero of the Fahrenheit scale is 459.69 R.

With metric units, temperatures are expressed in degrees Celsius (centigrade) or in degrees Kelvin absolute. The zero of the Celsius scale is 273.16 K. (See Par. I-3-18 for note on Celsius and centigrade scales.)

I-2-8 Units of Fluid Measurement. In the metering of liquids in volume units, as in the metering of water from city supply systems and in the retail sale of petroleum products, the *gallon* is the basic unit of measurement. The standard U.S. gallon is defined as a volume of 231 cu in. or 0.013368 cu ft. In the petroleum industry the (petroleum) barrel of 42 gal is the unit most used as a basis for bulk sales and in reports of operating activities. Another volume unit, used in irrigation and related activities, is the acre-foot, which is equivalent to 43,560 cu ft or 325,851 gal.

Note: The British Imperial gallon is equivalent to 277.420 cu in. Hence, 1 British Imperial gal = 1.20094 U.S. gal.

The pound (weight) is the unit of quantity most used in power plants, chemical plants, paper mills, and similar industries in the measurement of water and other liquids. It is used also in the measurement of fuel for many engines, especially aircraft engines, which use a liquid fuel.

Common units for rate of flow are: gallons per minute, (gpm); cubic feet per second, (cfs); pounds per second (pps); or pounds per hour, (pph); and million gallons per day, (mgd). In irrigation, the miner's inch is still used to a limited extent; however, the definition of this unit and its equivalent in terms of gallons per minute of cubic feet per second is not the same in all states [2, 3, 4].

I-2-9 For gases, especially commercial fuel gases, the *cubic foot* is the unit of measurement. However, a cubic foot of gas has no absolute or comparative value unless the pressure and temperature of the gas are specified when it fills a cubic foot space. A common unit used for evaluating rates of flow is "standard cu ft per hour" (scfh), which means at a temperature of 60 F, a pressure of 14.73 psia, and dry. In the transportation and bulk sale of fuel gases, it is customary to report gas quantities in terms of M (1000) cu ft. In recent years increased attention has been given to expressing wholesale quantities of fuel gases in terms of pounds or in energy units such as Btu or even kilowatt hours. In the sale of compressed gases (e.g., acetylene and carbon dioxide) and gases which are stored and transported in the liquid phase, the pound (weight) is the unit used most. Also, it is customary to express the quantity of air supplied for combustion to gas turbines and jet engines in terms of pounds per unit time, usually pounds per hour (pph) or pounds per second (pps).

I-2-10 Reference or Standard Conditions of Temperature and Pressure: Water. The reference temperature used in measurements of water varies with the purpose of the measurement and the industry. In scientific work the temperature of maximum density, 39.2 F (4 C), is generally used.

In commercial work and especially in pressure measurements by water column gages (manometer), 68 F (20 C) is used the most, although 60 F is used to a limited extent.

One atmosphere (14.696 psia) is the usual reference pressure for the measurement of water. In most commercial work, as well as hydraulic laboratory work, it is customary to neglect the changes in the density of water with pressure. This practice can lead to significant errors when the water is measured under high enough pressure and especially if also at an elevated temperature. The density of water for a wide range of temperatures and pressures is given in Part II.

I-2-11 Petroleum and Liquid Petroleum Products. The reference temperature for the measurement of these liquids is 60 F.

The reference pressure (absolute) is atmospheric pressure or the Reid vapor pressure, whichever is higher [5, 6].

Note: Reid vapor pressure is the absolute pressure of the hydrocarbon vapor, usually at 100 F, as determined with an apparatus in which the vapor space is approximately four times that of the liquid.

I-2-12 Gases. The reference temperature used in the measurement of commercial fuel gases is 60 F. In the measurement of other gases, and in laboratory work, both 32 F (0 C) and 68 F (20 C) are used also.

The reference pressures most generally recognized now in the metering of commercial fuel gases are: 30.00 in. Hg abs at the ice point, equivalent to 14.735 psia; and 14.73 psia, which is equivalent to 29.99 in. Hg abs at the ice point. The particular value used in any given case will be governed by the purpose of the measurement, the type of fuel gas being metered, the locality and parties involved.

In most scientific work and in tables of physical data on gases, the reference pressure is usually 760 mm Hg abs at 0 C (29.921 in. Hg abs).

A further element usually incorporated in the definition of the reference conditions for the measurement of gases is the amount of water vapor with the gas, which is referred to as the degree of water-vapor saturation or relative humidity. In the fuel-gas industry, if the gas is a manufactured or a mixed gas (i.e., a mixture of some type of manufactured and natural petroleum gases), the reference condition is almost always complete water-vapor saturation. With natural (petroleum) fuel gases, particularly in the transmission of the gas, the reference condition is always dry. The selection of the reference state for measurements of air and other gases will vary with the conditions and purposes of the measurements, but probably the dry state is used most often.

I-2-13 The reference or "standard" conditions which have been formally endorsed by the American Gas Association for the evaluation of fuel gas volumes are: a pressure of 14.73 psia; a temperature of 60 F; and dry (i.e., free of water vapor) [7].

Note 1: The American National Standard Institute has adopted 14.73 psia and 60 F as standard for natural fuel gas measurement, and designated it ANSI, Z132.1.

Note 2: The use of 60 F as a reference temperature for gas volumes and 68 F as the temperature of water for converting inches of water to psi is not inconsistent. Each one is entirely independent of the other.

I-2-14 In the measurement of gas pressures close to atmospheric pressure as well as in the measurement of differential pressures (see Par. I-3-13), the unit frequently used is "one inch of water column." Since the absolute value of this unit will depend upon the temperature of the water, that of 68 F is generally assumed. The density of distilled water at 68 F is 62.32 (or more exactly 62.316) lb_m per cu ft, so that 1 in. of water is equivalent to 0.03606 psi.

Note: The difference between the densities of distilled water and ordinary tap water usually will be less than 1 part in 10,000, so that the effect of using tap water in a manometer will be negligible for most fluid metering requirements.

I-2-15 **Letter Symbols.** In so far as conveniently possible, the letter symbols used in this edition have been selected in an effort to harmonize with: (1) some of the common commercial usages; (2) appropriate portions of Letter Symbols for Hydraulics and Letter Symbols for Heat and Thermodynamics approved by the American National Standards Institute (formerly the American Standards Association);" and (3) symbols common to other members of the Committee on Measurement of Fluid Flow of the International Organization for Standardization. In this effort, the general rules adopted by the ANSI as guides in the selection of letter symbols have had a considerable influence. For this reason, the more important of these rules will be stated and are:

(1) In general, the same symbol shall be used regardless of the system of units.

(2) In cases where more than one system of units is encountered, the same symbol should be used with the addition of suitable subscripts or superscripts to denote a unit other than the primary.

(3) The same symbol should be used for a given concept, regardless of the number of special values which occur, and subscripts or superscripts should be used to designate them.

(4) Where possible, capital letters denote total quantities, and small letters denote specific quantities or quantities per unit.

(5) Letter subscripts should be used to denote values under special conditions or in special states. (When possible, descriptive subscript letters are to be used.)

(6) Numeral subscripts should be used to denote values at designated points or sections in an apparatus, process, or cycle. In applying numerical subscripts to flowing fluids, the general practice is to number from the inlet to the outlet section.

In using numerical subscripts with differential pressure meters, certain conventions that are more or less generally used should be stated. The subscript "1" refers to the inlet or upstream section and usually to that particular section at which the inlet pressure is taken. The subscript "2" may refer to the section producing a change in the stream area or at which the stream section is a minimum, but in general it refers to the outlet or downstream section at which the outlet pressure is measured.

I-2-16 In the development of some of the equations, it has been necessary to use additional symbols to those given in the table. Where this has been done, these temporary symbols have been selected with little or no regard to their use elsewhere. Also, the definitions assigned these temporary symbols do not carry beyond the particular equations in the development of which they were used.

I-2-17 Numerous conversion factors are given in this and the following chapters. Additional factors and tables may be found in the many references which are listed at the end of each chapter [8].

In the rounding off of the results of computations, different rules have been followed. One such rule is that when the digit to be dropped is a 5 the last remaining digit is made an even number. With electronic computing equipment, the rule or instructions fed to the computer will depend on the programming method in use [9, 10].

Table I-2-1 Letter Symbols

Symbol	Symbol Description	Unit of Measurement	Dimensions
A	An area; specifically, cross-sectional area of a channel completely filled with fluid	sq in. (or sq ft) (see Notes 1 and 2)	L^2 (see Note 3)
a	Area of a Venturi throat, nozzle throat, or an orifice	sq in. (or sq ft)	L^2
D	Diameter; specifically, the diameter of a pipe at a specified section	in. (or ft) (see Notes 1 and 2)	L
d	Diameter of a Venturi throat, nozzle throat, or orifice	in. (or ft)	L
C	Coefficient of discharge	ratio	
c_p	Specific heat of a fluid at constant pressure	Btu per lb_m per $^\circ$R	
c_v	Specific heat of a fluid at constant volume	Btu per lb_m per $^\circ$R	
E	Velocity of approach factor $= 1/\sqrt{1 - \beta^4}$	ratio	
F	Flow function $= m\sqrt{T_t/a p_t}$	$lb_m \cdot {}^\circ$R per lb_f	
F_a	Area factor to account for the thermal expansion of a primary element	ratio	
f	Frequency	cycles per sec	T^{-1}
G	Specific gravity (see Pars. 1-3-21 to 1-3-24)	ratio	
g	Acceleration due to gravity (local) (see Note 4)	ft per sec^2	LT^{-2}
g_c	Proportionality constant in the force-mass-acceleration equation $= 32.1740$	factor	
g_o	International Standard acceleration of gravity $= 9.80665$ m/sec^2 $= 32.1740$ ft/sec^2	ft per sec^2	LT^{-2}
h	Effective differential pressure	ft of fluid	$ML^{-1}T^{-2}$

13

Table I-2-1 Letter Symbols (Continued)

Symbol	Symbol Description	Unit of Measurement	Dimensions
h_w	Effective differential pressure (see Note 5)	in. of water at 68 F	$ML^{-1}T^{-2}$
K	Flow coefficient $= CE$	ratio	
L	A length; specifically, the length of the flow channel included in a metering unit	ft	L
M	Mach number $= V/V_s$ (see Note 6)	ratio	
m	Mass rate of flow	lb_m per sec	MT^{-1}
N	Restriction factor $= a^*/a$	ratio	
p	Absolute pressure, the actual or observed pressure (see Notes 1 and 7)	lb_f per sq in. abs (psia)	$ML^{-1}T^{-2}$
p_t	Total or stagnation pressure $=$ static pressure $+$ the pressure equivalent of the kinetic energy due to the motion of the fluid	lb_f per sq in. abs (psia)	$ML^{-1}T^{-2}$
Q	A volume, e.g., volume of a container; specifically, a particular fluid volume or the volume of a particular quantity (mass) of fluid	cu ft	L^3
q	Volume rate of flow	cu ft per sec (cfs)	L^3T^{-1}
R	Gas constant in $pv = RT$ (here p is lb_f/ft^2, see Notes 1 and 7)	ft lb_f/lb_m °R	$L^2T^{-2}(R)^{-1}$
R_D	Reynolds number based on pipe diameter, D	ratio	
R_d	Reynolds number based on a throat diameter, d	ratio	
r	Ratio of outlet or throat static pressure to inlet static pressure $= p_2/p_1$	ratio	

Table I-2-1 Letter Symbols (Continued)

Symbol	Symbol Description	Unit of Measurement	Dimensions
T	Absolute temperature (static)	deg Rankine (R)	
T_t	Total or stagnation temperature of a flowing fluid	deg Rankine	
t	Time interval	sec	T
V	Velocity (see Note 6)	ft per sec	LT^{-1}
V_s	Velocity of sound in a fluid at the density under consideration	ft per sec	LT^{-1}
v	Specific volume $= 1/\rho$	cu ft per lb_m	$L^3 M^{-1}$
x	Ratio of differential pressure to inlet absolute static pressure $= (p_1 - p_2) / p_1 = \Delta p/p_1$	ratio	
Y	Expansion factor to account for effect of expansion of a gas flowing through a Venturi tube, flow nozzle, or orifice (see Par. I-5-30)	ratio	
Z	Compressibility factor to account for the departure from linearity of the compressibility of an actual gas	ratio	
β (beta)	Ratio of a throat or orifice diameter to the pipe diameter $= d/D$	ratio	
Γ (gamma)	Isentropic exponent, a function of p_1, p_2 and T		
γ (gamma)	Ratio of the specific heats of an ideal gas $= c_p/c_v$ (see Note 8)	ratio	
δ (delta)	Diameter of pressure tap holes	in.	L
Δ (delta)	A difference		

15

Table I-2-1 Letter Symbols (Continued)

Symbol	Symbol Description	Unit of Measurement	Dimensions
Δp	Differential pressure $= (p_1 - p_2)$	lb_f per sq in.	$ML^{-1}T^{-2}$
λ (lambda)	Reynolds number reciprocal, $\lambda = 1000/\sqrt{R_D} = 1000/\sqrt{\beta R_d}$	ratio	
μ (mu)	Absolute viscosity (see Note 9 and Par. I-3-35)	lb_m per ft sec	$ML^{-1}T^{-1}$
ν (nu)	Kinematic viscosity $= \mu/\rho$	ft² per sec	L^2T^{-1}
ρ (rho)	Density (see Pars. I-3-19 and I-3-20)	lb_m per cu ft	ML^{-3}
ψ (psi)	Function		
τ (tau)	Deflection of orifice plate by differential pressure	in.	L
σ (sigma)	Standard deviation		

Note 1: The units of measurement in which quantities are expressed, as given in this table and used throughout this report, are those used in other fields of engineering, and not fluid metering only. It must be remembered that in the English (engineering) system of units the *foot* is the basic unit of length and that the inch is a secondary unit. Hence, in developing and applying equations the relations between the inch and foot and between the square inch and square foot must be taken into consideration, especially when going from the unapplied to the working equations.

Note 2: The symbols A and D are used frequently as general symbols for area and diameter, respectively. In this report, when used in connection with differential pressure meters, it is to be understood that they apply specifically to that cross section of the inlet pipe at which the inlet pressure is measured. Likewise, the lower case letters a and d are used to designate the area and diameter, respectively, of a Venturi tube throat, of a nozzle throat, or of an orifice. This use of both upper and lower case letters to denote particular sections, or quantities, does not fully agree with the ANSI rules. However, it will probably prove to be a convenience to flow measurement engineers and is in agreement with the usages of other members of the International Standards Organization Committee No. 30 on Flow Measurement.

Note 3: In evaluating the dimensions of quantities in terms of the fundamental units of length, mass, and time, it is customary to use the upper case letters L, M and T, respectively, to represent the units. Each of these letters will be found in the list of letter symbols. However, this double use of these letters should cause no confusion, since the checking of an equation for dimensional equality is seldom presented concurrently with the development of the equation.

Note 4: For most engineering work it is not necessary to distinguish between the exact local value of gravity and the standard value, and it is generally sufficient to use $g = 32.17$ ft per sec². However, when for precise work it is desirable to take account of the effect of any difference between local gravity and standard gravity, the weighings made with a scale should be multiplied by g_o/g to give the weight at standard gravity. On the other hand, pressure measurements made with manometers are to be multiplied by g/g_o to give the pressures at standard gravity, as explained in Pars. I-2-5 and I-2-6.

Note 5: In connection with differential pressure meters, the use of h_w to represent the differential pressure in inches of water at 68 F (= 20 C) is in agreement with the use of this temperature as a datum. In the development of equations of flow, it may be desirable to express the differential pressure producing the flow as a column of the flowing fluid, measured in feet, the symbol for which is h.

Table I-2-1 Letter Symbols (Continued)

Note 6: As used in this report, V denotes the area average velocity. Hence, $V_1 = q/A_1$ and $V_2 = q/A_2$. V_s, as used in defining the Mach number, represents the velocity of sound in the fluid concerned at the density of that fluid under the conditions of pressure and temperature at the place of interest.

Note 7: It is general practice to express pressures in pounds (force) per square inch, (psi), and conforming to this practice the numerical factors associated with the working equations given in this report as based on the use of this unit. However, in the development·of these equations, especially where specific volumes and densities are involved, the pressure unit to be considered is pounds (force) per square *foot.*

Note 8: When $\gamma = c_p / c_v$ is used in relation to $pv^\gamma = $ constant, it is to be understood that this relation is for an ideal gas and a reversible system.

Note 9: There is no name for absolute viscosity in lb_m per ft-sec nor for kinematic viscosity in ft² per sec. The names of these fluid properties in metric units are, respectively, *poise* and *stoke.*

References

[1] "Systems of Units (Mass, Length, Time and Force)," ISO/TC-12 Draft Proposal.

[2] "ASTM-IP Petroleum Measurement Tables," ASTM D-1252-52.

[3] "Bibliography of Gas Meters and Metering," Report DMC-55-8; American Gas Association, 1955.

[4] "Water Works Practice: A Manual" American Water Works Association, p. 700; Williams and Wilkins, 1936.

[5] "Tests for Vapor Pressure of Petroleum Products: Reid Method" ASTM-D-1267-55.

[6] "Code for Installing, Proving and Operating Positive Displacement Meters in Liquid Hydrocarbon Service," API Code 1101, 1960, American Petroleum Institute.

[7] "Orifice Metering of Natural Gas," AGA Gas Measurement Committee Report No. 3, 1969.

[8] "Conversion Factors and Tables," Zimmerman and Lavine; Industrial Research Services, 1944.

[9] Rules for "Rounding Off Numerical Values," ASA Z25.1, 1940.

[10] "Numerical Methods and Fortran Programing," D. D. McCracken and W. S. Dorn; pp. 51 and 84, John Wiley & Sons, N.Y. 1964.

Chapter I-3
Special Terms

I-3-1 In most specialized branches of industry, some common terms have come to have particular meanings when used in discussing that special subject. This is an appropriate point at which to define some of the special terms used in connection with fluid meters and, also, to discuss the methods of measuring the quantities to which they are applied.

I-3-2 **Static Pressure and Its Measurement.** The static pressure in a stream of fluid is the pressure that would act on and be indicated by a pressure gage if it were moving along with the stream so as to be at rest or relatively "static" with respect to the fluid.

The measurement of static pressure in accordance with this definition, however, is impracticable so that some equivalent substitute method must be adopted. The customary procedure is to drill a small hole perpendicularly through the wall of the pipe in which the fluid is flowing. If a connection is made between this hole and a pressure gage, the pressure that will be observed is called the "static pressure" of the fluid that is moving past the hole [1, 2].

I-3-3 The idea is that if the hole is too small to cause any sensible disturbance in the motion of the fluid along the wall, neither suction nor impact will occur, and the motion of the fluid past the hole will have no effect on the pressure observed. This reasoning is not hard and fast, and the possibility of a slight suction always exists. Nevertheless, this mode of measuring static pressure is justified by its results, for, when this pressure is used in the theoretical reasoning about the behavior of streams of fluid as if it were measured in accordance with the

strict definition of static pressure, no conflict between theory and experiment which can be attributed to an error in the principle of measurement is found.

If the stream is turbulent and full of eddies, the pressure in the side hole will fluctuate irregularly; but, under all ordinary circumstances, the gage will indicate the average pressure at the side hole, provided there are no pulsations (see Par. I-3-44).

I-3-4 With a rough pipe, a very small hole might be placed with relation to the surrounding humps and hollows so that it is subjected to a more or less regular impact or suction. The gage reading would then be permanently too high or too low; or it might be high at one rate of flow and low at another. The obvious remedy is to make the diameter of the hole large in comparison with the average grain size of the roughness. But, on the other hand, the hole is a discontinuity in the wall and tends to disturb the regular tangential flow so that it should not be too large in relation to the pipe diameter.

In the case of a small pipe in which the grain size is a considerable fraction of the diameter, for instance a ¼-in. galvanized pipe, a side hole large enough to be safe from the effects of roughness would be liable to distort the flow too much. If the use of a side hole in such a pipe is unavoidable, the inside of the pipe should be made as smooth as practicable.

I-3-5 For smoothly finished surfaces, such as those of drawn brass pipe or the bronze throats of Venturi tubes, side holes can be very small, with the limit in this direction being set by considerations relating to possible obstruction of the hole by dirt, scale, etc., lag in transmission to the gage, or errors due to the leaks in the gage connections. A table in Part II gives the recommended maximum diameters of pressure holes for several sizes of pipes.

The edge of the hole where it breaks the inner surface of the pipe *must* be entirely free of burrs. In order to insure this condition, the edge of the hole may be dulled or rounded very slightly. However, the edge of the hole may be left square provided it is free of burrs.

I-3-6 **Variation of Static Pressure across a Stream.** In a straight parallel stream, the static pressure plus the elevation is constant or uniform all over any normal section. In a curved stream, the static pressure increases from the inside to the outside of the bend on account of centrifugal force. Hence, when speaking of the static pressure "at" a particular section of a stream, the flow through this section is implied and understood to be sensibly straight.

Eddies in a stream are merely curved currents superimposed on the general motion, and they cause local variations of static pressure. If they are carried along as turbulence, they cause rapid, irregular, local variations of the static pressure over any cross section of the stream, but they do not change the average pressure which can be correctly measured at a side hole.

A permanent eddy or crosscurrent due to some fixed obstruction or irregularity, such as a valve or bend upstream from the section under consideration, may vitiate the determination of pressure from a side hole. The usual method of eliminating this possibility is to have a sufficient length of straight pipe ahead of any section where the static pressure is to be determined. However, other means, such as the honeycomb type of straightening vane, radial vanes, or a disk orifice, can be employed to assist effectually in straightening the flow.

I-3-7 With the idea of reducing the chance of such an error, without using the above precautions, several similar side holes, uniformly spaced around the periphery of the section and all leading into a single, tight, ring-shaped pressure chamber, or "piezometer ring," from which a single gage connection is led off, have sometimes been used. Such a piezometer ring has been assumed to afford a reliable and accurate means of determining static pressures. But, if the pressures at the separate holes are appreciably different, some circulation through the ring must occur; and the pressure observed at the gage to the piezometer ring may not always be approximately the arithmetic average of

the pressures that would be observed at the separate holes.

If the pressures at the separate holes are appreciably and permanently different—assuming that the holes are all properly made—this condition demonstrates that the fluid is not flowing straight along the pipe but has permanent crosscurrents or eddies in it. Since the theoretical formulas to be used with differential pressure meters are based on the assumption of straight flow, they cannot be expected to represent correctly the facts in such a case as the one now under discussion.

The conclusion is that, when the flow is sufficiently straight and steady so that the static pressure is definite and sensibly uniform all over any normal section of the stream, this pressure can be found accurately from a single properly made side hole and can be used for determining the flow.

I-3-8 A uniform helical motion, such as follows two 90-deg bends in perpendicular planes, makes the static pressure increase from the axis outward, so that a side hole reading will always be higher than the average static pressure over the whole section. Such a twist can be straightened out by longitudinal vanes or will die of itself in a sufficient length of straight pipe, which may require several hundred pipe diameters.

I-3-9 **Velocity Pressure.** If a glass tube, bent at right angles, were held in a stream of water so that one leg points directly upstream while the other is vertical and projects above the free surface, the water in the vertical leg would stand above the level of the surrounding surface because water is driven into the open end of the horizontal leg by the impact of the stream. This is the original form of the Pitot tube for converting velocity energy into pressure energy.

Similarly, if the open end of a tube were to be held in any stream of fluid with its plane normal to the flow and the other end of the tube were to be connected to a pressure gage outside the stream, the pressure received by the open end of the tube would be found to exceed the static pressure at that section of the stream by an amount which is proportional to the density of the fluid and the square of its speed or, in other words, to its kinetic energy per unit volume. This pressure difference is known as the "velocity pressure" of the fluid at the place in question; and, if the density of the fluid is known, it can be used as a measure of the speed.

Velocity pressure can be measured directly on a differential gage with one side connected to the impact opening and the other to a static opening placed so as to give the static pressure close to the impact opening, or it can be determined as the difference in the readings of two separate gages connected to the static and impact openings, respectively.

So long as the plane of the impact opening is normal to the flow, the size, shape, and wall thickness can be varied within very wide limits without affecting the result. In general, all that is essential, if it is desired to find the impact pressure of the undisturbed stream, is that the entire "impact tube"—the opening and the tube which supports it and leads to the gage—shall not be so large as to obstruct the fluid passage unduly and so alter the speed of flow appreciably [1, 2].

The simplest form of impact tube is a piece of pipe bent at right angles and with the short end cut off square, so that, when this leg is parallel to the direction of flow, its mouth is perpendicular to this direction. The hole can also be made in some other symmetrical body such as a sphere, a small flat disk, or a long cylinder placed across the stream. In all such cases, the same maximum impact pressure is obtained when the plane of the hole is normal to the direction of the flow.

I-3-10 Effect of Changing the Plane of the Opening: Suction. With the ordinary impact opening formed by the end of a round tube cut off at right angles, the tube does not have to point exactly upstream, for it can be turned several degrees from this position before the pressure begins to fall off perceptibly. If the tube is turned more and more until it finally points directly downstream, the impact pressure gradually decreases to zero and then becomes negative, i.e., it changes into a suction.

The physical interpretation of this effect is that, when the opening is in the downstream or suction position, the current past its edge mixes with and entrains the "dead fluid" in the shielded space over the opening and so decreases the pressure there below the static pressure outside.

The rapidity of this mixing and the strength of the resulting entrainment will obviously depend on the character of the flow past the edge of the opening; and, since this is considerably influenced by the size and shape of the body in which the opening is made, the effect observed at a reversed impact or "suction" opening will depend on these, whereas, in the opposite or impact position, the flow up to the opening is unimpeded, and the resulting impact pressure is very much less affected by changes in the body or in the opening itself.

I-3-11 The Neutral Position: Measurement of Static Pressure in the Body of a Stream. Somewhere between the impact and suction positions, a neutral position is found where the pressure in the opening is the same as the static pressure measured at the standard side hole; but this position is not necessarily the one with the plane of the hole parallel to the general direction of flow. If the opening is a small hole in a sphere or in the side of a closed cylindrical tube reaching across the stream, the neutral position is only about 40 deg from the position of maximum impact [3, 4].

With an open-end tube, held with its axis normal to the direction of the fluid motion, the fluid does not simply move around the tube but also spills over the opening in convex lines which cause a centrifugal suction. Such a tube is not a reliable device for getting an observation of static pressure.

If this spilling is prevented by a thin sharp-edged disk forming a large flange set flush with the end of the tube and if the plane of this disk is set parallel to the flow, so that the fluid moves as smoothly over the face of the disk as it would along the wall of a smooth pipe, the pressure in the small hole at the center of the disk is the same as in a standard side hole.

I-3-12 Total or Stagnation Pressure. If, instead of measuring velocity pressure on a differential gage with one side connected to a static opening, an impact opening is connected to a gage adjusted to indicate absolute pressure, the pressure observed is the sum of the static pressure and the pressure equivalent to the kinetic energy due to the velocity of the fluid. This is known as the "total" or "stagnation" pressure and is an important quantity in the use of some rate-of-flow meters.

I-3-13 Differential Pressure and Its Measurement. As the term implies, "differential pressure" is the difference between the static pressures at two sections of the stream. To determine this difference of pressures, actually measuring the static pressures at the two sections is not necessary; in fact, the actual measurement can be made in various ways and in terms of any convenient units. In developing some of the equations it is convenient and helpful to express the differential pressure in terms of a column of h ft of the flowing fluid at the average density of the fluid at the section of measurement. This value of h can be obtained from the actual units of measurement by the use of appropriate conversion factors, as will be explained later.

Commercial instruments for indicating or recording the differential pressure and for converting it into cubic feet per second, gallons per hour, and similar units are not described in this report, but some general remarks on methods for the direct observation of differential pressure are in order here.

One very simple scheme is illustrated by Fig. I-3-1. The side holes at throat and entrance are connected to two vertical glass tubes which are set side by side so that the difference of level can be read directly on a scale. If both pressures are above atmospheric but not too high, the valve, B, can be left open, or the connection at the top can be omitted and the ends of the gage tubes left open. If the pressures are so high that this would require inconveniently long gage tubes, both columns can be equally depressed by introducing compressed air at B; or, if the throat pressure falls below atmospheric, the columns can be raised by applying suction at B. If the pressure (of air) is high enough, it may be necessary to apply a correction to the observed height, h_o, computed as described in Par. I-3-14 [5].

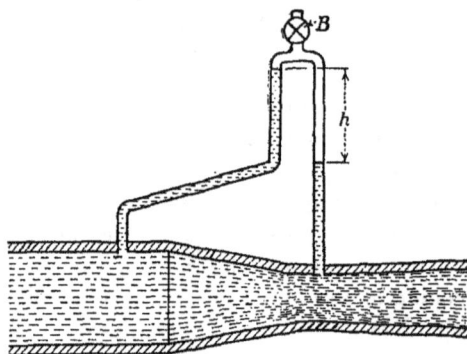

FIG. I-3-1 SIMPLE FORM OF DIFFERENTIAL PRESSURE GAGE (FLOW IS FROM LEFT TO RIGHT.)

Figure I-3-2 illustrates a modification of this arrangement which can be used when it is more convenient to have the gage tubes below the pipe. In this case, air pressure will have to be applied at B unless the entrance pressure is below atmospheric.

Any intermediate position of the gage tubes can be adopted also; and, if the primary element is properly made and installed, the levels on its circumference at which the pressure connections are made are immaterial, with the only requirement being that sufficient pressure or suction be applied at the air valve B to keep the tops of the columns of liquid in the gage tubes in sight and on the scale where their positions can be read.

If the differential pressure is so small that readings on a vertical scale are not sufficiently accurate, the tubes and scale can be inclined so as to magnify the readings. Another expedient for increasing the sensitiveness of the differential gage is to fill the

upper part of the inverted U with a second liquid that does not mix with the first and has a slightly lower density. If, on the other hand, the differential pressure is inconveniently large, it can be read on a reduced scale by using mercury in the gage, with the connections being made as shown in Fig. I-3-3 or in some equivalent manner.

I-3-14 The value of h to be used in equations for computing the rate of flow through differential-pressure meters depends upon the density of the liquid in the observed column (manometer) and of the fluid transmitting the pressure to the manometer or gage. To explain this, let

h = Pressure difference in terms of the height of a column of the fluid being metered

h_o = Height of the observed liquid column

ρ = Density of the fluid being metered

ρ_o = Density of the liquid in the observed column

ρ_b = Density of the fluid used to transmit the pressure

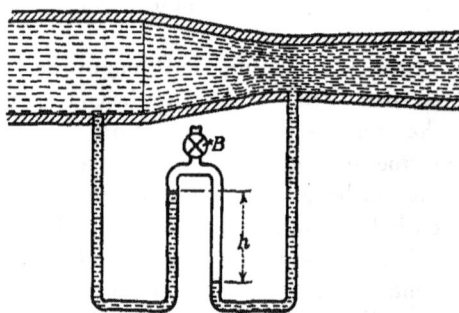

FIG. I-3-2 DIFFERENTIAL GAGE CONNECTED TO THE BOTTOM OF A HORIZONTAL PIPE (FLOW IS FROM LEFT TO RIGHT.)

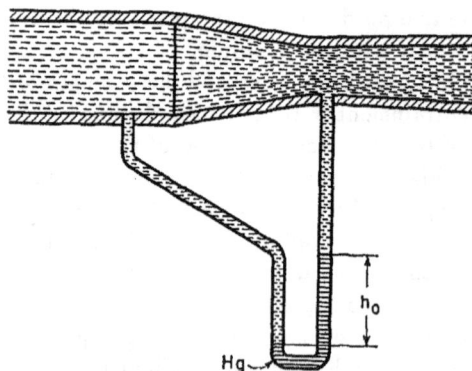

FIG. I-3-3 MERCURY DIFFERENTIAL GAGE (FLOW IS FROM LEFT TO RIGHT.)

Then

$$h\rho = (h_o\rho_o - h_o\rho_b)$$
$$= h_o(\rho_o - \rho_b) \qquad (\text{I-3-1})$$

or

$$h = h_o\left[\frac{(\rho_o - \rho_b)}{\rho}\right] \qquad (\text{I-3-2})$$

and if

$$\rho_b = \rho$$
$$h = h_o\left(\frac{\rho_o}{\rho} - 1\right) \qquad (\text{I-3-3})$$

In the development of the equations for differential-pressure meters, it is usual to assume that h is in feet. However, in actual practice the observed column height is generally expressed in inches. To take account of this difference in units, equations (I-3-2) and (I-3-3) may be modified to read

$$h = \frac{h_o}{12}\left(\frac{\rho_o - \rho_b}{\rho}\right) \qquad (\text{I-3-4})$$

and

$$h = \frac{h_o}{12}\left(\frac{\rho_o}{\rho} - 1\right) \qquad (\text{I-3-5})$$

I-3-15 For use in the final or working equations which would be used in computing the rate of flow through a differential pressure meter, it is desirable, and in the interest of exactness, to have the value of the pressure expressed in the equivalent inches of water at a definite temperature. This makes it possible to take account of the units of measurement with a fixed numeric and is the reason why, in the table of symbols, "h_w" is defined as the "effective differential pressure in inches of water at 68 F." To evaluate h and h_o in terms of h_w, let ρ_s = density of water at 68 F; then $h\rho = h_w\rho_s$ and

$$h = h_w\frac{\rho_s}{\rho} \qquad (\text{I-3-6a})$$

or

$$h = \frac{h_w\rho_s}{12\,\rho} \qquad (\text{I-3-6b})$$

Combining the first form of equation (I-3-6) with equation (I-3-2), or the second form with equation (I-3-4), gives

$$h_w = h_o\left(\frac{\rho_o - \rho_b}{\rho_s}\right) \qquad (\text{I-3-7})$$

I-3-16 Equations (I-3-2), (I-3-3) and (I-3-7) give the relations between the height of a column of the flowing fluid, the observed and equivalent column height of the same or a different liquid in a mano-

meter, and the equivalent value of h_w to be used in computing the flow rate with the use of a fixed numerical term. Of these, equation (I-3-7) is the most important, although the density of the flowing fluid, ρ, does not appear in it. Also, a broader definition than stated above should be given to ρ_b, namely, ρ_b is the density of the fluid in the low-column side of the gage tubes (manometer) regardless of whether or not this fluid transmits any pressure. For example, if in Fig. I-3-1 the columns were not joined at the top, and both open to the air, the right-hand would have a column of air, of height h_o, in excess of that in the left-hand column. To illustrate how much this may amount to, assume that $\rho_o = \rho_s = 62.316$ lb per cu ft, and for air $\rho_b = 0.076$ lb_m per cu ft. Hence,

$$h_w = h_o \frac{62.316 - 0.076}{62.316} \qquad (\text{I-3-8})$$
$$= 0.9988\,h_o$$

showing that the air in the low-liquid side of the manometer affects the value of h_w by over 0.10 per cent for each atmosphere of pressure.

Note: Sometimes the correction illustrated by equation (I-3-8) is referred to as gas column balancing effect or manometer factor [5].

Again, in Fig. I-3-3, the "observed" liquid is mercury, and at 68 F $\rho_o = 845.64$ lb_m per cu ft. If the other liquid is water, also at 68 F, so that $\rho_b = \rho_s = 62.316$, then

$$h_w = h_o \frac{845.64 - 62.316}{62.316} \qquad (\text{I-3-9})$$
$$= 12.570\,h_o$$

If in these two examples the manometers were of the single column and pot type, the relations developed in equations (I-3-8) and (I-3-9) would apply just the same.

Note: The correction for the air or other gas in one side of a manometer is omitted very frequently, even when under a pressure of several atmospheres. Also, especially in connection with water flow measurements, any difference between ρ, ρ_o, and ρ_s is generally neglected. This amounts to assuming that $h = h_o = h_w$.

I-3-17 Temperature of a Fluid. For the purposes of fluid measurement at velocities well below the sonic velocity, it is generally sufficient to use as the temperature of the fluid that which is indicated by a suitable thermometer placed in the fluid stream or held in a thermometer well which, in turn, extends well into the fluid stream. The quality of the thermometer used and the accuracy of its indications will naturally be governed by the nature of the work being done. In many instances, a thermometer with a scale graduated to 5-F intervals and which indicates temperatures correctly within ± 2 F will be all that is required. In other cases, as for example, laboratory tests and acceptance tests in power plants, a thermometer scale graduated to 0.5 F and correct within ± 0.2 F may be desirable. The type of thermometer used will depend on the location in which it is to be used. Most of the usual types can be had in scale and accuracy ranges suitable for fluid-measurement work.

When measuring the temperature of a flowing gas, there may be times when it will be necessary to take into consideration the effect of impact of the gas upon the thermometer or other sensing element. When the velocity is below 100 fps, this effect may be neglected; but, as the stream velocity is increased beyond 200 fps, this effect becomes of increasing importance. At these higher velocities the "stagnation" temperature, i.e., the temperature indicated by a thermometer if the impinging gas stream element came to zero velocity, becomes increasingly greater than the "static" temperature, i.e., the temperature that a perfect thermometer would indicate when moving with the stream and at the same velocity. The amount of this difference is given by $V^2/2g_cJc_p$, where V, g_c, and c_p have the meanings given in Table I-2-1 and J is the mechanical equivalent of heat (778.26 ft-lb$_f$/Btu). For a gas such as air at a velocity of 200 fps the difference would be about 4 F and about 83 F for a velocity of 1000 fps.

In general, a thermometer or other temperature-sensing element in a fluid stream "stops" only a part of the stream element impinging against it; hence, it "*indicates*" neither stagnation nor static temperature. That is, a part of the difference due to the velocity effect is "recovered" so that the actual relation between "static" and "indicated" temperature is: (static temperature) = (indicated temperature temperature) − $(r_c V^2/2g_cJc_p)$, and r_c is called the "recovery factor." All thermometric elements have a recovery factor; the numerical value is a function of both design and Mach number. The ordinary types of straight or tapered wells used extensively in fluid metering have recovery factors ranging between 0.5 and 0.9. Many of the temperature devices cited in the literature are not suitable to withstand the force of impingement and are used very seldom in flow measurement. Within practical limits stagnation temperatures may be obtained with special designs of thermometer wells (for which r_c = 1.0 approximately) that are independent of Mach number, yet are

rugged enough to be used in high-velocity gas streams [6, 7, 8, 9, 10].

I-3-18 Absolute Temperature. The temperature used in most equations for reducing observed gas volumes is not the observed thermometer reading in degrees Fahrenheit but the absolute temperature T. In engineering work, the absolute temperature is expressed in degrees Rankine, which is numerically very nearly 460 greater than the temperature observed on a Fahrenheit scale thermometer. The exact relation is given by

$$R = F + 459.69 \qquad (I-3-10)$$

Also, when using the Celsius scale for temperature measurements, the absolute temperature is referred to as degrees Kelvin. The relation between the Celsius and Kelvin scales is given by

$$K = C + 273.16 \qquad (I-3-11)$$

Note 1: The correct name of the temperature scale commonly called "centigrade" is "*Celsius*" in honor of its inventor. This name was adopted in 1948 at the 9th General Conference on Weights and Measures [11].

Note 2: At the 11th General Conference on Weights and Measures in 1960, the Kelvin thermodynamic scale was redefined by assigning to the triple point of water the value 273.16 K exactly. (The zero of the scale is understood to be the absolute zero of temperature.) Thus, on the Rankine scale the value of the triple point of water is 491.688 R. Also, the zero of the Celsius scale is defined as 0 C = 273.15 K exactly. Thus, the triple point of water on the Celsius scale is at 0.01 C. However, instead of using this exact relation most temperature conversion tables have used relations that give 273.15 K ≅ 0.00 C ≅ 32.00 F ≅ 491.67 R, whereas absolute temperatures in degrees K based on the above definition of the triple point of water are used in the computation and tabulation of the thermodynamic properties of fluids [12].

I-3-19 Density. Density is defined as the quantity (mass) of a substance per unit volume. In the system of units used in this report, densities are expressed in pounds-mass per cubic foot. In metric units densities usually are expressed in grams-mass per cubic centimeter (or per milliliter). Tables of densities of solids and fluids, such as given in the International Critical Tables and Smithsonian Physical Tables, are in these metric units.

I-3-20 At the General International Conference on Weights and Measures, October 1964, the definition of the liter as the volume occupied by one kilogram of water at its maximum density (at 4 C) was abolished. Thus, the liter becomes merely a special name of the cubic decimeter, that is, 1 liter = 1 cubic decimeter exactly (and *not* 1.000028 cubic decimeters as formerly). However, when using density data prepared prior to October 1964 in which density values are given in gram$_m$ per liter or per milliliter, the old relation *must* be used. The current relation *is* 1 liter = 1 cubic decimeter = 61.02374 cubic inches. The discarded relation *was* 1 liter = 61.0251 cubic inches [13, 14].

I-3-20 Specific Weight. Specific weight is defined as the weight, due to the gravitational pull of the Earth, of a substance per unit volume. In the engineering system of units, specific weight is expressed in pounds-force per cubic foot.

For a description of the methods of determining densities and specific weights and the corrections to be made to observed values, the reader is referred to any standard text on experimental physics for advanced schools.

I-3-21 Specific Gravity. In this country, the term "specific gravity" is used to mean the ratio of the density of one body to the density of a second or reference body, with the densities of both bodies having been determined under, or referred to, the same conditions of pressure, temperature, and gravity. A more appropriate name for this ratio is "relative density." (See note following Par. I-3-23.)

For liquids and solids, water is used as the reference body or substance, almost universally. The particular value of its density, or rather the temperature defining the density at standard gravity, is not the same for all substances and uses. For some work the maximum value is used, namely that at 4 C (39.2 F) and standard gravity. For other substances, particularly oils and industrial liquids, 60 F is the defining temperature. A third temperature which is used to a considerable extent, particularly in laboratory work, is 68 F (20 C). The density of distilled water at 60 F is 62.3707 lb$_m$ per cu ft and at 68 F, 62.3205 lb$_m$ per cu ft, both in vacuo and under standard gravity [15].

I-3-22 The determination of the specific gravity or relative density of a liquid can be made by comparing the weights of equal volumes of the liquid and water. If the quality of the work justifies it, these weighings may be corrected for the buoyancy of air as well as for temperature effects. For most commercial work, the specific gravities of liquids are obtained with hydrometers. The scales of hydrometers are graduated to read directly in specific gravities, in degrees Baumé or in degrees API (American Petroleum Institute) [16, 17]. The relations between specific gravity and degrees Baumé are given by the following formulas:

1. For liquids heavier than water

$$\text{Deg B} = 145 - \left(\frac{145}{\text{Sp gr } 60/60 \text{ F}} \right) \qquad \text{(I-3-12)}$$

2. For liquids lighter than water

$$\text{Deg B} = \left(\frac{140}{\text{Sp gr } 60/60 \text{ F}} \right) - 130 \qquad \text{(I-3-13)}$$

For use in the American petroleum industry, the following relation between degrees API and specific gravities is used:

$$\text{Deg API} = \left(\frac{141.5}{\text{Sp gr } 60/60 \text{ F}} \right) - 131.5 \qquad \text{(I-3-14)}$$

In the above equations the term "sp gr 60/60" means that the specific gravity value to be used or determined is that which prevails when the temperatures of the reference liquid (water) and of the oil or other liquid are *both* 60 F.

I-3-23 For gases air is used almost universally as the reference fluid. However, for use in this report, instead of a ratio of densities, the ideal specific gravity of a gas is defined as the ratio of the molecular wieght of the gas of interest to the molecular weight of air. The molecular weight of air is 28.9644 as given in Table I-3-1 [18].

The reason for not using the ratio of the densities is that the effects of pressure and temperature on the densities of gases vary from one gas or gas mixture to another. Thus, even though the densities may be determined at very nearly identical ambient conditions and the resulting values adjusted to a common basis of pressure and temperature, when the resulting ratio is used at a state differing from the common basis, an error may be incurred. The magnitude of this error is likely to increase as the state of use departs further and further from the common starting basis. On the other hand, so long as the composition of the gas used undergoes no change, the ratio of molecular weights will remain the same regardless of changes of pressure, temperature, and location.

The differences between the specific gravities of gases by the ratio of molecular weights and by the ratio of densities for conditions close to the normal ambient values of pressure and temperature will depend upon the closeness of the ambient temperature to the critical temperature of the gas. Assuming an ambient temperature of 68 F and atmospheric pressure, the difference in question with methane would be less than 0.1 per cent whereas with ethylene it might be over 0.5 per cent. In commercial activities involving gases, the specific gravity balance or a recording gravitometer will continue to be used for some time [19]. However, as the conditions of pressure and temperature at which gases are measured and used are extended to both higher and lower values, the importance of using the ratio of molecular weights will increase.

Note: The term *specific gravity* has been used by some writers in Europe to mean *specific weight* as defined in Par. I-3-20, and the term *relative density* is used in place of specific gravity. Such differences in terminology should cause no misunderstandings provided that writers are careful to define the meanings of the terms used.

I-3-24 Since specific gravity is a ratio, the numerical value is independent of the system of units of measurement being used. Also, the ratio of two specific gravities often can be used in place of the ratio of the corresponding densities of liquids or solid materials. For example, let the subscripts m, o, and w refer to mercury, oil, and water, respectively; then,

$$\frac{\rho_m - \rho_o}{\rho_o} = \frac{\rho_m/\rho_w - \rho_o/\rho_w}{\rho_o/\rho_w}$$

$$= \frac{G_m - G_o}{G_o} \qquad \text{(I-3-15)}$$

Obviously, a similar substitution cannot be made in cases where a gas and a liquid are involved, since their specific gravities are not referred to the same substance as unity.

I-3-25 The Relations between Pressures, Temperatures, Volumes, and Densities of an Ideal Gas. Unlike a liquid, any quantity of gas completely fills the space in which it is confined, and the pressure that it exerts upon the confining walls depends upon the temperature. Therefore, to specify a particular quantity of gas, three variables, namely, pressure, temperature, and volume, must be stated. The relation between these three variables may be represented by the equation

$$\frac{p_1 v_1}{T_1} = \frac{p_2 v_2}{T_2} = \text{a constant} \qquad \text{(I-3-16)}$$

or, since by definition $\rho = 1/v$,

$$\rho_2 = \rho_1 \frac{p_2 T_1}{p_1 T_2} \qquad \text{(I-3-17)}$$

Instead of using a pound mass as the unit quantity of gas, it frequently is more convenient to use that quantity which has a mass in pounds numerically equal to the molecular weight of the gas under consideration. This quantity is known as a "pound mol"

or simply a "mol" of gas. When the mol is used as the unit of quantity, equation (I-3-16) can be written in the form

$$p v_m = RT \qquad \text{(I-3-18)}$$

or

$$pQ = NRT \qquad \text{(I-3-19)}$$

in which N is the number of mols of gas in any volume Q, R is known as the gas constant, and v_m is the molar volume. The numerical value of R will depend upon the units in which p, T, and v_m are expressed. For example, when $p = 14.696$ psia and $T = 491.7$ R, the volume v_m of 1 mol of any gas has been found experimentally to be very nearly 359 cu ft. Using these values in equation (I-3-19) and solving for R gives $R = 10.73$. Also, if p (lb_f per sq ft abs) is used, $R = 144 \times 10.73 = 1545$. For an ideal gas these values are, respectively, 10.7313 and 1545.33 [12, 18, 20].

Equations (I-3-16), (I-3-17), (I-3-18), and (I-3-19) are "equations of state" of an ideal gas.

Note: Using the above value of R in equation (I-3-19) gives $v_m = 359.04$. The reciprocal of v_m, i.e., ρ_m, is the density in lb_m-mol/ft³. In order to have ρ from equation (I-3-18) in the conventional units of lb_m/ft³ it will be necessary to multiply ρ_m by the molecular weight of the gas or to divide the universal gas constant, R, by the molecular weight, thus giving the individual gas constant $R_g = R/\text{molecular weight}$.

The equation of state for an ideal gas, in one of the forms given above, or one derived therefrom, is used very extensively and will give rather accurate values of the density of a gas for any state for which the pressure is relatively low and the temperature is relatively high (i.e., relative to the critical values). This accuracy improves significantly as the temperature is increased above the critical temperature and the pressure is decreased below the critical pressure of the particular gas.

I-3-26 Equation of State of Real Gases: Compressibility. All real gases deviate by varying amounts from the equation of state for an ideal gas. In a general way, such deviations may be described as the result of molecular attractive and repulsive forces, and the molecules effectively occupy a finite volume. Hence, the internal pressure of a gas can differ from that felt on a surface (e.g., a wall or pressure gage), and the actual volume available for the free movement of the molecules is somewhat less than the volume of the container.

It has been found convenient to account for such deviations by introducing into the ideal-gas equation a new variable factor. The most common form of the thus modified equation is

$$\rho = \frac{p}{ZRT} \text{ or } pv = ZRT \qquad \text{(I-3-20)}$$

in which Z is called the "compressibility factor."

Equation (I-3-20) may be written in a form similar to (I-3-17), namely,

$$\rho = \rho_o \frac{P\,T_o\,Z_o}{p_o\,T\,Z} \qquad \text{(I-3-21)}$$

in which the subscript o refers to the reference state defined by $p_o = 14.696$ psia and $T_o = 491.69$ R. At this state for air $\rho_o = 0.0807\,223$ lb$_m$/ft^3 and $Z_o = 0.999\,41$ so that the equation for the density of dry air becomes, with p in psia,

$$\rho = 2.6991\,\frac{p}{T\,Z} \qquad \text{(I-3-22)}$$

Since specific volume, v, is volume per unit mass, that is, the reciprocal of density, then from equation (I-3-21)

$$v = v_o \frac{p_o\,T\,Z}{p\,T_o\,Z_o} \qquad \text{(I-3-23)}$$

For gases in general, equation (I-3-20) may be written

$$\rho = \frac{p}{R_g T\,Z} \qquad \text{(I-3-24)}$$

or using equation (I-3-22)

$$\rho = 2.6991\,G\,\frac{p}{T\,Z} \qquad \text{(I-3-25)}$$

where R_g is the gas constant for the particular gas, Z is the compressibility of that gas at the conditions p and T, and G is the ratio of molecular weights as defined above. However, there may be times when the composition of a gas mixture and therefore its molecular weight are not known. In such cases the value of G will have to be determined by other methods, e.g., a specific gravity balance. Such a determination should be made at conditions of p and T as close to 14.696 psia and 32 F as possible.

Note 1: It has been stated in many texts that close to the usual atmospheric conditions of p and T, i.e., 14.7 psia and 60 F, most gases follow the simple relation, $pv/T = $ constant, rather closely. However, by reference to compressibility tables it will be found that for a 30 F change of temperature, as from 32 F to 62 F, there may be about a 1 per cent change in the compressibility factor for some gases, e.g., CO_2, whereas for other gases, notably air, the resultant change may be less than 0.1 per cent. It is these differences in the compressibilities of gases that can cause errors in specific gravity determinations, especially those made with balance-type instruments, unless correction is made with the aid of compressibility tables.

Note 2: For many years in the fuel gas industry the departures from the ideal-gas equation have been taken in-

to account by using the equivalent of equation (I-3-20) in the form

$$v_o = \frac{p\,s\,v\,T_o}{p_o\,T} \qquad \text{(I-3-26)}$$

in which s is the "supercompressibility ratio." In routine computations of gas measurements by orifice meters, the industry uses the supercompressibility factor, F_{pv}, which is defined by

$$F_{pv} = \sqrt{s} = 1/\sqrt{Z} \qquad \text{(I-3-27)}$$

It is important to note that the factors F_{pv}, s, and Z all serve the same purpose, and each may always be evaluated in terms of the others by use of (I-3-27). Extensive tables of F_{pv} for methane and natural fuel gases have been prepared by the American Gas Association [21].

The preceding discussion and equations for the densities of gases apply to dry water-vapor-free gas. Equations for taking account of water vapor are given later.

Values of the molecular weights, densities, and other constants of some of the common commercial gases are given in Table I-3-1.

I-3-27 Theorem of Corresponding States. The following terms will be used in this discussion:

Critical Temperature, T_c, is that temperature of a gas above which the gas cannot be liquefied by the application of pressure alone, regardless of the amount of pressure.

Critical Pressure, p_c, is the saturation pressure of the gas at the critical temperature.

Critical Volume, v_c, is the volume of a unit mass of the gas at the critical temperature and pressure, i.e., the specific volume at T_c and p_c. In the units used here, v_c is in ft^3/lb$_m$.

Critical Density, ρ_c, is the density of the gas at T_c and p_c in lb$_m$/ft^3.

Reduced Temperature, T_r, *Reduced Pressure*, p_r, and *Reduced Volume*, v_r, are the ratios of the *actual* temperature, pressure, and specific volume to the critical temperature, critical pressure, and critical volume, respectively. That is,

$$T_r = T/T_c \qquad p_r = p/p_c \qquad v_r = v/v_c$$
$$(\text{or } \rho_r = \rho/\rho_c) \qquad \text{(I-3-28)}$$

All equations of state describe the relationship of pressure, temperature, density (or specific volume), and composition (molecular weight) of a gas; and, for any given composition, such a relationship describes a geometrical surface in the space coordinates $(p, T, v \text{ (or } \rho))$. Possibly the most frequently used equation of state is that of Van der Waals, which may be written

$$(p + a\rho^2)\,(1 - b\rho) = \rho RT \qquad \text{(I-3-29a)}$$

or

$$p = \frac{R\,T}{v - b} - \frac{a}{v^2} \qquad \text{(I-3-29b)}$$

This is an approximation of the more extensive virial equation of state, namely,

$$\frac{p \, v}{R \, T} = 1 + \frac{B_2(T)}{v} + \frac{B_3(T)}{v^2} + - - - - = Z \quad \text{(I-3-30)}$$

where the temperature functions $B_n(T)$ are called the virial coefficients and are listed in various handbooks.

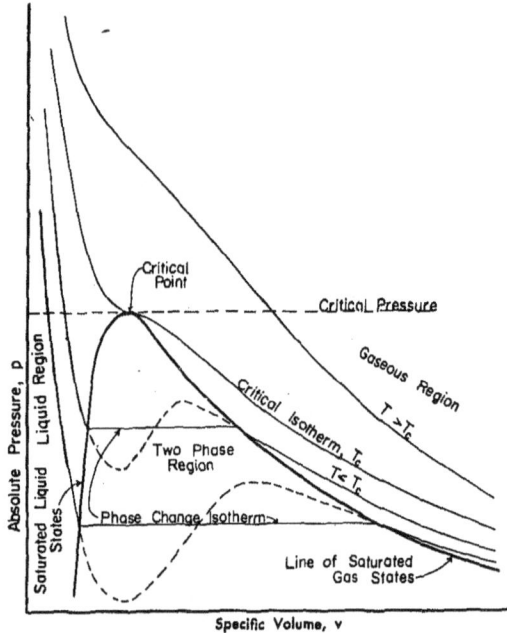

FIG. I-3-4 END OR SIDE VIEW OF THE p-T-v SURFACE DESCRIBING LIQUID, GAS AND LIQUID-GAS EQUILIBRIUM STATES

The Van der Waals equation is an empirical cubic equation in specific volume, and the factors a and b are constants for any given fluid. It represents both the gaseous and liquid phases and the common critical point of a fluid substance. A contour representation of the surface described by the equation is shown in Fig. I-3-4. The heavy line identifies all saturation states, liquid to the left and gaseous to the right of the line. The common point is the *critical point*. At this point the isothermal lines have a horizontal tangent and a point of inflection; therefore,

$$\frac{\partial p}{\partial v} = 0 \quad \text{and} \quad \frac{\partial^2 p}{\partial v^2} = 0 \quad \text{(I-3-31)}$$

Equating the first and second derivatives of equation (I-3-29) to zero and solving for the critical point values of p, T, and v gives

$$p_c = a/27 \, b^2 \quad T_c = 8a/27 \, b R \quad v_c = 3b \quad \text{(I-3-32)}$$

I-3-28 Substituting these in equation (I-3-29) and using the reduced coordinates as defined by equation I-3-28) gives

$$\left(p_r + \frac{3}{v_r^2}\right)\left(3 - \frac{1}{v_r}\right) = 8\frac{T_r}{v_r} \quad \text{(I-3-33)}$$

With equation (I-3-33) it can be shown that practically all fluids have the same equation of state when the pressure, temperature, and density or specific volume are expressed in the reduced coordinates. Values of two of the reduced coordinates fixes the value of the third. This is the principle of cooresponding states. It has numerous practical uses, one of the most common being the evaluation of the compressibility of actual gases. Thus, for a nearly ideal gas to which the Van der Waals equation (I-3-29) applies, the value of Z_c is 3/8. (For most real gases, the value of Z_c is nearer 3/10 than 3/8.)

I-3-29 For a mixture of gases not combined chemically, the critical pressure and temperature may be calculated on the assumption that each gas contributes to the mixture critical pressure and temperature in proportion to its fraction by *volume* (not weight). Thus,

$$p_c \text{(mix)} = p_{c_1} V_1 + p_{c_2} V_2 + p_{c_3} V_3 + - - -$$
$$T_c \text{(mix)} = T_{c_1} V_1 + T_{c_2} V_2 + T_{c_3} V_3 + - - - \quad \text{(I-3-34)}$$

where p_c (mix) and T_c (mix) are the critical pressure and temperature of the mixture and subscripts 1, 2, and 3 indicate the corresponding values of the gases 1, 2, and 3. V_1, V_2, and V_3 are volumes (not specific volumes) of each gas per cu ft of the mixture, so that

$$V_1 + V_2 + V_3 = 1 \text{ ft}^3 \text{ of mixture} \quad \text{(I-3-35)}$$

It should be noted that this method is generally applicable only for conditions somewhat well removed from the critical values of the constituents.

I-3-30 Humidity. The term "humidity" is used largely in connection with gases to indicate the presence of perceptible moisture or water vapor.

Absolute Humidity. This the weight of water vapor per cubic foot of the moist gas (i.e., mixture of gas and water vapor) at the existing conditions of pressure and temperature.

Relative Humidity. It is the ratio, expressed as a per cent, of the actual quantity of water vapor present to the maximum quantity that would be present at complete saturation at the given temperature. Alternately, it is the ratio of the actual partial pressure of the water vapor to the maximum partial pressure that would be exerted at complete saturation at the given temperature. When, at a given temperature, no more water vapor can be added to a given space without causing condensation, the point of water vapor saturation has been reached, and the relative humidity is 100 per cent.

Thus, relative humidity is equal to $(p_v/p_s)_T$, where p_s is the absolute saturation pressure of steam at temperature T and p_v is the partial absolute pressure of the water vapor in the given space or atmosphere at T. $p_v \lessgtr p_s$.

Note: The presence of water vapor in a given space is independent of the presence or absence of other gases in the same space, assuming, of course, no chemical action between the water vapor and the other gases occurs. The quantity of saturated water vapor that can exist in any given space depends entirely upon the temperature. This saturated water vapor will exert a certain pressure, termed the "pressure of saturated water vapor" or steam, which varies with temperature. If in a given space the water vapor is not saturated, the pressure which it will exert will be proportionally less. For example, at 70 F, the pressure of saturated steam is 0.74 in. Hg. At the same temperature and 50 per cent relative humidity, the vapor pressure would be one-half of that at saturation, or 0.37 in. Hg. Similarly, the saturation pressure of the vapor of other fluids depends on the temperature.

I-3-31 Psychrometry—Determination of the Pressure of the Water Vapor in a Moist Gas.
Several methods can be used to determine the quantity of water vapor in a moist gas or the partial pressure of this vapor. The first is to collect, with a suitable

Table I-3-1 Physical Data on Some Common Commercial Gases

Gas Name	Formula	Molecular Weight (basic C^{12})	Density at 32 F and 14.696 psia (lb_m/ft^3)	Specific Gravity (see Par. I-3-23)	Ratio Specific Heats [see Note (e)]	Boiling Point at 14.696 psia (°R)	Critical Temperature (°R)	Critical Pressure (psia)	Critical Volume (ft^3/lb_m)
			ρ_o	G	γ	T_b	T_c	P_c	v_c
Air		28.9644(a)	0.0807223(b)	1.00000	1.41	142.0	238.4	547.	0.0517
Argon	Ar	39.948		1.3792	1.67	157.4	272.08	705.4	0.0301
Acetylene	C_2H_2	26.0382	0.06860	0.89897	1.24	340.7	557.1	905.	0.0661
Ammonia	NH_3	17.0306	0.0452	0.58798	1.31	431.6	731.1	1657.	0.0684
Benzene	C_6H_6	78.11	0.0548	2.6967		635.9	1010.9	700.9	0.0527
Butane-n	C_4H_{10}	58.1243	0.15805	2.0068	1.09	490.8	765.3	550.7	0.0704
Butane-iso	C_4H_{10}	58.1243	0.15788	2.0068	1.10	470.6	734.6	529.1	0.0725
Carbon dioxide	CO_2	44.00995	0.12342(b)	1.5194	1.30	350.4	547.7	1073.	0.0348
Carbon monoxide	CO	28.01055	0.078065(b)	0.96707	1.40	143.0	241.7	510.	0.0515
Ethane	C_2H_6	30.0701	0.07987	1.0382	1.19	332.2	549.8	708.3	0.0787
Ethylene	C_2H_4	28.0542	0.07391	0.96858	1.24	305.0	509.5	742.1	0.0705
Ethyl alcohol	C_2H_5OH	46.07	49.2759 (liq)	1.5905	1.13	632.75	929.3	927.3	0.0581
Helium	He	4.0026	0.011143(c)	0.13819	1.66	7.669	672.41	1306.4	
Hydrogen	H_2	2.0159	0.0056114(b)	0.069599	1.41	36.8	59.9	188.	0.5168
Methyl alcohol	CH_3OH	32.04	49.6942 (liq)	1.1061	1.203	608.06	923.7	1156.6	0.0588
Hydrogen sulphide	H_2S	34.0799	0.09050	1.1766	1.32	383.2	672.4	1306.	
Methane	CH_4	16.0430	0.042355	0.55389	1.31	201.0	343.2	673.1	0.0993
n-Octane	C_8H_{18}	114.23	43.9257 (liq)	3.9438		715.968	1024.5	361.5	0.0684
Nitrogen	N_2	28.0134	0.078064(b)	0.96717	1.40	139.3	226.9	492.	0.0515
Oxygen	O_2	31.9988	0.0892102(b)	1.1047	1.40	162.3	277.9	730.	0.0373
Propane	C_3H_8	44.0972	0.11806	1.5225	1.33	416.0	666.	617.4	0.0730
Sulphur dioxide	SO_2	64.07	0.1826	2.212		473.7	774.6	1141.9	0.0308
Water (steam, dry)	H_2O	18.0153		0.62198	1.30	671.7	1165.1(d)	3208.2(d)	0.05078(d)

(a) "U.S. Standard Atmosphere," U.S. Government Printing Office, 1962, p. 9. [18]
(b) National Bureau of Standards Cir. 564, Tables of Thermodynamic Properties of Gases, 1960. [20]
    (Interpolation within these tables should be by a method of second differences.)
(c) *U.S. Bureau of Mines Journal of Chemistry and Engineering Data*, vol. 5, Jan. 1960, p. 51.
(d) "1967 ASME Steam Tables." [15]
(e) Values for an ideal gas and a reversible system.

drier, the water vapor in a known or measured volume of the mixture, the temperature of which is also taken. From the weight of water collected, the pounds of water vapor per cubic foot of the mixture, that is, the absolute humidity, can be calculated. From steam tables, the weight per cubic foot and the partial pressure of saturated steam at the observed temperature of the mixture can be found, and, thus, the relative humidity and partial pressure of the water vapor in the mixture calculated. If carefully done, this method is probably the most accurate.

I-3-32 The Wet- and Dry-Bulb Psychrometer. Another method, which is usually more convenient, uses the wet- and dry-bulb psychrometer. This consists of two thermometers that are separated by 1 or 2 in. with the bulb of one covered with thin muslin or some other suitable material which can be wetted. To make an observation, the bulbs of both thermometers are held in the gas stream, the velocity of which should be at least 15 fps, and their readings noted after the wet-bulb thermometer ceases to fall. For determining atmospheric humidity, the thermometers are often attached to a sling that can be whirled gently to provide the desired ventilation.

The theory of calculating the partial vapor pressure from the wet- and dry-bulb thermometer readings will not be developed and discussed here [22, 23]. It will suffice to give an approximate working equation, since, even for experimental work, it will seldom be necessary to determine the value of p_v closer then about ±0.05 psi (i.e., about ±0.1 in. of mercury). Thus, the error in the density of the moist gas mixture from this factor should not exceed and will probably be less than about ±0.1 per cent. An equation that will fulfill this requirement, especially within the temperature range of 32 to 104 F, is

$$p_v = p_s - [0.000\ 095\ c_{mp}\ (T_d - T_w)p] \qquad \text{(I-3-36)}$$

in which

p_v = Actual partial pressure of the water vapor at the conditions defined by T_d and T_w psia

p_s = Saturation pressure of steam (i.e., water vapor) at T_w psia

p = Total pressure of the gas-water vapor mixture   psia

T_d = Dry-bulb temperature, i.e., the gas stream temperature   F or R

T_w = Wet-bulb temperature   F or R

c_{mp} = Mean molecular heat per deg F at constant pressure of the gas and T_d and T_w are within the range

suggested above, as given by Table I-3-2, or computed on the basis of the volume composition of the gas

Table I-3-2 Mean Molecular Heats of Some Gases

Gas	Air, H_2, N_2, CO	CO_2	CH_4	C_2H_2	C_2H_4	C_2H_6
c_{mp}	3.87	4.95	4.72	5.84	5.67	6.89

I-3-33 Modern moisture monitoring equipment measures the specific humidity of a gas stream directly by measuring the amount of electric current necessary to keep a viscous film of a hygroscopic salt at equilibrium with a metered gas flow. Examples of such salts are phosphorus pent oxide and lithium chloride. With recorders of this type, the measured specific humidity must be expressed in dimensionless units, and the implicit value of p_v in equation (I-3-37) below is solved from S, G, and p data.

I-3-34 Density of Moist Gas. The following special symbols will be used, in addition to G, the specific gravity:

p_d = Partial absolute pressure of the *dry* gas in the moist gas mixture   psia

p_v = Partial absolute pressure of the water vapor in the moist gas mixture   psia

p = Total absolute pressure of the moist gas = $p_d + p_v$ psia

ρ_d = Density of the dry gas at p_d and T lb_m/ft^3

ρ_v = Density of the water vapor at p_v and T lb_m/ft^3

ρ_m = Density of the moist gas at p and T lb_m/ft^3

Assuming the temperature remains constant,

$$\frac{\rho_m}{\rho_d} = \frac{\rho_d + \rho_v}{\rho_d} = 1 + \frac{18.0153}{28.9644G}\frac{p_v}{p_d}$$

$$= 1 + \frac{0.622}{G}\frac{p_v}{p_d} = 1 + \frac{0.622}{G}\frac{p_v}{p-p_v} = 1 + S \quad \text{(I-3-37)}$$

where
S = Specific humidity

$$= \frac{\rho_v}{\rho_d} = \frac{0.622}{G}\frac{p_v}{p-p_v} \qquad \text{(I-3-38)}$$

By equation (I-3-25)

$$\rho_d = 2.6991\ \frac{p - p_v}{T\ Z}\ G \qquad \text{(I-3-39)}$$

and

$$\rho_m = 2.6991\ (1 + S)\ \frac{p - p_v}{T\ Z}\ G \qquad \text{(I-3-40)}$$

I-3-35 Converting the volume of a moist gas mixture, that is, a mixture of gas and water vapor, from one state to another is done, ordinarily, by applying the general pressure-temperature-volume relation to the mixture, taking account of the partial pressures of the gas and the water vapor. Using the relation expressed by equation (I-3-23)

$$Q_2 = Q_1 \left[\frac{p_1 - (p_v)_1}{p_2 - (p_v)_2} \right] \frac{T_2}{T_1} \frac{Z_2}{Z_1} \qquad \text{(I-3-41)}$$

where Q is the total volume of the mixture and Z is the compressibility factor of the primary gas. The values of p_v are obtained from steam tables or psychrometric determinations.

The application of equation (I-3-41) is based on the assumptions (1) that the general relation for gases as expressed by equation (I-3-20) holds for the gas-water vapor mixture and (2) that both the gas and water vapor have the same coefficient of thermal expansion. Furthermore, within the range of usual commercial application, the pressure and temperatures may have any values whatever.

As the total pressure of the mixture is changed, the partial pressures of the constituents will change proportionally. To illustrate, let it be assumed that the temperature T remains constant, and let p_s be the saturation pressure of steam at T. Then, as the pressure p is increased, the partial pressure, p_v, increases proportionally; and a condition may be reached where $p\,(p_s/p_v) = p_c$, the total absolute pressure of the mixture at which the water vapor is just saturated, that is, $p_v = p_s$. Above this pressure, p_c, condensation of the water vapor will take place. Any water vapor which is condensed to water particles is no longer considered a part of the mixture and, accordingly, does not appear in further computations of volume or density. Conversely, if at some elevated pressure p, $p_v = p_s$, then, as the total pressure is decreased, p_v will decrease proportionally. In this case, the mixture becomes relatively "drier" as the pressure decreases.

In general, the temperature will change along with changes of pressure, and both p_s and p_v are temperature dependent.

At times it may be required to compute the portion of *dry* gas in a gas-water vapor mixture. Using the same symbols as above and assuming that there is no change of temperature,

$$Q_d = Q \frac{p - p_v}{p} = Q \frac{p_d}{p} \qquad \text{(I-3-42)}$$

I-3-36 Viscosity. The coefficient of viscosity measures the temporary rigidity of a fluid. The frictional resistance which a fluid offers to a change of shape, that is, to shearing, is directly proportional to the rapidity with which the change is accomplished, namely, to the shear per unit time. This shearing can be regarded as the relative sliding of parallel planes without change of their mutual distance, and the tangential force per unit area of one of the planes (i.e., the intensity of the shearing stress) is the proper measure of the frictional resistance of the fluid at the actual rate of shearing. Thus, the quotient of

$$\frac{\text{tangential force per unit area}}{\text{shear per unit time}}$$

is the measure of that quality of the fluid, by virtue of which it resists distortion. It is called the "coefficient of viscosity" or simply the "viscosity" of the fluid. The omission of "per unit time" from the denominator would convert the definition into that of simple rigidity.

The classical method of determining the viscosity of a fluid is by measuring both the rate of flow of the fluid through a capillary tube and the pressure drop causing the flow. The length of the tube should be very great in comparison to its diameter. The coefficient of viscosity determined by this method, or by some equivalent method (e.g., the oscillating disk), is called the "absolute viscosity" [24, 25]. *Poise* is the name of the unit of viscosity in the metric system, as thus determined, and is expressed in dyne-sec per sq cm (or $\text{gram}_m/\text{cm sec}$). The centipoise = poise/100 is used extensively in reporting and using viscosity values. In the ft-lb_m-sec system of units, viscosity values are given in lb_m per ft-sec ($= 32.174[\text{lb}_f\text{-sec}/\text{ft}^2]$), for which there is no name.

$$\mu_{\text{poise}} \times 0.067197 = \mu_{\text{lb}_m}/\text{ft sec} \qquad \text{(I-3-43)}$$

It has become a common practice in this country to use viscosity values in poise or centipoise and to incorporate the appropriate conversion factor along with other factors in whatever computation is involved. In presenting equations in which viscosity is a factor, the appropriate numerics are given for use with both the lb_m/ft-sec unit and the poise (or centipoise).

I-3-37 Kinematic Viscosity. This is known also as the kinetic viscosity and is the absolute viscosity divided by the density at the same temperature. That is,

$$\nu = \mu/\rho \qquad \text{(I-3-44)}$$

In the metric system kinematic viscosities are given in cm² per sec, called the *stoke*. There is no name for kinematic viscosities given in ft² per sec.

$$\nu_{stoke} \times 0.001076 = \nu_{ft^2/sec} \qquad (I\text{-}3\text{-}45)$$

Kinematic viscosity is the quantity derived with a viscometer, such as the Saybolt universal. The equations used with this viscometer, which give the approximate value of ν in stokes, are, when $32 < t < 100$,

$$\nu = 0.00226\ t - \frac{1.95}{t} \qquad (I\text{-}3\text{-}46)$$

and when $100 < t$

$$\nu = 0.00220\ t - \frac{1.35}{t} \qquad (I\text{-}3\text{-}47)$$

t = outflow time in sec [27].

I-3-38 Ratio of Specific Heats. In developing the theoretical equations for the flow of compressible fluids through differential pressure meters, it is customary to use the relation

$$p\upsilon^\gamma = constant \qquad (I\text{-}3\text{-}48a)$$

or

$$\frac{\upsilon_1}{\upsilon_2} = \left(\frac{p_2}{p_1}\right)^\gamma = r^\gamma \qquad (I\text{-}3\text{-}48b)$$

in which $\gamma = c_p/c_\upsilon$, the ratio of the specific heat at constant pressure to that at constant volume of a gas. It is important to note that the relation given in equation (I-3-48) and the definition of γ are rigorously true only for an ideal gas and a reversible system. No real gas follows these relations exactly; and, although a few do not depart from them but slightly, others depart noticeably therefrom. However, for the purposes of metering compressible fluids, it is more convenient to use the above relations as valid and to incorporate the effects of departures therefrom in other experimentally determined factors. Values of γ are given in Table I-3-1.

Note 1: The definition of γ with equation (I-3-48) and the values in Table I-3-1 are special values of the exponent n in the general relation $p\upsilon^n = constant$, which fits all polytropic changes. The values of γ and n are related by the equation

$$\frac{c_p/c_\upsilon}{n} = \frac{\gamma}{n} = 1 - \left(\frac{\partial Z/Z}{\partial p/p}\right)_T \qquad (I\text{-}3\text{-}49)$$

As the value of the partial derivative term approaches zero, the values of γ and n approach identity.

Note 2: The specific heat of solids and liquids is always understood to be that at constant pressure, i.e., c_p, unless specifically stated otherwise. For a complete discussion of specific heat and the values of γ and n, reference should be made to a current textbook on thermodynamics.

I-3-39 Similitude and Dimensional Analysis. In many problems of experimental and applied physics, it is helpful to compare the behavior of two machines of the same kind but of very different size or to predict the behavior of a proposed machine from the behavior of a model. In order that such comparisons may be made, the two machines or the model and proposed original must necessarily be geometrically similar in all respects. When this comparison involves the flow of a fluid or of different fluids, it is necessary, although not so evident, that the forces (consequently the pressures), due to the inertias and viscosities of the fluids, be in a constant ratio, if the flows are to be similar in direction and distribution at corresponding points and sections. That is, in the case of fluid flows, "kinetic" similarity must exist. This comparison of the behavior of machines and fluids on the basis of geometric and kinetic similarity is generally referred to as "dimensional analysis."

In applying dimensional analysis to any problem, such as fluid measurement, *all* the quantities on which the result may in any way depend must be specified at the start. Later, one or more of the quantities that have been found to have no bearing on the result can be dropped. No prospect exists for determining a numerical coefficient from the principle of similarity alone. Such coefficients must be obtained, if found at all, by further calculation or by experiment. Thus, although dimensional analysis is a useful tool in analyzing and comparing the results of physical experiments, it is not a solution of a problem.

This subject will be referred to again in the discussion of coefficients of differential pressure meters. However, in this report, the methods of developing dimensionless criteria will not be discussed in detail [28, 29, 30].

I-3-40 In the absolute system of units, the primary units are *length, mass,* and *time,* represented in dimensional formulas by L, M, and T. In the gravitational system the primary units are *force, length,* and *time,* represented by F, L, and T. The dimensional formulas for quantities of interest in fluid metering are listed in Table I-3-3 [31].

Note: The use of F, L, M, and T in dimensional formulas should cause no confusion with their use as defined in Table I-2-1. They are never used with more than one meaning in an equation, and the meaning in each use is always evident from the context.

I-3-41 The Reynolds Number. In studying the flow of a fluid along a channel, especially a closed one, it is helpful to use a dimensionless criterion to aid in comparing results of observations made under different conditions. The quantities upon which this flow will depend are: the size of the channel which may be represented by any transverse dimension, D (with circular channels, D is, of course, the diameter); the rate of flow which may be represented by V, the mean speed; the density of the fluid, ρ; and the viscosity, μ. The combination of these four quantities that will be dimensionless is $DV\rho/\mu$. This product is known as Reynolds criterion, or Reynolds number. Since this product is dimensionless, the numerical value will be the same for any given case, so long as the separate factors are all expressed in a

Table I-3-3   Dimensional Formulas

Quantity	Symbol from Table I-2-1	Absolute System	Gravitational System
Acceleration, linear		$L\,T^{-2}$	$L\,T^{-2}$
Acceleration, angular		T^{-2}	T^{-2}
Density	ρ	$M\,L^{-3}$	$F\,L^{-4}T^2$
Energy		$M\,L^2T^{-2}$	$F\,L$
Frequency	f	T^{-1}	T^{-1}
Force		$M\,L\,T^{-2}$	F
Mass		M	$F\,L^{-1}T^2$
Momentum, linear		$M\,L\,T^{-1}$	$F\,T$
Momentum, angular		$M\,L^2T^{-1}$	$F\,L\,T$
Pressure	p	$M\,L^{-1}T^{-2}$	$F\,L^{-2}$
Specific volume	v	$M^{-1}L^3$	$F^{-1}L^4T^{-2}$
Velocity	V	$L\,T^{-1}$	$L\,T^{-1}$
Viscosity	μ	$M\,L^{-1}T^{-1}$	$F\,L^{-2}T$
Viscosity, kinematic	ν	L^2T^{-1}	L^2T^{-1}
Work		$M\,L^2T^{-2}$	$F\,L$

consistent system of units, regardless of what system of units is used.

As listed in Table I-2-1, R_D represents the Reynolds number when the pipe diameter, D, is the length dimension used in the ratio, and R_d when the diameter, d, of a throat section or an orifice is used. For flows in an open channel and also in a pipe when the flow of a liquid does not fill the cross section considered, the length dimension to use is the hydraulic radius. The hydraulic radius is the quotient of the area of that portion of the channel cross section that is filled with the liquid, divided by the wetted perimeter.

Although the combination $DV\rho/\mu$ is the classical expression for the Reynolds number, there are several other equivalent combinations. First, the ratio ρ/μ may be replaced by $1/\nu$, thus giving DV/ν.

Again, the volume rate of flow is $q = \pi(D^2/4)V$, thus a second alternate combination is $4\,q\rho/\pi D\mu$. Also $q\rho = m$, the mass rate of flow, so that a third alternate combination is $4m/\pi D\mu$. It should be remembered that when D or d is given in inches it is necessary to divide them by 12. Thus, for example, the last combination becomes $48m/\pi D\mu$. Also, $R_D = \beta\,R_d$. If μ is in poise, i.e., μ_p, this last combination for R_D becomes $227.37\,(m/D\mu_p)$.

I-3-42 Mach Number. Another dimensionless ratio of interest in fluid metering is the Mach number. It is the ratio of the velocity of the fluid to the velocity at which sound will travel in the fluid at its flowing density. Thus, if ρ is the density of the flowing fluid at the section of interest, p is its absolute pressure in pounds$_f$ per square foot, and γ the ratio of specific heats, the velocity of sound in the fluid is

$$V_s = \sqrt{\frac{p}{\rho}\gamma} \qquad (I\text{-}3\text{-}50)$$

Then, since the velocity of the fluid is V, the Mach number is

$$M = V/V_s = V\sqrt{\frac{\rho}{p\,\gamma}} \qquad (I\text{-}3\text{-}51)$$

I-3-43 The Acoustic Ratio. When compressible fluids (gases) are discharged through a throttling device, the mass rate of discharge has been observed to increase less rapidly than that of an incompressible fluid (liquid) for a like increase of the differential pressure (drop) across the throttling device. For any one gas, this difference between the gas and liquid rates of discharge has been found to be a function of the pressure ratio $r = p_2/p_1$ or the differential ratio $x = (p_1 - p_2)/p_1 = 1 - r$. When the mass rates of flow of different gases are compared, the differential-pressure ratio x alone is no longer sufficient. Since the flow of a compressible fluid through a differential pressure meter is usually assumed to be isentropic, for which $pv^\gamma = $ a constant, the specific heat ratio, γ, must be included, also. Such a criterion, involving both x and γ, is x/γ, which has been called the "acoustic ratio" [32].

A more precise criterion would be $r^{1/\gamma}$, but the ratio x/γ is simpler and has been found to be suitable for all ordinary cases, even when a high degree of accuracy is involved.

The acoustic ratio is closely related to the Mach number. The ratio of the stagnation pressure, p_t, to the free-stream static pressure is a function of the Mach number, as expressed by the equation

$$\left(\frac{p_t}{p}\right)^{\frac{\gamma - 1}{\gamma}} = 1 + \frac{\gamma - 1}{2} M^2 \qquad \text{(I-3-52)}$$

This can be written

$$M^2 = \frac{2}{\gamma - 1}\left[\left(\frac{p}{p_t}\right)^{\frac{1-\gamma}{\gamma}} - 1\right]$$

$$= \frac{2}{\gamma - 1}\left[\left(1 - \frac{\Delta p}{p_t}\right)^{\frac{1-\gamma}{\gamma}} - 1\right] \qquad \text{(I-3-53)}$$

When $p/p_t < 1$ and by using a binomial expansion

$$M^2 = \frac{2}{\gamma - 1}\left(1 + \frac{\gamma - 1}{\gamma}\frac{\Delta p}{p_t} - 1\right)$$

$$= \frac{2}{\gamma}\frac{\Delta p}{p_t} \qquad \text{(I-3-54)}$$

$$= \frac{2x}{\gamma}$$

hence,

$$\frac{x}{\gamma} = \frac{M^2}{2} \qquad \text{(I-3-55)}$$

The acoustic ratio is a useful parameter for plotting theoretical and experimental values of the expansion factor, Y, is easier to compute than exponential values of the pressure ratio, and introduces negligible error when it is used as an index of compressible flow effects.

I-3-44 Kinds of Flow. The character of the flow of a fluid along a channel can be classed as *laminar* or *turbulent*. Consider the flow of a fluid along a straight, smooth-walled channel of uniform and regular cross section. If the paths of all fluid particles are exactly parallel to the axis of the channel, the flow would be called "laminar" or "viscous." That the velocity of all particles shall be equal is not essential, but no particle can have any transverse component of motion. Experiments have indicated that *laminar* flow occurs at very low Reynolds numbers, which may be obtained by low average velocities, in channels of very small dimensions, or with fluids of high viscosities.

When the paths described by a few or all of the particles are not parallel to the axis of the channel, the flow is called "turbulent." That is, in their movement along the channel, some or all of the particles have a transverse as well as a longitudinal component of motion. The degree of this transverse movement may vary from a slight wavering to violent swirls and eddies, often resulting in the formation of continuous swirls so that the fluid particles describe helical paths. This latter condition of turbulent flow is frequently referred to as swirling or helical flow to distinguish it from the more simple forms of turbulent flow. Turbulent flow is by far the most common and, hence, the most important.

I-3-45 Steady Flow and Pulsating Flow. If all of the variables associated with fluid flow, such as pressure, velocity, density, and mass or volume rate do not change or change only slowly with respect to time, the flow would be characterized as *steady*. Such a truly steady flow may be attained or nearly attained if the flow is in the laminar or viscous regime. However, under the usual turbulent flow conditions, it is doubtful if true steady flow is ever attained, even when the time interval of observation is short.

If any or all of the flow variables, such as pressure, velocity, or density, change cyclically with time at one location, the flow is called *pulsating*. Noncyclical variations or transients are not included, and neither are those cyclical variations which take place so slowly that quasi-equilibrium conditions exist at all times in the flow system. Excluded also are those random fluctuations which exist in almost every actual flow system due to turbulence and which usually are so small that they do not seriously affect the accuracy of flow measurement.

What is meant by *slowly*, as used above, depends primarily on the frequency response of the flow-measuring system. Flow oscillations slower than about one-fifth of the undamped natural frequency of the measuring system are not classed as pulsating, as that term is used here.

I-3-46 In most practical cases where pulsating flow exists, it is caused by reciprocating or rotating machinery, such as compressors, pumps, or turbines. Also, pulsations may be produced by sirens, chattering check valves, liquid which surges back and forth in a low portion of a gas or vapor line, or anything else which produces cyclic, intermittent flows or pressures in a measuring system. Pulsations may occur in a self-resonating flow system without any moving parts [33].

Pulsating flow is usually detected as differential or total pressure fluctuations, but it may exist without visible oscillations of the flow recorder pen or pressure gage. Under pulsating conditions, flow-meter indications are likely to be in error. Although a pulsating flow can affect the indications of meters of the volumetric class, this section on pulsation applies particularly to meters of the differential-pressure group such as the orifice, flow nozzle, and

Venturi tube. With few exceptions, the error is always such as to indicate a flow greater than actual. A completely satisfactory method to predict the size of the error is not yet available. The best that can be offered to calculate the extent of the error is the Hodgson number, (N_H), which is a measure of pulsation attenuation between the pulsation source and the flow meter. The Hodgson number is defined by

$$N_H = \frac{Qf}{q}\frac{\Delta p}{p} \qquad (I\text{-}3\text{-}56)$$

where

Q = Volume of the flow system (piping) between the pulsation source and the meter, ft^3

f = Pulsation frequency, cycles per sec

q = Average volume rate of flow, cfs

Δp = Average pressure drop in the system from the pulsation source to the meter, psi

p = Average absolute pressure at the meter, psia

The units should be consistent so that N_H is dimensionless. If N_H is larger than 2.0, the pulsation error is almost always less than 1 per cent [29].

I-3-47 For corrective action for the effects of pulsating flows, a distinction must be made between incompressible fluids (liquids) and compressible fluids (gases or vapors). For liquids, a correction for pulsating flow error may be possible. For gases or vapors it is necessary to reduce the pulsations at or near the source.

The average flow rate in pulsating liquid flow is computed with good accuracy by the steady-flow equation using a *time average of the square root* of the instantaneous differential pressure measured at the meter. The square root of the average differential pressure, a value which most secondary elements (recording gages) give, will be in error. This error is commonly called the *square root error.*

If the frequency and amplitude of a cyclic or pulsating flow of a liquid are known, steady-flow equations can be used to compute the actual flow rate. This, along with other conclusions given below, were determined from a research program on pulsating flow conducted by the Southwest Research Institute under the sponsorship of the American Gas Association. In the first part of the program, the fluid was water; and studies were made of the error in flow rate as determined with sharp-edged orifices subjected to various pulsation frequencies up to about 10 cps and amplitudes up to nearly reversing movement. From these tests and within the range of the variables covered, the following statements are drawn [34, 35]:

1. Water flow computed from the steady component of the differential pressure across an orifice as measured from flange taps is in error when the flow is pulsating in character.

2. The error in the thus computed water flow is always positive, i.e., the computed flow is greater than the actual.

3. The error is independent of the direction from which the pulsations eminate, that is, whether the source is upstream or downstream of the orifice.

4. The square root relationship

$$(1 + E)^2 = 1 + \tfrac{1}{2}\left(\frac{u}{V}\right)^2 \qquad (I\text{-}3\text{-}57)$$

correctly expresses the error within experimental uncertainty. E is the error in the flow rate as from an orifice under pulsating flow, u is the zero-to-peak amplitude of the sinusoidal component of the axial velocity through the orifice, and V is the magnitude of the steady axial velocity through the orifice.

5. The certification of error from the extremes of the total instantaneous differential pressure is correct within experimental uncertainty.

6. The effect of the Strouhal number, (N_S), up to 0.52 is smaller than the experimental uncertainties. $N_S = \omega d/V$ where ω is the angular frequency in radians per sec, d is the orifice diameter (ft), and V is the magnitude of the steady axial velocity through the orifice. N_S is dimensionless.

I-3-48 In the second part of the research program on pulsating flows, air was the fluid used. From that work as well as from other sources, it is evident that the error with pulsating compressible fluids is far too complex to be evaluated by the square root error, equation (I-3-57). No relationship has been found to exist between the average velocity of the jet through an orifice meter and the observed differential pressure across the orifice when pulsations are present, and no consistent explanation for this discrepency has yet been given [36]. Therefore, with compressible fluids, it is recommended that the error be eliminated by eliminating or at least reducing pulsations at or near the source. (This recommendation may be followed to advantage with liquid flows.) This can be accomplished with surge bottles or pulsation dampeners, and criteria for the sizing and design of these are available [37]. (Surge bottles may be used with liquid flows to advantage, but their design is more difficult because a compressible medium such as a gas cushion or spring loaded piston must be provided [38].)

References

[1] "Measurement of Air Velocity and Pressure," A. F. Zahm; *Physical Review*, series 1, vol. 17, Dec. 1903, p. 410.

[2] "Einfluss der Manometermundung auf die Druckablesung" (Influence of the (pressure) Aperture on Manometer Readings), K. Buchner; *Zeitshrift des Verines Deutscher Ingenieure*, vol. 48-II, July 23, 1904, p. 1100.

[3] "A Three-Dimensional Spherical Pitot Probe," J. C. Lee and J. E. Ash; *Trans. ASME*, vol. 78, Apr. 1956, p. 603.

[4] "The Pitot Tube in Current Practice," E. S. Cole; *Civil Engineering*, vol. 5, Apr. 1935, p. 220.

[5] "Correction Factors for the Balancing Effect of a Gas in One Leg of a Manometer," H. S. Bean and F. C. Morey; *Instruments*, vol. 24, May 1951, p. 528; *Oil & Gas Journal*, July 1951, p. 106; *Gas*, vol. XXVII, Aug. 1951, p. 48.

[6] "Review of Practical Thermometry," R. P. Benedict; ASME paper 57-A-203.

[7] "Measurement of Temperatures in High-Velocity Steam," J. W. Murdock and E. F. Fiock; *Trans. ASME*, vol. 72, Nov. 1950, p. 1155.

[8] "Temperature Measurements in High-Velocity Air Streams," H. C. Hottel and A. Kaladinsky; *Journal of Applied Mechanics*, vol. 12, Mar. 1945, p. A-25; *Trans. ASME*, vol. 67, 1945.

[9] "Measurement of High Temperature in High-Velocity Gas Streams," W. J. King; *Trans. ASME*, vol. 65, July 1943, p. 421.

[10] "Performance of Three High Recovery-Factor Thermocouple Probes for Room Temperature Operation," M. D. Scadron, C. C. Gettelman and G. J. Pack; NACA Research Memorandum E-50129.

[11] "Celsius versus Centigrade," H. F. Stimson; *Science*, vol. 136, Apr. 20, 1962, p. 254.

[12] "Tables of Thermodynamic and Transport Properties of Air, Argon, Carbon Dioxide, Carbon Monoxide, Hydrogen, Nitrogen, Oxygen and Steam;" National Bureau of Standards Circular 564; also, Pergamon Press Ltd., London, 1960.

[13] "The Twelfth General Conference on Weights and Measures," *National Bureau of Standards Technical News Bulletin*, vol. 48-12, Dec. 1964, p. 207.

[14] "Units of Weights and Measures," *National Bureau of Standards Miscellaneous Publication*, No. 287, May 1967.

[15] "1967 ASME Steam Tables," ASME, New York, 1967.

[16] "Standard Density and Volumetric Tables," National Bureau of Standards Circular 19. (Out of print, reference copies in many libraries.)

[17] "ASTM-IP Petroleum Measurement Tables," ASTM D-1250-52.

[18] "U.S. Standard Atmosphere," 1962; U.S. Government Printing Office, Washington, D.C.

[19] "Gas Measurement Manual," American Gas Association, 1963, Ch. VII.

[20] "Relations between the Temperatures, Pressures and Densities of Gases," National Bureau of Standards Circular 279. (Out of print, reference copies in many libraries.)

[21] "Manual for Determination of Supercompressibility Factors for Natural Gas," American Gas Association, 1962.

[22] "Rational Psychrometric Formulas," W. H. Carrier; *Trans. ASME*, vol. 33, 1911, p. 1005.

[23] "Theory of the Psychrometer," J. H. Arnold; *Physics*, vol. 4, 1933, pp. 255, 334.

[24] "Determination of Viscosity of Exhaust-Gas Mixtures at Elevated Temperatures," J. C. Westmoreland; NACA Technical Note 318, June 1954.

[25] "Measurement of the Viscosity of Five Gases at Elevated Pressures by the Oscillating Disk Method," J. Kestin and K. Pilarczyk; *Trans. ASME*, vol. 76, Aug. 1954, p. 987.

[26] "Units and Conversion Factors for Absolute Viscosity," G. A. Hawkins, H. L. Solberg and W. L. Sibbitt; *Power Plant Engineering*, vol. 45, Nov. 1941, p. 62.

[27] "Standard Method of Conversion of Kinematic Viscosity to Saybolt Universal Viscosity," ASTM D446-53.

[28] "Model Experiments and the Forms of Empirical Equations," E. Buckingham; *Trans. ASME*, vol. 37, 1915, p. 263.

[29] "The Laws of Similarity for Orifice and Flow Nozzle," J. L. Hodgson; *Trans. ASME*, vol. 51, 1929, FSP-51-42.

[30] "Dimensional Analysis," H. E. Huntley; Richard & Co., New York, 1951.

[31] "Mechanical Engineers' Handbook," Marks; McGraw-Hill Book Co., New York, 6th ed., pp. 3-88.

[32] "Quantity-Rate Fluid Meters," E. S. Smith, Jr.; *Trans. ASME*, vol. 52, 1930.

[33] "Pulsating-Flow Measurement, a Literature Survey," E. G. Chilton; *Trans. ASME*, vol. 77, Feb. 1955, p. 231.

[34] "Analysis of the Effect of Pulsations on the Response of Mercurial-Type Differential-Pressure Recorders," R. J. Martin and D. S. Moseley; *Trans. ASME*, vol. 80, Oct. 1958, p. 1343.

[35] "Measurement Error in the Orifice Meter on Pulsating Water Flow," D. S. Moseley; Flow Measurement Symposium, ASME, Sept. 26-28, 1966, p. 103.

[36] "A Study of Pulsation Effects on Orifice Metering of Compressible Flow," C. R. Sparks; Flow Measurement Symposium, ASME, Sept. 26-28, 1966, p. 124.

[37] "Pulsations in Gas Compressor Systems," E. G. Chilton and L. R. Handley; *Trans. ASME*, vol. 74, Aug. 1952, p. 931.

[38] "Pulsation Absorbers for Reciprocating Pumps," E. G. Chilton and L. R. Handley; *Trans. ASME*, vol. 77, Feb. 1955, p. 225.

Elements of Quantity Meters

I-4-1 The distinctive features of quantity meters is that all of the fluid passes through the primary element in completely or almost completely isolated quantities. The number of these quantities is counted and indicated in terms of weight or volume units by the secondary element or register.

Before taking up the descriptions of quantity meters it will be convenient to define three terms used in connection with these meters, especially those for liquids. These are: positive seal, capillary seal and slip.

I-4-2 Positive Seal is the closing off of a measuring chamber so tightly as to prevent (momentarily) any fluid from entering or leaving the chamber when under any pressure difference to which the meter would normally be subjected. An example is the water seal of a wet gas meter. In the case of liquid meters an absolute positive seal is difficult to attain as in general it depends on the closeness of contact between two metal surfaces or a metal and a nonmetalic material. It may be approached as between a piston and cylinder by the use of close-fitting piston rings; but, as the number and stiffness of the rings are increased, the force (pressure difference) required to move the piston increases to an extent which is impracticable.

I-4-3 Capillary Seal attains a practical degree of measuring-chamber tightness by virtue of the strength of the surface tension of a film of the fluid between two of the chamber surfaces which do not actually touch. The distance (clearance) between these surfaces must be very small compared to the length of the shortest path the fluid would travel between them in escaping. As an example, in Fig. I-4-12 the distance (clearance) between the circumference of the rotor and the case may be as little as 0.002 in. while the arc length of a rotor tip may be as little as 0.125 in. Since the capillary action depends on the properties of the fluid and the condition of the surfaces in addition to the dimensions of the passage, it is not practical to specify any definite limits of what constitutes a capillary seal. Sometimes the effect of the capillary passage between a piston and cylinder is supplemented by circumferential grooves around the piston without applying piston rings.

I-4-4 Slip or flow through clearances is the difference between the volume of fluid which passes through a meter per cycle and the meter displacement per cycle as determined from geometrical measurements. The difficulties of accurately determining the amount of the slip are due in part to those of measuring the true displacement of the meter and in part to the fact that the magnitude of the slip is affected by the character of the fluid, the condition of the surfaces of the meter parts, the amount of wear, the rate of flow, and the pressure drop across the meter which in turn is affected by the load imposed by the register.

I-4-5 Weighing Meters. In general, weighing-type meters are applicable only to the metering of liquids, and especially those with a low vapor pressure. The simplest form of a weigher would be a tank mounted on a scale. However, with a single tank the measurement of a fluid flow would be intermittent, since the flow would have to be stopped or diverted while the tank is emptied. Obviously, two such tanks may be mounted close together with a diverter and outlet-valve operating mechanism so connected that one tank is filling while the other empties. The outlet valve and connections must be such as to permit the rate of outflow to equal, and preferably exceed, the rate of inflow.

I-4-6 A Tilting Tank Meter is diagrammatically illustrated by Fig. I-4-1. Two tanks, A and A′, are supported on the pivots (knife edges), K and K′, and counterbalanced by the weights, W and W′. When the gross weight of the liquid and tank reached a certain value, the tank tips on its pivot to the position of A and thereby causes the liquid to discharge through the syphon, S. This tipping of the tank tilts the diverting trough about its axis, C, diverting the liquid into the other tank. When a part of the liquid has drained out of the tank which tilted, the counterweight, W, will cause the tank to rock back to an upright position, but liquid will continue to discharge through the syphon until the tank is emptied down to the inlet end of the syphon tube. As the second tank is filled, it tilts, shifting the diverting trough to the position for refilling the first tank. Obviously, the rate of emptying the tanks must equal or exceed the rate of inflow.

There is a variation of this design in which the two tanks are mounted on a common shaft in such a way that the tilting of one tank brings the other into position for filling. In this design the syphon may be eliminated.

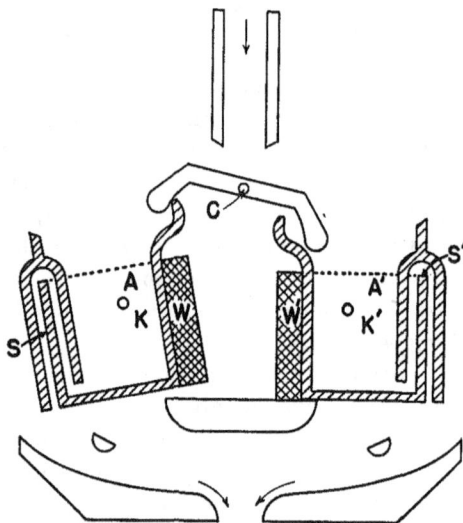

FIG. I-4-1 WEIGHING METER TILTING TRAP
TYPE — SCHEMATIC

I-4-7 Positive Displacement Meters: Volumetric Types.

General Features. Positive displacement meters, or simply displacement meters, measure the quantity or volume of fluid by filling repeatedly a given container. The total quantity of fluid flowing through the meter in a given time is the product of the volume of the container and the number of fillings. Usually, the number of fillings is obtained by a counter, or register, which is operated by the meter. The registers of most meters for liquids are geared to indicate quantities in gallons, cubic feet or barrels (42 gal). The registers of gas meters are usually geared to indicate quantities in cubic feet.

I-4-8 A displacement meter may be considered as a fluid motor operating with a high volumetric efficiency under a very light load. This load is made up of two parts — the internal load, or that due to the friction within the metering element; and the external, or that due to the registering mechanism. As in all fluid motors, work done against a load results in an expenditure of pressure, i.e., there will be a drop in pressure from inlet to outlet. The main factors influencing the amount of this pressure drop are: the type of seal, the power required to drive the register, the viscosity of the liquid, and the rate of flow.

I-4-9 Displacement meters may be divided into three classes according to the method used to close or seal the measuring compartments of the meter. The first class is that in which one wall of a measuring chamber incorporates a flexible material which allows the wall to move and thereby displaces increments of volume with no leakage to another chamber. In the second class a mechanical or pack seal is used between the movable and stationary walls of a measuring chamber. The third class is the film or capillary-seal type, in which the clearances between moving and stationary parts are very small and carefully controlled. The amount of slippage in this third class is controlled by keeping the clearances small and the pressure drop across the seal as low as possible.

There will be a marked difference in the pressure drop characteristics of the three classes. The pack seal class, which includes the reciprocating-piston-type meters, will have a much higher pressure drop because of the friction between the fixed and moving parts. These meters can carry the high load imposed by a register with two or more sets of dials. Meters of the first class with a flexible wall can also carry a high-register load without materially changing their displacement. The film-class meter must have a relatively low pressure drop across the clearances, so the load from the bearings and register must be as low as practicable.

I-4-10 Metering Tank. The metering tank is the most elementary form of volumeter and embodies an open tank which is repeatedly filled, to a fixed depth, and emptied. The number of fillings may be counted by visual observation or by some form of mechanical counter. Further elaboration of the design could include two or more tanks filled and emptied in succession by a suitable mechanism.

I-4-11 Liquid Sealed Drum. The measuring element of these meters is a drum or cylinder divided into four chambers by approximately helical partitions. These partitions do not extend to the central shaft on which the drum is mounted, and there are narrow radial openings into each compartment in each end of the cylinder. A fifth chamber is formed at one end of the cylinder by a second end of slightly convex shape and having a central opening sufficiently large to accommodate the shaft and a tube through which gas can enter (or leave) this chamber and the measuring compartments. Figure I-4-2 shows one of these drums with part of the cylinder wall rolled back. The drum is mounted within a case with the end of the shaft which carries the pointer, extending through a bearing and packing gland. The case is filled with water (or some other liquid) until both openings of a compartment are below the water surface when the center of the compartment is uppermost. This usually requires that the drum be about two-thirds submerged. Thus, as the drum rotates, due to the pressure of the gas against the compartment partitions, each compartment when filled is momentarily sealed off. The capacity of a compartment when sealed will depend therefore upon the height of the water surface. Thus, the calibration of one of these meters consists in determining the height to which the meter should be filled with water to cause the meter to register the correct amount. When this water height has been determined, it can be marked by a suitable gage mark. In subsequent use, the height of the water surface is adjusted to this gage mark, after the meter has been leveled. When adjusting the water gage or setting the water surface to it, both inlet and outlet should be open to the atmosphere, or to a common pressure.

I-4-12 Bellows Meters. The common name for these meters is "dry positive displacement meter," or simply "dry meter," and they are used only for the metering of gases. The principal elements of these meters are the flexible partitions or diaphragms of the measuring compartments, valves for controlling and

FIG. I-4-2 METERING DRUM OF A SMALL WET-GAS METER, PARTLY OPENED

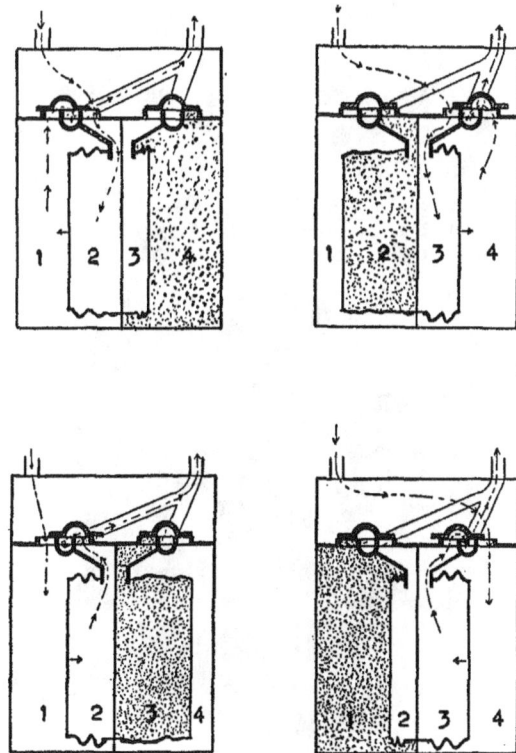

FIG. I-4-3 FOUR STAGES IN THE CYCLE OF OPERATIONS OF A DRY-GAS METER

directing the gas flow in filling and emptying the
measuring compartments, appropriate linkage to keep
the diaphragms and valves in synchronism, and a
register for counting the number of cycles. To obtain
continuous flow and power for operating the register,
it is necessary to have three or more measuring compartments or chambers, with two or more movable
walls which are sealed with a flexible material that
is impervious to gas. The movements of the walls or
diaphragms are so regulated that the total cubical
displacement on successive cycles is the same. The
amount of travel or stroke of the diaphragms is
regulated in most meters by the radial position of the
crank pin to which the diaphragm linkage arms are
attached. Figure I-4-3 shows diagrammatically the
sequence of filling and emptying of the four
measuring chambers of a two-diaphragm meter,
while Figure I-4-4 shows some of the interior of such
a meter.

I-4-13 Reciprocating Piston, or Piston, Meters.

Meters of this type are essentially reciprocating
piston engines; however, whereas an engine is
designed to extract as much energy as possible from
the fluid passing through it, the meter extracts only
enough energy to overcome the frictional resistance
in valves and register. A section through a typical
reciprocating piston meter is shown in Fig. I-4-5,
from which the sequence of its operations may be
traced. Through a suitable linkage and gear system
(not shown), the reciprocations of the pistons
operate the meter register.

There are a number of variations to be found in
the design of reciprocating piston meters. For example, the valves may be built integrally with the
pistons. In another design only one cylinder and
piston is used, with an auxiliary piston to assist in
operating the valve. Again, there may be a number of
cylinders (five or more) placed radially in a common
plane, with the pistons connected to a single common
crank. Or the cylinders may be arranged in a circle
with axes parallel (e.g., vertical) and the pistons
linked together by a nutating disk. Meters of this
design may have a single valve driven from the
crank or disk.

In reciprocating piston meters, sealing rings of
metal, leather or other material may be used to

FIG. I-4-4 DIAPHRAGMS AND VALVE MECH-
ANISM OF A DRY-GAS METER

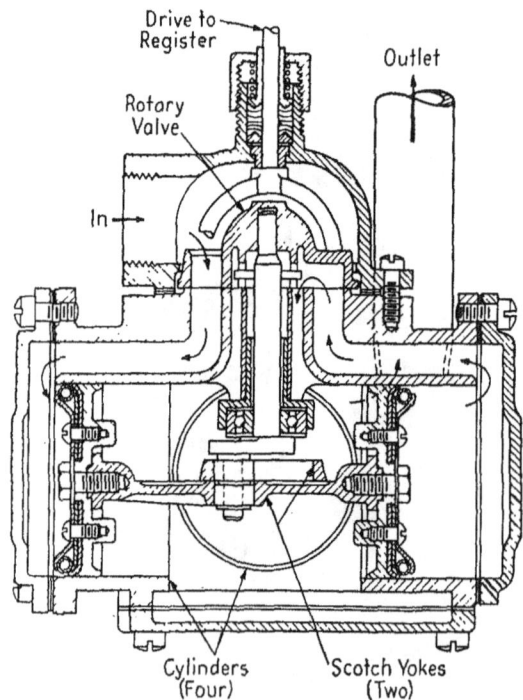

FIG. I-4-5 CROSS SECTION OF A RECIPRO-
CATING PISTON METER

greatly reduce, if not practically eliminate, any slippage past the piston. Another method is to use a longer piston without any sealing rings and to depend on capillary sealing to reduce slippage.

In some designs, adjustment of the capacity per cycle is made by regulating the stroke of one or more of the pistons or by altering the stroke of all pistons simultaneously by changing the crank radius or the nutating-yoke movement.

I-4-14 Ring Piston Meters. (Also called oscillating and rotary piston meters.) In these meters the piston and cylinder are the same length (except for a small operating clearance), but the diameter of the piston is much less than that of the cylinder. Moreover, the piston is hollow, and the cylinder chamber is annular, as shown in Fig. I-4-6. Between the outer and inner walls of the cylinder there is a radial partition, A, and in the sidewall of the piston there is a slot wide enough to fit over the partition without binding. It is this partition and slot that prevent rotation of the piston. Mounted in the center of the cylinder base is the post, B, and in the center of the closed upper end of the piston is the post, C. The outer and inner diameters of the hollow piston, the major and minor diameters of the annular cylinders, the inner diameter of the inner cylinder wall, and the diameters of the posts B and C are all so related that the outer surface of the piston is al-

ways in contact with the inner surface of the outer wall of the cylinder, while the inner surface of the piston is always in contact with the outer surface of the inner wall of the cylinder. The sealing of the measuring chambers depends on sliding contact between the cylinder base and the lower open end of the piston and on a combination of rolling and sliding line contact between the cylinder and piston sidewalls.

The irregular shaped areas, D and E, are the inlet and outlet ports, respectively, and are located in the base of the cylinder. The motion of the piston and its functioning as a valve, as well as a piston, may be seen by a study of the four positions shown by Fig. I-4-6 a-d. The circular motion of the extension of post C is used to drive the meter register.

In each cycle there is measured out the volume of fluid required to fill the chambers 1 and 2 of Fig. I-4-6a and 3 and 4 of Fig. I-4-6c. Since the displacement per cycle is fixed, the relation between the displacement and the register can be adjusted only by changing the gearing ratio to the register. Then, having adjusted the gearing ratio to give the desired relation between displacement and register, as nearly as possible, the establishment of the exact relationship will depend upon a calibration or "proving" of the meter. This is especially important if fluids of different viscosities are to be metered or if the meter is to be operated at different rates within its range. In such cases the meter should be calibrated or "proved" with a fluid of each grade or viscosity to be metered and over the range of rates at which it will be used [1].

I-4-15 Nutating Disk Meters. Other, and possibly more appropriate, names for these meters are *precession disk* and *wabble plate meters*. Figure I-4-7 shows a vertical section through such a meter.

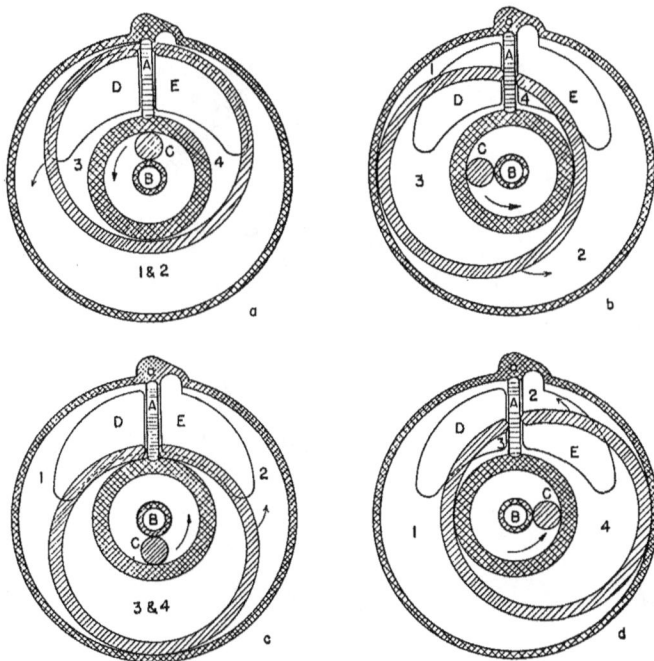

FIG. I-4-6 CROSS SECTION ILLUSTRATING THE OPERATION OF A RING PISTON METER

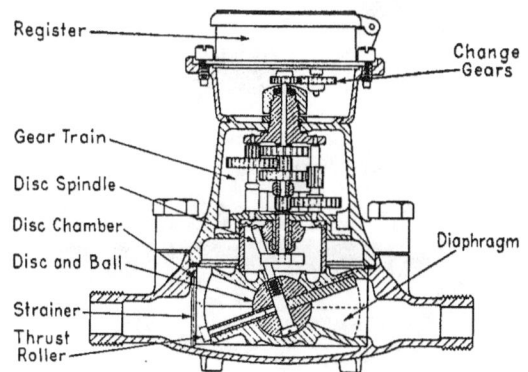

FIG. I-4-7 SECTION OF A NUTATING DISK METER

The top and bottom surfaces of the measuring chamber are conical, extending inward, and the chamber sidewall is spherical. The movable element is the disk mounted on the central ball, from the top of which a shaft extends that is perpendicular to the disk. This shaft is held in an inclined position by a cam roller so that the disk is in contact with the chamber bottom along a radial element on one side of the ball and in contact with the top in the same radial plane on the other side of the ball. The disk is prevented from rotating about its own axis by a radial partition, which is more clearly shown in Fig. I-4-8. The inlet and outlet ports are in the sidewall of the case, adjacent to, and on opposite sides of the radial partition. Liquid enters the measuring chamber alternately above and below the disk and passes around the conical measuring chamber to the outlet port. This movement of the liquid around the measuring chamber first above and then below the disk imparts a nutating motion to the disk (nodding in a circular path without revolving about its own axis). As the disk nutates, the top of the shaft moves in a circular path and, by engaging a crank, operates the meter register.

Between different makes of nutating disk meters there may be minor variations in design. For example, instead of a flat disk, as shown in Figs. I-4-7 and I-4-8, some meters have a coned disk. In such a case the cone angles of the top and bottom will not be the same.

Since the displacement per cycle is fixed, the relation between displacement and registration is adjusted by means of a set of change gears in the drive system between the metering chamber and register.

Because of the simplicity of its construction this type of meter can be produced very economically. Partly for this reason, and also because it will

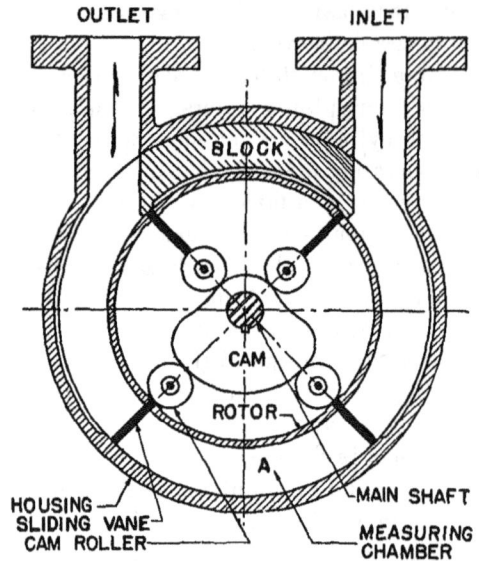

FIG. I-4-9 METERING CHAMBER OF A SLIDING VANE METER

FIG. I-4-8 MEASURING CHAMBER AND DISK OF A NUTATING DISK METER

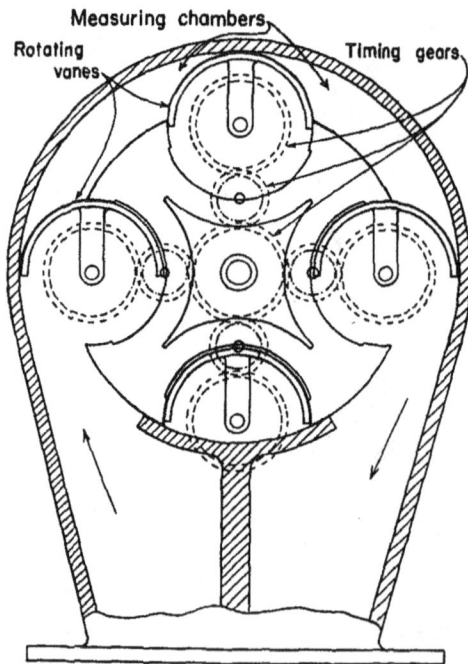

FIG. I-4-10 SCHEMATIC REPRESENTATION OF THE OPERATING FEATURES OF A ROTATING VANE METER

maintain a satisfactory degree of accuracy over a long period of time, it is used extensively as a water meter [2].

I-4-16 **Sliding and Rotating Vane Meters.** Sliding and rotating vane meters may be considered as continuous piston meters having two or more measuring compartments, as indicated by *A* in Figs. I-4-9, I-4-10, and I-4-11. The design of these meters must incorporate some feature by which to "gate" the rotating vanes from the outlet back to the inlet side. This may be with a cam and follower system as in Fig. I-4-9 or with a set of timing gears as in Fig. I-4-10. With any one fluid, the amount of slip depends on the thinness of the clearances and the

pressure drop; therefore the clearances between rotating and stationary parts are made as small as practicable. In addition, low friction type bearings are used. The operating range of these meters, for acceptable accuracy, may be relatively wide. Normally, meters such as represented schematically by Figs. I-4-9 and I-4-10 are used with liquids, while that represented by Fig. I-4-11 is used for the metering of gas.

I-4-17 **Lobed Impeller and Gear Meters.** Figures I-4-12, I-4-13, and I-4-14 illustrate the general features of this group of meters. The measuring chambers are the spaces between the gears, or rotor lobes, and the outside case. When there are only two or three lobes on each rotor, it is necessary to have equal-toothed spur gears on the rotor shafts at one or both ends of the rotor to keep them in the correct rotational relationship with each other. These timing

FIG. I-4-11 A ROTATING TWO-VANE METER USED FOR METERING GAS

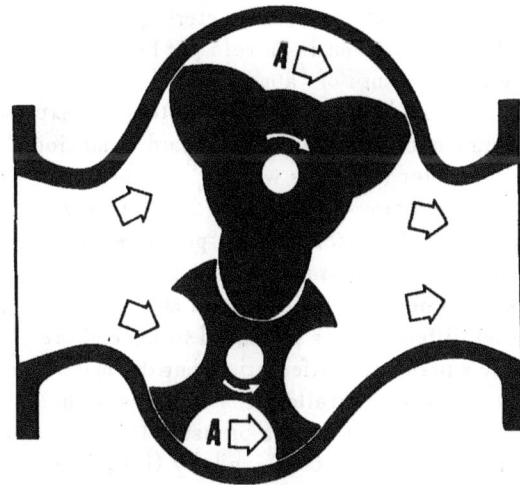

FIG. I-4-13 A COMBINATION ROTARY DISPLACEMENT METER FOR LIQUIDS

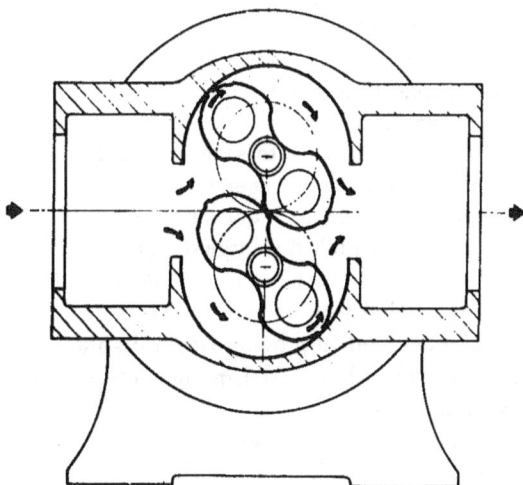

FIG. I-4-12 A TWO-LOBED ROTARY DISPLACEMENT METER FOR GAS

FIG. I-4-14 A GEAR DISPLACEMENT METER FOR LIQUIDS

gears may be either inside or outside the end walls of the meter case.

By careful machining, the clearances between the rotors and case and between the rotors themselves may be made very small, thereby obtaining an effective degree of capillary sealing of the measuring chamber. Also, by the use of precision bearings, the overall pressure drop can be kept to a low value.

Normally, the two-lobed meter is used for gas measurement, while the three-lobed and gear meters are used for metering liquids. The volumetric capacity of the impeller meter at line conditions depends upon the maximum safe gear tip speed, or rpm of the meter. This safe gear tip speed is independent of the operating pressure of the line to which the meter is connected.

I-4-18 Meter Capacities. The capacities of displacement meters for liquids are usually expressed in terms of gallons per minute, gpm, or gallons per hour, gph. In the case of meters for metering petroleum, the capacity may be given in barrels per hour, bph.

The low pressure or "atmospheric" capacities of diaphragm-type displacement gas meters are expressed in terms of cu ft per hr at "standard conditions" which the meter will pass when operating with a pressure drop across it of 1/2 or 2 in. of water (depending on the size). At high pressure, it is customary to make a practical, but arbitrary, compromise between meter differential and meter speed; the meter differential is permitted to exceed the basic (low pressure) differential, but the meter speed is simultaneously curtailed. For fuel gases the "standard conditions" are: a pressure (inlet) of 14.73 psia, a temperature of 60 F, and dry (i.e., free of water vapor).

With the rotary-type meters the rated capacity is that when the meter is operating at rated speed (rpm) and with an inlet line pressure less than 1 psig. The capacity of a meter at other line conditions may be evaluated, approximately, by use of the general pressure-temperature-volume relationship for gases [3].

I-4-19 Calibration, or "Proving," Quantity Meters. Practically all types of quantity meters discussed in this chapter are calibrated or "proved" to insure that they register correctly the quantity of fluid being metered. This applies to meters for both liquids and gases regardless of size. In the case of meters used in the sale of fluids, the methods to be used, the procedures to be followed and the indicated accuracy limits to be attained are prescribed in considerable detail by the appropriate commissions of most states. For meters not involved in sales, the testing pro-

cedures will vary from one company to another and with the purpose of the measurement. Hence, none of these procedures will be given here. However, testing or proving equipment will be described briefly, with comments on tolerances and operating.

I-4-20 Gas Meter Provers. All displacement-type gas meters measure gas quantities by volume increments. By far the most generally used prover for testing small- to medium-sized meters is an inverted bell liquid-sealed type gasometer. Most such provers have a working capacity of 2, 5 or 10 cu ft; however, provers with capacities up to 100 cu ft are obtainable. Ordinarily these provers are operated at the ambient conditions of barometric pressure and the temperature existing in the shop where used, or very close to those conditions. These shop conditions are usually sufficiently close to the standard conditions so that no adjustments are made for any difference there may be [4, 5].

For the larger-capacity gas meters a secondary or transfer prover is used very often. Such a transfer prover may be another displacement meter that has been carefully tested and adjusted against a primary prover. However, a more common type of transfer prover is a rate-of-flow meter, such as a specially prepared orifice or flow-nozzle meter. Transfer provers of this kind are used frequently for routine checking in the field of large-sized gas meters. Adjustments of these field tests to standard conditions are not always made unless the pressure and temperature conditions are such as to make them necessary. In such cases the simple pressure-temperature-volume relations for an ideal gas are normally used [5].

I-4-21 Liquid Displacement Meter Proving. The types of provers used are: weigh tanks; volumetric tanks, open and closed; another meter as a transfer prover; and mechanical displacement provers. The selection of the type and size will include consideration of the capacities of the meters to be tested, the liquid to be metered or used in the proving, and the location of the meters. As to size of liquid capacity of a prover, the usual recommendation is: the volume of the prover should be at least sufficient to hold the quantity of liquid which the largest meter to be tested will pass in one minute at its maximum rated capacity, and twice this volume is desirable. Other items to be observed in proving liquid displacement meters, especially those used in connection with the sales of liquids, are:

1. The meter should be tested with the liquid it is to meter, or one having similar density and viscosity characteristics, at a temperature as close as possible to that at which it is used.

2. The meter should be tested over the range of rates of flow at which it is used.

3. The meter should be tested under a pressure equal to that at which it is operated.

4. The liquid must be free of entrained air or vapor. This will require that the pressure be appreciably above the vapor pressure of the liquid. Since this will apply to the liquid when in the prover, it may be necessary to have a closed prover so that it may be under pressure also, and operated by liquid or gas displacement.

5. There should be little or no difference in the temperature of the fluid while passing through a meter and while in a volumetric prover. With some liquids, a temperature difference of 5 F will cause a volume change of 0.5 per cent, while with other liquids the effect is insignificant.

When a meter that registers in volume units is proved with a weigh tank, the weight of the liquid must be converted to the volume it would occupy when at the temperature at which it passed through the meter [6, 7].

When a hot water meter is geared to register in pounds, it will register the correct quantity if the water is at the temperature for which the meter gearing is set. If the water is at a lower temperature, the meter will under-register slightly, that is, it will register fewer pounds than were actually discharged. Similarly, the meter will tend to over-register on water at a higher temperature.

I-4-22 A common weakness in the proving of displacement meters (gas as well as liquid) with either volumetric or gravimetric-type provers is the necessity to start and stop the meter at the beginning and end of each prover filling or discharge. In general, this is due to the need of having the test hand, or dial, of the meter index start and stop at the same mark. This makes it difficult to test a meter up to as high a rate as would be desirable, and the rate during any test is not uniform. This weakness may be overcome more or less completely by using supplemental equipment to provide high speed reading and even recording of meter, prover or both. Alternately, dual volumetric tanks may be used and the fluid stream diverted from one to the other without interruption or loss.

A third way of overcoming the start and stop weakness is the use of a mechanical displacement prover. The volume chamber of such a prover is a section of pipe, the inner surface being bored, honed and possibly surface treated to improve surface smoothness and retard corrosion. In some designs there is an enlarged chamber at each end of the

special section from which and into which a displacer, such as a spheroid or a piston with flexible wiper rings, may be launched and received. In other designs additional lengths of pipe serve as the launching and receiving chambers. In either design, especially the latter, the displacer may be moved in either direction, i.e., a bidirectional prover. The piping connections, including valves, are usually such that the displacer may be launched and received without interrupting or altering the fluid flow. Near the ends of the special pipe section, detector units or switches are located to provide signals as the displacer passes. The volume of the pipe section between the cross sections in which the detectors are located must be determined carefully. With a bidirectional prover, since the displacer may trigger the detector units slightly differently when moving in one direction than the other, the volume should be determined for the movement of the displacer in both directions. Also, in the testing of meters, one or more "round trips" of the displacer should be used in each test.

Note: Mechanical displacement provers are being used for testing gas meters, also [8].

In order to divide the last unit of the usual meter index into smaller subdivisions, it is a common practice to attach supplemental index equipment to the meter index.

I-4-23 Accuracy of Provers. The accuracy of all types of provers should be established by appropriate procedures. The scales used with a weigh tank may be checked with master weights [9]. The displacement volumes of gas-meter provers may be determined by dimensional measurements or by comparison with a volumetric standard, such as a cubic foot bottle [10, 11]. The capacities of volumetric-type provers for liquid meters may be established by dimensional measurements, by weighing water into or out of the prover, and by filling or emptying with a standard measure, such as a 5- or 10-gallon test can [1]. The displacement volume of a mechanical displacement prover may be determined by similar methods.

I-4-24 Tolerances. The accuracy requirements or tolerances applied to displacement meters depends upon the meter service, size, legal requirements, and local or company practices. With gas meters used in the sale of fuel gas, most state and local commissions require that before a meter is placed in service it shall register correctly within ± 1.0 per cent, and some commissions and companies may use ± 0.5 per cent. However, before any adjustment may be considered—because of a meter inaccuracy, it is usual to require that the meter be over 2 per cent in error.

With liquid meters it is common practice to use a maintenance tolerance of twice the acceptance tolerance. For example, in the retail sale of gasoline, on a 5-gallon delivery the acceptance tolerance is ± 3.5 cu in., or ± 0.3 per cent; but adjustment may not be required so long as the inaccuracy does not exceed ± 7.0 cu in., or ± 0.6 per cent. With a larger meter, used for wholesale or bulk deliveries of 50 gallons or more, the acceptance tolerance is 0.5 cu in. per gallon or ± 0.22 per cent [12]. A common requirement for new domestic and small commercial water meters is that they register the actual delivery correctly within ± 1.0 per cent.

I-4-25 Installation and Operation. With most of the small meters for both gas and liquids it is common practice to support them entirely by the piping to which they are connected. As the meter size increases, the need for a substantial foundation increases. With the larger meters a firm concrete foundation is desirable, and the piping connections to the meter should be well supported and carefully aligned so there will be no strain placed on the meter.

If the temperature of the fluid to be metered is likely to be appreciably different from the normal ambient range (e.g., 50 F either below or above a 20 to 80 F range), consideration should be given to the effect of temperature on some of the materials in the meter to be used. On this question the recommendations of the meter manufacturer should be obtained.

In actual gas-meter service conditions, both the pressure and temperature may be far from the reference or standard values. For such conditions there are available temperature compensating elements which will adjust, automatically, the meter registration for the effects of the flowing temperature of the gas. While similar elements can be used to compensate for the effects of line pressure, a common practice is to record the pressure with a recording gage that may be driven by the meter index; or, a volume indicating pen is added to a clock driven pressure gage. From the record on a chart from such a pressure gage the volume recorded by the meter

register may be adjusted to the standard or other contract conditions, including any factor for the effects of compressibility [5].

Regardless of what the liquid is that is being metered, it is necessary that the pressure be maintained well above the vapor pressure of the liquid at the metering temperature. This is necessary to prevent the liquid from flashing to vapor, since the vapor would be registered as liquid.

Most displacement meters may be operated at rates of flow up to about 125 per cent of rated capacity for short intervals. However, if it is found that a meter is operating close to its rated capacity a considerable part of the time, it should be replaced with a larger meter.

References

[1] "Code for Installing, Proving and Operating Positive-Displacement Meters in Liquid Hydrocarbon Service," API Standard 1101, 1960; and Mechanical Displacement Provers, API Standard 2531, First Ed., 1963, American Petroleum Institute.

[2] "Standard Specifications for Cold Water Meters, Displacement Type," AWWA-C700, 1964, American Water Works Association.

[3] "Gas Measurement Committee Report No. 3," 1969; American Gas Association.

[4] "Testing Domestic and Large Capacity Displacement Meters with Bell Type Provers," R. A. Seifert; Proc. 9th Appalachian Gas Measurement Short Course, Univ. of West Virginia, 1949, p. 350.

[5] "Gas Measurement Manual," 1963, Ch. V; American Gas Association.

[6] "ASTM-IP Petroleum Measurement Tables," American Edition; American Society for Testing Materials, 1952.

[7] Part II, Table II-I-4: The Density of Water.

[8] Mechanical Displacement Meter Prover for Gas Meters, A. W. Jasek; Proc. 26th Appalachian Gas Measurement Short Course, Univ. of West Virginia, 1966, p. 191.

[9] PTC 19, 5, 1- 1964: Chapter 1, Weighing Scales, "Supplement to Performance Test Codes," ASME.

[10] "Calibration of Bell Provers," H. V. Beck; American Gas Association Monthly, vol. 46, No. 9, Sept. 1964, p. 30.

[11] "Calibration of Bell Provers by Dimensional Analysis and by a Cubic Foot Standard," C. T. Collett; Proc. 24th Appalachian Gas Measurement Short Course, Univ. of West Virginia, 1964, p. 59.

[12] "National Bureau of Standards Handbook 44," 2nd Ed., 1955, Specifications, Tolerances and Regulations for Commercial Weighing and Measuring Devices, p. 92a.

Differential Pressure Meters: Theory of Fluid Flow in Terms of Differential Pressures and Equations for Differential Pressure Meters

1-5-1 Principal primary elements. In the differential pressure group these are the Venturi tube, the flow nozzle, and the thin-plate square-edged orifice. Other primary elements in this group, which are discussed in this chapter, are the nozzle-Venturi and other modifications of the Venturi tube, the quadrant-edged orifice, and eccentric and segmental orifices. In addition, there are the centrifugal (elbows), linear resistance (capillary tube and porous plug), and frictional resistance (pipe sections). The distinctive feature of this group of meters is that there is a marked pressure difference or pressure drop associated with the flow of a fluid through the primary element and that this pressure difference can be measured and related to the mass or volume rate of flow. Hence, the designation *differential pressure* meters. The theoretical considerations used as a basis for computing the rate of flow from the pressure measurements are the same for all meters in the group except for the last three. While the characteristics of fluid flow through all of the primary elements have been studied to some extent, such studies and tests with the Venturi tube (Fig. I-5-1), the flow nozzle (Fig. I-5-2), and square-edged orifice (Fig. I-5-3), have been very extensive. From such extensive studies the discharge coefficients and expansion factors for these three primary elements have been so well established that these meters are used extensively, even for important measurements, without calibration. Instructions on the construction, installation and operation of these primary elements are given in Part II.

I-5-2 Theory of the Flow of Fluids in Terms of Pressure Differences. In the following discussion and development of equations, a primary assumption is: the mass flow rate is constant with respect to a considerable period of time (e.g., 5 to 10 minutes or more), and the flow is steady. In the past, when the Bourdon gage or a liquid manometer were the principal pressure-indicating instruments, the adjective, "steady," implied that there were no noticeable periodic or cyclic pressure variations. Any momentary movements of the gages were entirely random and transitory, and thus the readings of the gages could be taken as sensibly steady and represented correctly the movement of the fluid. Today, with the high-speed pressure transducers and recorders that lack the inertia-dampening characteristics of the older gages, the sense of steadiness may be obscured. For this reason, a better statement of the requirement is that the flow is not subject to *pulsations* as that term is defined in Chapter I-3, Par. I-3-45.

I-5-3 The following letter symbols will be used throughout this chapter:

A = Area of first or upstream section — sq ft or sq in.

a = Area of second or down-stream section — sq ft or sq in.

C = Coefficient of discharge — ratio

D = Diameter of pipe at up-stream section — ft or in.

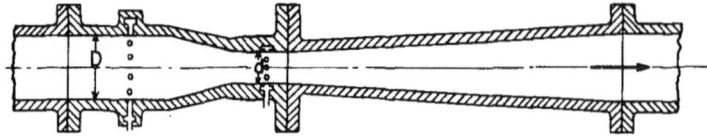

FIG. I-5-1 VENTURI TUBE—HERSCHEL TYPE

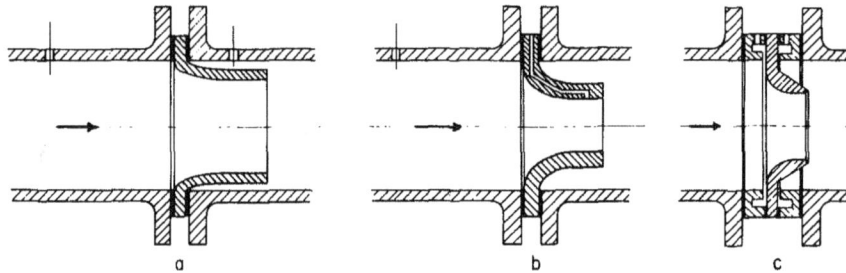

FIG. I-5-2 THREE FLOW-NOZZLE SHAPES AND
LOCATIONS OF PRESSURE TAPS

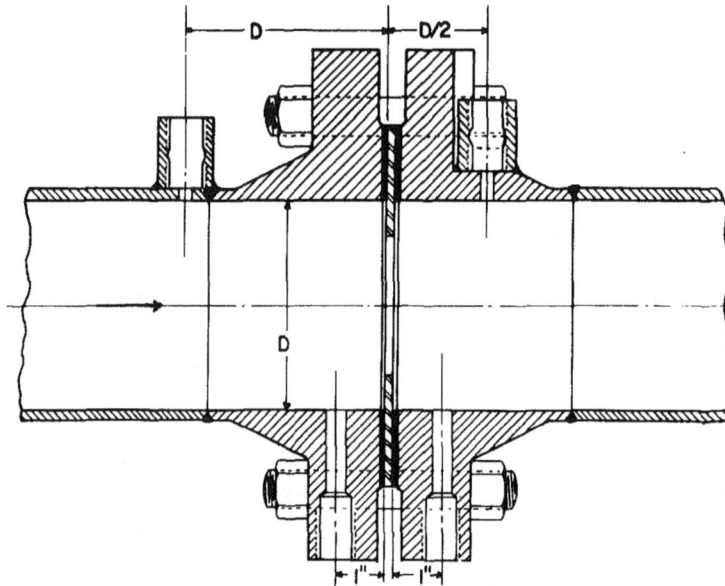

FIG. I-5-3 THIN-PLATE SQUARE-EDGED ORIFICE MOUNTED BETWEEN
FLANGES. (TWO PAIRS OF PRESSURE TAP LOCATIONS ARE
SHOWN; EITHER PAIR MAY BE USED, AND THE OTHER PAIR
OMITTED OR PLUGGED. ALSO, THERE ARE SEVERAL DE-
SIGNS OF SPECIAL FITTINGS IN WHICH ORIFICE PLATES
ARE MOUNTED.)

d = Diameter of primary element throat — ft or in.

E = Velocity of approach factor = $1/\sqrt{1-\beta^4}$ — ratio

F_a = Thermal expansion factor (of metals) — ratio

G = Specific gravity — ratio

g = Local acceleration of gravity — ft/sec^2

g_c = Proportionality constant = 32:174

H = Enthalpy = $p/\rho + u_i$ — ft lb$_f$/lb$_m$

h = Effective differential pressure — ft of fluid

J = Mechanical equivalent of heat — 778.16 ft lb$_f$/Btu

K = Flow coefficient = CE — ratio

M = Mach number — ratio

MW = Molecular weight

m = Mass rate of flow — lb$_m$/sec

p = Pressure, absolute pressure unless stated otherwise — psia or psfa

q = Volume rate of flow — cfs

q_H = Heat transferred to or from the fluid — ft lb$_f$/lb$_m$

R = Universal gas constant (see Note) — 1545.32 ft.lb$_f$/°R mole lb$_m$

R_g = Individual gas constant = R/MW — ft lb$_f$/lb$_m$°R

R_D = Reynolds number based on D — ratio

R_d = Reynolds number based on d — ratio

r = Ratio of pressures, p_2/p_1 — ratio

T = Absolute temperature — °R

u_i = Internal energy of the fluid — ft lb$_f$/lb$_m$

u_k = Kinetic energy of the fluid — ft lb$_f$/lb$_m$

V = Velocity — ft/sec

V_s = Velocity of sound in a fluid — ft/sec

v = Specific volume — ft^3/lb$_m$

x = Ratio of differential pressure to inlet pressure = $1 - r$ — ratio

Y = Expansion factor — ratio

Z = Compressibility factor of a gas — ratio

β = Ratio of diameters, d/D — ratio

γ = Ratio of specific heats of an ideal gas, c_p/c_v — ratio

Γ = Isentropic exponent of a real gas

Δp = Differential pressure, $p_1 - p_2$ — psi

λ = $1000/\sqrt{R_D}$ = $1000/\sqrt{\beta R_d}$ ratio

Λ = Height of a section above a datum — ft

μ = Absolute viscosity — lb$_m$/ft sec

ν = Kinematic viscosity — ft^2/sec

ρ = Density of fluid — lb$_m$/ft^3

σ = Standard deviation — per cent

ϕ = Flow function — ratio

Subscript 1 refers to the first or upstream section.

Subscript 2 refers to the second or throat section.

Subscript i refers to an ideal gas or a property thereof.

Subscript T refers to the theoretical rate or condition.

Subscript t refers to the total or stagnation pressure or temperature and may follow another subscript; thus, p_{1t} = total pressure at first section.

* refers to that section or to a fluid property or flow function, where or when the fluid speed is equal to the speed of sound.

Note: R = 10.7314 lb$_f$ft^3/in.^{2} °R mole lb$_m$ [21] Table 2-a and 10.7314 × 144 = 1545.32 lb$_f$ ft/°R mole lb$_m$, so that for air R_g = 1545.32/28.9644 = 53.3525 lb$_f$ft/lb$_m$°R. For other gases R_g = 1545.32/MW.

Concerning the units for A, a, D, d and p, refer to Notes 1 and 7 following Table I-2-1.

I-5-4 Let the values of p_1, u_{i_1}, u_{k_1}, V_1, v_1, and ρ_1 be arithmetical mean values obtained by averaging over the whole section, A, and, if the fluid motion is not strictly steady (laminar) but turbulent, a time average over the section. Likewise, let $p_2, \ldots \rho_2$ be the corresponding values over a second section, a. Let the area of a be considerably less than A. Then, since the flow is constant and steady, the same number of molecules must pass through section a as through A; but they must travel faster through section a than section A. In order to produce this change in velocity, which gives an increase in kinetic energy, there must be a decrease in other kinds of energy, particularly potential energy as evidenced by a decrease of static pressure. Also, in general, there will be a decrease in the internal energy of the fluid represented by a decrease in temperature. The decrease in pressure and temperature between sections A and a results in a change in the fluid density. With liquids this change in density is generally negligible, whereas with gases it must be taken into consideration.

I-5-5   The energy exchange represented by the decrease in static pressure between sections A and a, and which can be measured, is used in evaluating the difference in the velocities at the two sections and thence the rate of flow. So far as the analytical considerations are concerned, it is immaterial whether the area change from A to a takes place gradually or abruptly. However, the manner in which the area changes has a significant effect upon the magnitude of the static pressure difference and the positions and manner of measuring the static pressures. If the area change is gradual, or relatively so, so that the stream cross section is more or less well guided in changing from A to a, the minimum cross section of the stream appears to coincide with a. However, if the area change is abrupt and there is no guidance to the stream and if the axial length of a is very short compared to the diameter of the pipe, the cross section of the stream continues to decrease for a short distance downstream of a. That section at which this cross-sectional area is a minimum is called the "vena contracta." The distance of the vena contracta from a and its area depend upon the relation of a to A and the characteristics of the flow.

If downstream of section a the channel area returns to the same cross-sectional area as at section A, either gradually or abruptly, the static pressure, temperature and velocity of the fluid tend to return to nearly the same values that occurred at section A.

I-5-6   The basic physical concepts or equations from which are developed the equations for the flow of fluids through differential pressure meters are the equations of continuity and energy. The equation of continuity for steady flow is a special case of the general physical law of the conservation of matter. According to this equation the mass of fluid passing any section, A, per unit time is not only constant but is equal to that passing a second section, a, per unit time; thus,

$$AV_1 \rho_1 = aV_2 \rho_2 \qquad (I\text{-}5\text{-}1)$$

I-5-7   The Energy Equation.   The general energy equation is simply an energy accounting, and a statement of it is: as each pound of fluid passes from A to a, the increase of its total energy, kinetic plus internal, is equal to the work done on it plus the heat added to it. The work done upon the fluid due to the pressure change is $(p_1 v_1 - p_2 v_2)$ and that due to gravity or the change in gravitational potential (i.e., elevation) is $(\Lambda_1 - \Lambda_2)$. Thus, the general energy equation becomes

$$(u_{k_2} + u_{i_2}) - (u_{k_1} + u_{i_1}) = (p_1 v_1 - p_2 v_2) \\ + (\Lambda_1 - \Lambda_2) + q_H \qquad (I\text{-}5\text{-}2)$$

Particular attention should be paid to the fact that no assumption has been made about resistance to flow. Work done against resistance at the walls may be dissipated into thermal energy which stays in the fluid and increases its internal energy. In such a case, kinetic energy is lost, but the sum of the kinetic and internal energy is unaffected.

Since nothing has been said about the nature of the fluid, equation (I-5-2) is equally valid for liquids and gases; but the further developments are simpler for liquids because the compressibility of gases is an additional complication which requires the use of thermodynamics. Liquids will, therefore, be considered first.

I-5-8   For the development of the theoretical equations of flow it is both conventional and convenient to make the following limitations and assumptions concerning the section of pipe and the fluid flow through it.

1.   The pipe section is horizontal so that the effect of gravity is the same at all sections, and $\Lambda_1 = \Lambda_2$.

2.   In flowing from section A to section a, the fluid performs no external work.

3.   The flow is steady and axial, and the velocity profile at each section is relatively flat and normal to the pipe axis.

4.   There is no transfer of heat between the fluid and the pipe, i.e., q_H is zero.

5. With liquids *only*, there is no change of temperature between sections A and a, which implies there is no change in the internal energy of the fluid.

I-5-9 Theoretical Equations for Liquids. Since with an incompressible fluid, i.e., a liquid, the temperature does not change, the density is constant, so that $\rho_1 = \rho_2 = \rho$. Thus, the continuity equation, (I-5-1), becomes

$$AV_1 = aV_2 \qquad \text{(I-5-3a)}$$

or

$$V_1 = \frac{a}{A}V_2 \qquad \text{(I-5-3b)}$$

Under the conditions imposed by the preceding assumptions, the energy equation reduces to

$$u_{k1} + p_1v_1 = u_{k2} + p_2v_2 \qquad \text{(I-5-4)}$$

Because a *flat* velocity profile is assumed, the kinetic energy per pound (mass) is

$$u_k = \frac{V^2}{2g_c} \qquad \text{(I-5-5)}$$

and the general energy equation may be written

$$\frac{p_1}{\rho} + \frac{V_1^2}{2g_c} = \frac{p_2}{\rho} + \frac{V_2^2}{2g_c} \quad , \qquad \text{(I-5-6)}$$

Using the value of V_1 from equation (I-5-3) in equation (I-5-6) and rearranging gives

$$V_2^2 = 2g_c \left(\frac{p_1 - p_2}{\rho}\right)\left[\frac{1}{1 - \left(\frac{a}{A}\right)^2}\right] \qquad \text{(I-5-7)}$$

Attention is directed to two of the factors in equation (I-5-7). First, the quantity, $(p_1 - p_2)/\rho$, is equal to the difference between the static pressures at A and a if measured by a column of the flowing liquid h ft in height. This column is referred to generally as a "head" of the liquid and gives us, hence, the term, "differential head meter," which has been used in fluid-metering literature. The second is $1/[1 - (a/A)^2]$, the square root of which is called the "velocity of approach" factor, since it resulted from substituting the expression for V_1 into equation (I-5-6). If the areas are circular, which is by far the most common condition, the diameters are the

dimensions measured and known; hence, $(a/A)^2$ may be and usually is replaced with $(d/D)^4 = \beta^4$.

Note 1: A column of liquid when used as a measurement of pressure has the dimensions of force per unit area and not simply length.

Note 2: A table of values of the velocity of approach factor, $\sqrt{1/1 - \beta^4}$, is given in Part II.

In general, the user of a differential pressure meter is interested in knowing the rate of flow in terms of either mass (weight) or volume for some unit of time, such as second, minute, hour, or day. The theoretical equation for the mass rate of flow is

$$m_T = \rho\, a\, V_2$$
$$= a\sqrt{2g_c\rho\,(p_1 - p_2)}\,\sqrt{\frac{1}{1 - \beta^4}} \quad \text{lb}_m \text{ per sec} \qquad \text{(I-5-8)}$$

with the use of equation (I-5-7). The theoretical equation for volume rate of flow is

$$q_T = a\, V_2$$
$$= a\sqrt{\frac{2g_c\,(p_1 - p_2)}{\rho}}\,\sqrt{\frac{1}{1 - \beta^4}} \quad \text{cfs} \qquad \text{(I-5-9)}$$

$$= a\sqrt{2g_c h}\,\sqrt{\frac{1}{1 - \beta^4}} \quad \text{cfs} \qquad \text{(I-5-10)}$$

where

$$h = \frac{(p_1 - p_2)}{\rho} \quad \text{ft of flowing fluid} \qquad \text{(I-5-11)}$$

The subscript, T, indicates a theoretical value in contrast to actual rates of flow. Frequently equations (I-5-8), (I-5-9) and (I-5-10) are called the theoretical "hydraulic" equation for the flow of a fluid through orifices, flow nozzles and Venturi tubes.

I-5-10 Theoretical Equations for Compressible Fluids, Gases. The assumption of no transfer of heat between the fluid and the pipe, which implies no friction, permits assuming that any change of state between sections A and a is a reversible isentropic (adiabatic) change for which

$$p_1 v_1^\gamma = p_2 v_2^\gamma = p v^\gamma = \text{a constant, } c' \qquad \text{(I-5-12)}$$

If it is assumed that it is an ideal gas for which the general equation of state is

$$p/\rho = R/T \qquad \text{(I-5-13)}$$

then the energy equation, (I-5-2), may be written in the form

$$p_1/\rho_1 + \frac{V_1^2}{2g_c} + u_{i_1} = p_2/\rho_2 + \frac{V_2^2}{2g_c} + u_{i_2} \quad \text{(I-5-14)}$$

or, applying the definition for enthalpy,

$$\frac{V_2^2}{2g_c} - \frac{V_1^2}{2g_c} = H_1 - H_2 \quad \text{(I-5-15)}$$

For an ideal gas

$$H_1 - H_2 = \int_{p_1}^{p_2} v\,(dp) \quad \text{(I-5-16)}$$

Using equation (I-5-12) in the form, $v = c'/p^{1/\gamma}$,

$$H_1 - H_2 = c' \int_{p_1}^{p_2} p^{1/\gamma}\,(dp) \quad \text{(I-5-17)}$$

the integrand of which is

$$H_1 - H_2 = c'\,p_1^{\left(\frac{\gamma-1}{\gamma}\right)} \frac{\gamma}{\gamma-1}\left(1 - r^{\frac{\gamma-1}{\gamma}}\right) \quad \text{(I-5-18)}$$

Since $c' = v_1 p_1^{1/\gamma}$ and with equation (I-5-15),

$$\frac{V_2^2}{2g_c} - \frac{V_1^2}{2g_c} = p_1 v_1 \left(\frac{\gamma}{\gamma-1}\right)\left(1 - r^{\frac{\gamma-1}{\gamma}}\right) \quad \text{(I-5-19)}$$

Since the mass rate of flow is the same at sections A and a,

$$m = AV_1\rho_1 = aV_2\rho_2 \quad \text{(I-5-20)}$$

Using equation (I-5-12) in the form $\rho_2/\rho_1 = r^{1/\gamma}$

$$V_1 = V_2\left(\frac{a}{A}\right)r^{1/\gamma} \quad \text{(I-5-21)}$$

Substituting the relation from equation (I-5-21) in (I-5-19) and solving for V_2 gives

$$V_2 = \left[\frac{\dfrac{2g_c p_1}{\rho_1}\left(\dfrac{\gamma}{\gamma-1}\right)\left(1 - r^{\frac{r-1}{\gamma}}\right)}{r - \left(\dfrac{a}{A}\right)^2 r^{2/\gamma}}\right]^{1/2} \quad \text{(I-5-22)}$$

Using this value of V_2 in equation (I-5-20) and noting that $\rho_2 = \rho_1\, r^{1/\gamma}$ gives

$$m_T = a\left[\frac{2g_c p_1 \rho_1\, r^{2/\gamma}\left(\dfrac{\gamma}{\gamma-1}\right)\left(1 - r^{\frac{\gamma-1}{\gamma}}\right)}{1 - \left(\dfrac{a}{A}\right)^2 r^{2/\gamma}}\right]^{1/2} \quad \text{(I-5-23)}$$

This equation may be modified by using $p_1 = (p_1 - p_2)/(1 - r)$ and $\beta^2 = (a/A)$ so that it can be written

$$m_T = a\left[\frac{2g_c (p_1 - p_2)\,\rho_1\, r^{2/\gamma}\left(\dfrac{\gamma}{\gamma-1}\right)\left(\dfrac{1 - r^{\frac{\gamma-1}{\gamma}}}{1-r}\right)}{1 - \beta^4 r^{2/\gamma}}\right]^{1/2} \quad \text{(I-5-24)}$$

In contrast to equation (I-5-8), equation (I-5-24) is called the "theoretical adiabatic" equation for the mass rate of flow of an ideal compressible fluid across section a in terms of the initial pressure or the pressure difference and the density, ρ_1. Equation (I-5-24) may be written in the form

$$m_T = a\sqrt{\frac{2g_c \rho_1 (p_1 - p_2)}{1 - \beta^4}}$$

$$\left[r^{2/\gamma}\left(\frac{\gamma}{\gamma-1}\right)\left(\frac{1 - r^{\frac{\gamma-1}{\gamma}}}{1-r}\right)\right]^{1/2}\left(\frac{1-\beta^4}{1-\beta^4 r^{2/\gamma}}\right)^{1/2} \quad \text{(I-5-25)}$$

This amounts to using the hydraulic equation, (I-5-8), modified by the expansion factor,

$$Y = \left[r^{2/\gamma}\left(\frac{\gamma}{\gamma-1}\right)\left(\frac{1 - r^{\frac{\gamma-1}{\gamma}}}{1-r}\right)\left(\frac{1-\beta^4}{1-\beta^4 r^{2/\gamma}}\right)\right]^{1/2} \quad \text{(I-5-26)}$$

The value of Y depends upon the diameter ratio, β, the pressure ratio, r, and the ratio of specific heats, γ. For routine computations it will be found convenient to prepare curves or tables from which the values may be read or to use the curves or tables in Part II.

The values of ρ_1 in equations (I-5-24) and (I-5-25) should be computed with the general equation of state for an actual gas which, for this purpose, may be written

$$\rho_1 = \frac{p_1}{Z_1 R_g T_1} \qquad \text{(I-5-27)}$$

in which Z_1 is the compressibility factor for the particular gas being metered, corresponding to the conditions defined by p_1 and T_1.

I-5-11 Since most materials expand or contract as their temperature increases or decreases, a factor, F_a, must be introduced to take account of any change of the area of section a of the primary element when the operating temperature differs appreciably (e.g., more than 50 F) from the ambient temperature at which the device was manufactured and measured. If the meter is to be used under temperature conditions within the range of ordinary atmospheric temperatures, any difference between the thermal expansion of the pipe and the primary element may be ignored; and the diameter ratio, β, may be considered to be unaffected by temperature. However, if the meter is to be used at temperatures outside the ordinary range, then the material for the throat liner of a Venturi tube, a flow nozzle or an orifice plate should have a coefficient of thermal expansion as close as possible to that of the pipe. (An exception to this last statement could be where an orifice plate is mounted in a special fitting such that its outer rim is not clamped rigidly between flanges.) Values of F_a are given in Part II, Fig. II-1-2.

I-5-12 *Definition of Discharge Coefficient*. The *actual* rate of flow through a differential pressure meter is very seldom, if ever, exactly equal to the theoretical rate of flow indicated by the particular theoretical equation used. In general, the actual rate of flow is *less* than the indicated theoretical rate. Hence, to obtain the actual flow from the theoretical equation, an additional factor, called the "discharge coefficient," must be introduced. This coefficient is represented by C and defined by the equation:

$$C = \frac{\text{actual rate of flow}}{\text{theoretical rate of flow}} \qquad \text{(I-5-28)}$$

The rate of flow may be in terms of mass (or weight) or volume per unit of time. When volume is used, it is necessary that both the actual and the theoretical volumes be at or be referred to the same conditions of pressure and temperature. Thus, the actual rate of flow through a Venturi tube, flow nozzle or orifice,

when the general hydraulic equation is used, will be

$$m = \left(\frac{\pi d^2 F_a}{4}\right)\left(\frac{C}{\sqrt{1-\beta^4}}\right)\sqrt{2g_c \rho_1 (p_1 - p_2)} \qquad \text{(I-5-29)}$$

or

$$q = \left(\frac{\pi d^2 F_a}{4}\right)\left(\frac{C}{\sqrt{1-\beta^4}}\right)\sqrt{2g_c h} \qquad \text{(I-5-30)}$$

Using the adiabatic equation, (I-5-25), would require multiplying the right-hand side of equations (I-5-29) or (I-5-30) by the expansion term, equation (I-5-26).

The factor, $C/\sqrt{1-\beta^4}$, may be and frequently is replaced by the flow coefficient K, that is

$$K = C/\sqrt{1-\beta^4} \qquad \text{(I-5-31)}$$

Note: By using equation (I-5-31) and the relation given by equation (I-2-7) in equation (I-5-30) gives

$$q = a F_a K \sqrt{2gh} \qquad \text{(I-5-32)}$$

The values of C and K will be different for each different type of primary element, Venturi tube, flow nozzle and orifice. Also, with flow nozzles and orifices, the values will depend upon the locations of the pressure taps; and, with the orifice, the values will differ with the type of inlet edge, whether square and thin or rounded, i.e., the so-called quadrant-edge orifice. Since both C and K are ratios, their numerical values are independent of the system of units in which the various quantities are measured. Values of C for these several types and modifications of primary elements are given in Part II of this report in connection with the application and use of these meters.

I-5-13 Equations for Computing Actual Rates of Flow: Foot-Pound-Second-Fahrenheit Units. In most of the metering of fluids with differential pressure meters as used in this country, diameters are measured in *inches* (instead of feet); static pressures and differential pressures, in pounds (force) per sq in., inches of mercury or inches of water; densities, in pounds (mass) per cu ft; and temperatures, in degrees F. The rate of flow may be given in pounds per second (pps), cubic feet per second (cfs), or gallons per second (gps). Of course, these rates of flow may be given in terms of other time intervals such as a minute, hour or day. When in cubic feet or gallons, the temperature of the fluid should be obtained; and, with gases, the static pressures and, possibly, also the relative humidity will be needed in addition. For any combination of units of measure that may be selected, the necessary conversion factors can be combined with some of the other terms into a single numerical factor. Some of the more common combinations of units used in commercial work with the corresponding numerical factors are given below.

With p_1 and p_2 in psia, T_1 in deg R, and ρ_1 in lb_m per cu ft at the conditions p_1 and T_1, and using Y computed from equation (I-5-26),

$$m \text{ (lb}_m \text{ per sec)} = \frac{d^2 CY}{576}\left(\frac{F_a}{\sqrt{1-\beta^4}}\right)$$

$$\sqrt{2 \times 144 \times 32.174\,\rho_1\,(p_1 - p_2)} \quad \text{(I-5-33)}$$

$$= 0.52502\left(\frac{CYd^2 F_a}{\sqrt{1-\beta^4}}\right)\sqrt{\rho_1\,(p_1 - p_2)}$$

If the differential pressure is measured in inches of water at 68 F, then

$$(p_1 - p_2) = h_w\,\frac{62.3164}{1728} \quad \text{(I-5-34)}$$

Applying this relation to equation (I-5-33) gives

$$m \text{ (lb}_m \text{per sec)} = 0.099702$$

$$\left(\frac{CYd^2 F_a}{\sqrt{1-\beta^4}}\right)\sqrt{\rho_1 h_w} \quad \text{(I-5-35)}$$

or

$$q_1 \text{ (cfs at } p_1,\ T_1) = 0.099702$$

$$\left(\frac{CYd^2 F_a}{\sqrt{1-\beta^4}}\right)\sqrt{\frac{h_w}{\rho_1}} \quad \text{(I-5-36)}$$

Also,

$$m \text{ (lb}_m \text{ per hr)} = 358.93\left(\frac{CYd^2 F_a}{\sqrt{1-\beta^4}}\right)\sqrt{\rho_1\,h_w} \quad \text{(I-5-37)}$$

and

$$q_1 \text{ (cfh at } p_1,\ T_1) = 358.93\left(\frac{CYd^2 F_a}{\sqrt{1-\beta^4}}\right)\sqrt{\frac{h_w}{\rho_1}} \quad \text{(I-5-38)}$$

I-5-14 As stated in Par. I-2-11 of Chapter I-2, the reference condition for fuel-gas volumes is 14.73 psia, 60 F and dry and is indicated by the subscript,

b. Thus, for fuel gas the volume rate at the reference state is

$$q_b \text{ (scfh)} = q_1\,\frac{519.69\ p_1 Z_b}{14.73\ T_1 Z_1} \quad \text{(I-5-39)}$$

For many years it has been the practice in the fuel-gas industry to evaluate the density of the gas at the metering conditions in terms of its specific gravity referred to air. Usually this specific gravity has been based on a determination with a specific gravity balance or the indication of a recording gravitometer, rather than the ratio of molecular weights (see Par. I-3-29). Applying equation (I-3-22), the density of the gas at p_1, T_1 will be

$$\rho_1 = 2.6991\,\frac{p_1 G}{T_1 Z_1} \quad \text{(I-5-40)}$$

Combining equations (I-5-31), (I-5-39), and (I-5-40) with (I-5-38) gives

$$q_b \text{ (scfh)} = 7708\,KYd^2 F_a Z_b \sqrt{\frac{h_w p_1}{GT_1 Z_1}} \quad \text{(I-5-41)}$$

The use of these equations, or equivalents thereto, generally provides acceptable results as long as the metering conditions as represented by p_1, T_1 are not too far from normal ambient conditions (e.g., 0 psig $< p_1 <$ 200 psig and -20 F $< T_1 <$ 140 F). As the metering conditions depart further and further from normal ambient values, the use of equation (I-5-40) instead of (I-5-27) may give to rise to significant errors. This is because the behavior of *air* and *fuel gases*, as represented by the compressibility factor, Z, differ more and more as the metering conditions are extended further and further in either direction from the normal ambient air conditions. For this reason it is suggested that the use of equation (I-5-27) in conjunction with equation (I-5-33) and equations derived rigorously from these are to be preferred.

I-5-15 Strictly, equation (I-5-40) is applicable to a dry gas only. If the gas is wet, i.e., if there is water vapor mixed with it, then equation (I-3-39) should be used, which may be written in the form

$$\rho_1 \text{ (of wet gas)} = 2.6991\,\frac{G}{T_1 Z_1}$$

$$\left[p_1 - p_w\left(1 - \frac{0.622}{G}\right)\right] \quad \text{(I-5-42)}$$

Hence, when a higher degree of accuracy is desired in metering a wet gas, the effect of moisture upon the density can be accounted for by replacing p_1 in equation (I-5-41) with $[p_1 - p_w(1 - 0.622/G)]$.

Also, average room temperatures are usually close enough to 68 F that the observed differential pressure in inches of water at room temperature may be assumed to be the same as if measured at 68 F without introducing any appreciable error as far as the solutions of many practical problems are concerned. However, there will be cases in which the requirements of the problem will justify taking account of any difference between the room and reference temperatures, and the method of doing this is illustrated in one of the typical problems in Part II.

I-5-16 Equations for Actual Rates of Flow: Kilogram-Meter-Second-Celsius Units. Let the several dimensions and quantities be expressed as follows:

D = Centimeters (cm)

d = Centimeters (cm)

g_c = 980.652

h_w = Centimeters of water at 20 C

p = Kilogram$_f$ per square centimeter, (kg$_f$/sq cm), or gram$_f$ per sq cm, (gm$_f$/sq cm)

Δp = Gram$_f$ per square centimeter, (gm$_f$/sq cm) or (gm$_m$/cm-sec^2)

m = Gram$_m$ per second, (gm$_m$/sec), or kilogram$_m$ per second, (kg$_m$/sec)

q = Cubic decimeter per second, (cu dm/sec), or cubic meter per second, m^3/sec)

ρ = Gram$_m$ per cubic centimeter, (gm$_m$/cc)

The basic equation for differential pressure meters, as represented by equation (I-5-29), may be expressed in the manner

$$m \ (\text{gm}_m/\text{sec}) = \frac{\pi}{4} \ \text{cm}^2$$

$$\left\{ 2 \left[g_c \left(\frac{\text{gm}_m}{\text{cm}-\text{sec}^2} \right) \right] \left(\frac{\text{gm}_m}{\text{cm}^3} \right) \right\}^{\frac{1}{2}} \quad \text{(I-5-43)}$$

Using $K = C/\sqrt{1 - \beta^4}$, the numerical value of g_c, and the appropriate letter symbols,

$$m \ (\text{gm}_m/\text{sec}) = 34.783 \, KYd^2 F_a \sqrt{\Delta p \rho_1} \quad \text{(I-5-44)}$$

or

$$m \ (\text{kg}_m/\text{sec}) = 0.034783 \, KYd^2 F_a \sqrt{\Delta p \rho_1} \quad \text{(I-5-45)}$$

At 20 C (68 F), the density of water is 0.9982336 gm$_m$/cc; thus,

$$\Delta p = 0.99823 \, h_w$$

and

$$m \ (\text{kg}_m/\text{sec}) = 0.034752 \, KYd^2 F_a \sqrt{h_w \rho_1} \quad \text{(I-5-46)}$$

Also,

$$q_1 \ (\text{m}^3/\text{sec at } p_1, \, T_1) = 0.000034752 \cdot$$

$$KYd^2 F_a \sqrt{\frac{h_w}{\rho_1}} \quad \text{(I-5-47)}$$

A committee on units of the International Gas Union recommends using as the reference or "standard" conditions for the evaluation of fuel gas volumes a pressure of 1013 millibars, a temperature of 15 C, and dry (i.e., free of water vapor).

Note 1: These conditions are equivalent to 14.696 psia, 59 F and dry. Thus a cu ft of gas at the AGA reference conditions (Par. I-2-11, Chapter I-2) would be 1.000385 cu ft at the IGU conditions.

Note 2: 1 atmosphere = 1013 millibars
= 760 mm Hg at 0 C
= 1.033226 kg$_f$/cm^2

The density of dry air at 1 atm and 0 C (273.16 K) is 0.00129304 gm$_m$/cc; thus

$$\rho_1 \ (\text{gm}_m/\text{cc}) = 0.34185 \frac{p_1 G}{T_1 Z_1} \quad \text{(I-5-48)}$$

Combining equation (I-5-48) with (I-5-45) gives

$$m \ (\text{kg}_m/\text{sec}) = 0.020339 \, KYd^2 F_a \sqrt{\frac{p_1 G}{T_1 Z_1} \Delta p} \quad \text{(I-5-49)}$$

and, with equation (I-5-47),

$$q_1 \ (\text{m}^3/\text{sec}) = 0.000059431 \, KYd^2 F_a \sqrt{h_w \frac{T_1 Z_1}{p_1 G}} \quad \text{(I-5-50)}$$

Applying the reference conditions given above,

$$q_b = q_1 \frac{288.16}{1.033226} \frac{p_1}{T_1 Z_1} \quad \text{(I-5-51)}$$

and, combined with equation (I-5-50),

$$q_b \ (\text{m}^3/\text{sec}) = 0.016575 \, KYd^2 F_a \sqrt{\frac{h_w \rho_1}{G T_1 Z_1}} \quad (\text{I-5-52})$$

The rates of flow per minute and per hour can be obtained, of course, by multiplying the above equations for m or q by 60 and 3600, respectively.

I-5-17 Determination of Discharge Coefficients. The basic procedure in determining discharge coefficients of differential pressure meters is to discharge the flow from the meter into a weighing tank, volumetric tank or holder. By noting the increase in the scale reading or the change in the content of the tank or holder for a measured interval of time, the *actual* mass rate of flow (corrected for air buoyancy) or volume rate of flow is determined. Simultaneously, the necessary indications of the meter under test are observed; and by substitution of these values in one of the equations given above, letting $C = 1$, the "theoretical" rate of flow is computed. The ratio of the actual rate of flow to the calculated theoretical rate is the discharge coefficient. Making such comparisons over even a limited range of conditions requires the use of much special equipment and involves considerable tedious computation if done manually. Fortunately, individual calibration is not now necessary in many cases since recognized coefficient values are available for use with certain definite and reproducible forms of differential pressure meters, which will be specified in later sections.

A considerable number of tests have been made to determine the discharge coefficients of certain differential pressure meters. As would be expected, the conditions under which these tests were made and their relative self-consistency differ rather widely, so that it is difficult to form an altogether satisfactory estimate of the proper discharge coefficient to be used in any given case. However, it is possible to refer the results of some of these different groups of tests to a common basis for comparison, and, to the extent that this is possible, composite discharge coefficient values can be obtained which are reliable within the limits of the experimental observations. Before a discussion of the correlation of coefficients of discharge, some comments on the correlation tools or ratios may be helpful.

I-5-18 Kinematic Similarity of Fluid Flows. Much can be learned about the flow of a fluid through a particular (large) channel by careful study of the flow through a small model of the original or prototype, *provided* that there is complete kinematic similarity between the model and the prototype. This condition may be satisfied if two conditions are fulfilled:

1. The model channel must be geometrically similar to the prototype channel. For example, if the prototype channel is a Venturi tube of certain shape, then the model should be a geometrically similar Venturi tube.

2. The flow pattern or pattern of the streamlines in the model must be similar to that in the prototype. The flow pattern of the streamlines, in turn, is determined by all the forces acting.

Among the forces just referred to are those arising from conditions preceding and following the special section. In other words, the flow pattern in such a section as the Venturi tube has been influenced by what preceded the tube, and what follows the tube may have some slight influence also. Therefore, the second condition cannot be fulfilled unless under the first condition the upstream and downstream configurations of the model are geometrically similar to those of the prototype. Although in principle this similarity of configuration should extend indefinitely, both preceding and following the particular section, in practice an approximation to such complete geometric similarity may have to suffice. No rules are at hand to be a guide as to what may constitute "approximate" geometric similarity, but possibly the requirements given in Fig. II-II-1 of Part II could be considered as minimums. However, it must be recognized that results of tests made with a model will be of doubtful value when applied to the prototype unless there is both geometric and dynamic similarity between model and prototype. Moreover, the correlation of test data, as discussed below, depends on the same degree of complete kinematic similarity.

I-5-19 Types of Forces Acting. Various specific laws of similarity or similitude could be devised, depending upon the type of forces acting. The types of forces are inertia, viscous, pressure, and elastic or compressible.

By inertia force is meant the resistance of an inert mass to acceleration. The magnitude of the inertia force is proportional to the product of particle mass and particle acceleration.

It is customary to consider separately two cases, or combinations, each with only three forces; thus: (1) viscous, inertia, and pressure and (2) elastic, inertia, and pressure. The first combination is characterized by the Reynolds number and the second combination by the Mach number. In each case, specifying two of the forces automatically specifies the third force because the three forces are in equilibrium. Therefore, in case (1) a significant pair of forces can be taken as viscous and inertia, whereas in case (2) a significant pair of forces can be taken as inertia and elastic.

I-5-20 Reynolds Number. Assume that the fluid is incompressible and that the flow takes place within a completely enclosing channel or that bodies having motion with respect to the stream are fully immersed in the fluid so that free surfaces do not enter into consideration and gravity forces are balanced by buoyant forces. For such a flow, the inertia and viscous forces are the only ones which need to be taken into account. Mechanical similarity exists if, at points similarly located with respect to the bodies, the ratios of the inertia forces to the viscous forces are the same.

Since the product of mass multiplied by acceleration is proportional to volume × density × (velocity/time), the inertia force is proportional to

$$\frac{L^3 \rho V}{T} = \frac{L^3 \rho V}{L/V} = L^2 \rho V^2 \qquad \text{(I-5-53)}$$

where V is some characteristic velocity (for example, the average velocity over a fixed cross section of pipe), L is some characteristic length (such as the internal diameter of a pipe), ρ is density, and T is time.

The magnitude of the viscous or laminar internal friction force is proportional to the viscous shear stress, S, times some length squared, or SL^2. Since, for laminar flow, the shear stress equals the dynamic viscosity, μ, times the velocity gradient, the shear stress, S, is proportional to $\mu V/L$. The shear force, SL^2, is thus proportional to μVL. Then the dimensionless ratio, (inertia force)/(viscous force), is proportional to

$$\frac{L^2 \rho V^2}{\mu VL} = \frac{\rho VL}{\mu} \qquad \text{(I-5-54)}$$

The name, Reynolds number, has been given to the ratio, $\rho VL/\mu$.

I-5-21 In the product composing the Reynolds number, L denotes any linear dimension of the section of the channel; and, in its use with differential pressure meters, the custom is to replace L with either D or d. Also, for brevity and convenience R_D is used to represent $DV_1\rho/\mu$ and R_d to represent $dV_2\rho/\mu$.

If viscous, inertia, and pressure forces determine the flow of an incompressible fluid for a prototype, then mechanical similarity between model and prototype is realized when the Reynolds number of the model equals the Reynolds number of the prototype.

With incompressible fluids, i.e., liquids, any change in the temperature as the fluid moves from section 1 to section 2 is in general so slight as to be negligible. Thus, the general practice is to assume that the density, ρ, and viscosity, μ, are the same at the two sections, so that $R_d = dV_2\rho/\mu_1$. On the other hand, with compressible fluids, gases, there may be an appreciable decrease in temperature accompanying the decrease in pressure from p_1 to p_2. These changes affect both ρ and μ, so that the correct evaluation of R_d is

$$R_d = \frac{dV_2\rho_2}{\mu_2} \qquad \text{(I-5-55)}$$

However, as T_2 is not readily determined by direct measurement, it is convenient to note that $V_2 = 4m/\pi d^2\rho_2$ and, thus, $R_d = 4m/\pi d\mu_2$. Since, in the great majority of cases the effect on the viscosity of the temperature change from T_1 to T_2 is small, it is customary to assume $\mu_2 = \mu_1 = \mu$. This results in arbitrarily using for evaluating the Reynolds number the relations,

$$R_d = \frac{dV_2\rho_2}{\mu} = \frac{4m}{\pi d\mu} \qquad \text{(I-5-56)}$$

and

$$R_D = \frac{DV_1\rho_1}{\mu} = \frac{4m}{\pi D\mu} \qquad \text{(I-5-57)}$$

In the preceding equations for R_d and R_D, the normal unit in which d and D would be expressed is the foot. As stated in Par. I-5-13, it is the general practice to express these diameters in inches; and, to keep d and D in inches, it is necessary to use

$$R_d = \frac{dV_2\rho_2}{12\mu} = \frac{48m}{\pi d\mu} \qquad \text{(I-5-58)}$$

and

$$R_D = \frac{DV_1\rho_1}{12\mu} = \frac{48m}{\pi D\mu} \qquad \text{(I-5-59)}$$

Note 1: By dividing equation (I-5-59) by equation (I-5-58), it will be seen that $R_D = \beta R_d$.

Note 2: The normal metric units for the Reynolds number are: D and d, in centimeters; V_1 and V_2, in cm/sec; ρ_1 and ρ_2, in gm_m/cm^3; m, in gm_m'/sec; and μ, in poise.

I-5-22 Mach Number. Consider a flow of a compressible fluid through two geometrically similar channels in which the only forces involved are inertia, pressure and elastic. A significant ratio is (inertia force)/(elastic force). The inertia force is proportional to $\rho L^2 V^2$. The elastic or compressible force is proportional to EL^2, where E is the bulk modulus of the fluid. The bulk modulus, E, in turn, equals ρV_s^2, where V_s is the acoustic velocity in the fluid. Then the ratio, (inertia force)/(elastic force), is proportional to

$$\frac{\rho L^2 V^2}{EL^2} = \frac{\rho L^2 V^2}{\rho V_s^2 L_2} = \frac{V^2}{V_s^2} \qquad (I\text{-}5\text{-}60)$$

If inertia, pressure and elastic forces determine the flow for a prototype, then mechanical similarity between model and prototype is realized when the ratio, V^2/V_s^2, for the model equals the corresponding ratio, V^2/V_s^2, for the prototype.

For purposes of flow similarity, the ratio, V/V_s, could be used as well as the ratio, V^2/V_s^2. The name Mach number has been given to the ratio, V/V_s.

I-5-23 Application of Similitude. The use of some parameter, as Reynolds number or Mach number, for correlating data depends upon the forces involved in the flow.

For example, consider the flow of a liquid through a Venturi meter. The Reynolds number could be a very useful parameter for correlating data because the forces in the number are those involved in the flow. For flow over a weir, however, other forces may exist, such as gravity, surface tension and capillarity. The Reynolds number does not involve these additional forces and, therefore, may not be adequate for a complete correlation of data.

As another example, consider the flow of a gas through a meter. At low velocities the gas may behave as an incompressible fluid, and the Reynolds number may provide a suitable parameter for correlating data. At high velocities, however, compressibility effects may be present, and it may be necessary to use the Mach number in order to correlate the data.

I-5-24 Fluid-Flow Characteristics: Jet Contraction with an Orifice. With both the Venturi and the flow nozzle, the section used in all computations is the one of minimum cross-sectional area of the tube or the nozzle; and the fluid stream completely fills this section, being guided by the walls of the tube or the nozzle. With an orifice, the fluid stream is not so guided; and, as shown in Fig. I-5-4, the cross section of the stream continues to decrease after passing through the orifice. In strict analogy to the Venturi tube, the area of the minimum jet section, known as the vena contracta, which corresponds to the throat of a Venturi tube, should be used in the flow equation. However, no satisfactory method of actually measuring this minimum jet area is known, whereas measuring the diameter of an orifice and thus determining its area is a relatively simple matter. Some

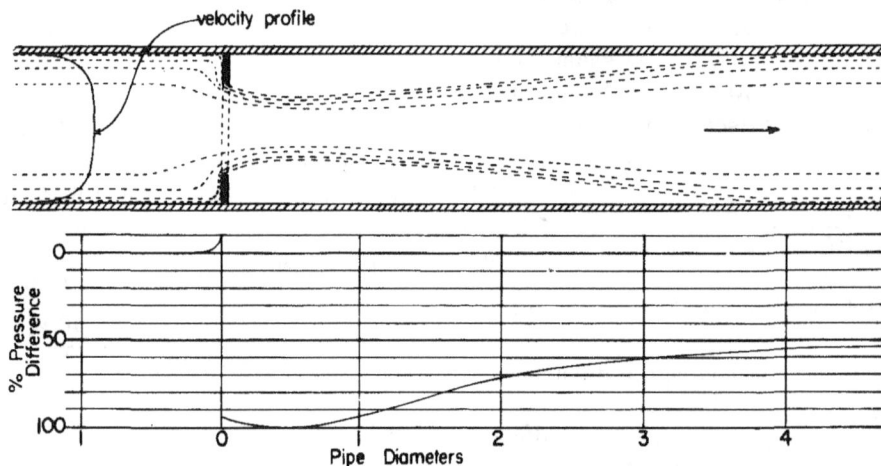

FIG. I-5-4 DIAGRAMMATIC REPRESENTATION OF FLUID FLOWING
THROUGH THIN-PLATE SQUARE-EDGED ORIFICE AND
RELATIVE PRESSURE CHANGES

investigators have represented the ratio of the minimum jet area to the orifice area by a separate factor, which then is called a contraction coefficient. Doing this, however, is of no practical advantage, and including the effects of contraction in the orifice discharge coefficient is more convenient. By experiment, the amount of contraction has been found to depend primarily upon the diameter ratio, β, and the properties of the fluid; hence, the discharge coefficient will also vary with these factors when it includes the effects of jet contraction.

With any one fluid and the same differential pressure, the relative amount of jet contraction increases as the diameter ratio, β, decreases. That is, the ratio of the jet area, at the vena contracta, to the area of the orifice decreases as β decreases. This is exactly what would be expected, because, as the fluid particles near the wall of the channel converge toward the orifice, they attain greater radial velocity inward when β is small than when β is large. Thus, the discharge coefficients for orifices will increase as β increases [1, 2, 3].

At ordinary rates of flow, the fluid properties that have the most influence on jet contraction are compressibility or its reciprocal, expansibility, and viscosity. Between the plane of the orifice and the vena contracta, a radial pressure gradient outward is present; and, if the fluid is a gas, it expands transversely as well as longitudinally, whereas in the case of a liquid, no expansion occurs. The cross section of the vena contracta is, therefore, larger with a gas than with a liquid [1]. Hence, if the hydraulic equation is used, as is the general practice, the discharge coefficient corresponding to a given jet velocity, computed from the use of the inlet density, will be numerically lower for a gas than for a liquid like water.

As the viscosity of the fluid increases, jet contraction decreases. The reason is that, as the viscosity increases, the effects of friction against the surface of the channel extend farther toward the center and the radial velocity of the stream filaments from near the wall is small in proportion to the axial velocity of the central filaments, thus making the vena contracta increase. Hence, for the same jet velocity, the discharge coefficient for such a fluid as a medium oil will be higher than for water [4, 5].

I-5-25 Static-Pressure Changes Close to an Orifice. All studies of the static-pressure changes close to a thin-plate square-edged orifice have shown that on the inlet side, within the last $0.1\,D$, the static pressure increases to a maximum at the corner of the plate and pipe wall. On the outlet side, the

static pressure continues to decrease until a minimum is reached somewhere between about $1/4\,D$ and $1\,D$. For any one orifice, the distance from the plane of the inlet side of the orifice plate to that where the static pressure at the wall of the pipe is the minimum is nearly independent of the rate of flow if the fluid is a liquid. Regardless of the nature of the fluid, this distance is a function of the orifice-to-pipe diameter ratio, β, becoming less as the value of β becomes larger. Numerous experiments by different investigators over a wide range of pipe sizes have shown that the relation between the position of minimum static pressure and the diameter ratio, β, is independent of the actual pipe size for pipes 2 in. or more in diameter. The average maximum and minimum values of the distance of the plane of minimum static pressure from the plane of the orifice, as reported by different experiments, are shown in Fig. I-5-5.

With the flow of compressible fluids, that is, gases, through an orifice, the position of minimum static pressure has been found to depend upon the rate of flow as indicated by the pressure ratio, $p_2/p_1 = r$, as well as upon the diameter ratio, β. If the value of r is very close to unity, the position of the minimum pressure will be indistinguishable from that observed with liquids [6]. However, as r decreases, that is, as the rate of flow increases, the position of the minimum static pressure moves farther away from the orifice. The fact that the plane of minimum static pressure is not at the orifice indicates that the area of the jet at the plane of minimum pressure is less than the area of the orifice. The general assumption is that this plane of minimum pressure coincides with that of the vena contracta. A few observations

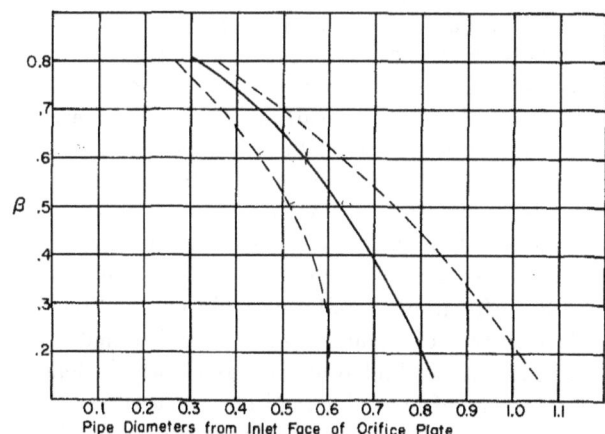

FIG. I-5-5 LOCATION OF VENA CONTRACTA OUTLET PRESSURE TAP WITH CONCENTRIC SQUARE-EDGED ORIFICES (BROKEN LINES SHOW MAXIMUM VARIATION LIMITS.)

on a stream of fluid passing at low rates of flow through an orifice held in a large glass tube indicated that, under those conditions, the assumption is correct. However, this assumption has not been positively established as valid for all conditions, although no experimental evidence to the contrary is known. Therefore, the coincidence of the plane of the vena contracta and that of minimum static pressure must be considered as an assumption even though it may be used as if it were a fact.

I-5-26 Location of the Pressure Taps. The locations of pressure taps used in test programs and commercial work will be named and defined as follows:

1. *Flange Taps.* The centers of the pressure holes are respectively 1 in. from the upstream, or inlet, and downstream, or outlet, faces of the orifice plate. Allowing for a 1/16-in. gasket, the centers of the holes will be 15/16 in. from the bearing faces of the flanges.

2. *Taps at One D and One-Half D.* The center of the inlet pressure tap is located one pipe diam preceding the inlet face of the orifice plate. The center of the outlet pressure tap is placed one-half pipe diam following the inlet face of the orifice plate, regardless of the value of the diameter ratio, β.

3. *Vena Contracta Taps.* The center of the inlet pressure tap is located between one half and two pipe diam from the upstream face of the orifice plate; usually a distance of one pipe diam is used. The center of the outlet pressure tap is placed at the position of minimum pressure (which is assumed to be the plane of the vena contracta) as given in Fig. I-5-5.

Note: Due to the flatness of the pressure gradient in the region of $\frac{1}{2}D$, the differential pressures observed with D and $\frac{1}{2}D$ taps and with vena contracta taps are almost the same over the central part of the range of β ratios.

4. *Corner Taps.* The pressure holes open in the corner formed by the pipe wall and the orifice plate. The method of doing this and the widths of the openings, either single holes or ring slits, is shown in Fig. I-5-2 (c). The axial width of the slits or openings should be $0.02\,D$. The same corner-tap proportions are used with both nozzles and orifices.

I-5-27 Effects of Installation and Construction. In the development of equations such as (I-5-8) and (I-5-29), a uniform fluid velocity was assumed, thus neglecting any effects of normal stream turbulence. This normal turbulence can be greatly increased, by the configuration of the channel or obstructions in the channel, so that the distribution of velocity becomes very irregular or a pronounced spiral motion may be set up. For convenience in this discussion,

this exaggerated turbulence will be termed "disturbed flow." Obviously, the character of this disturbed flow will depend in large measure upon the installation conditions, that is, upon the type of fittings and their relative distances, on the inlet side particularly, from the primary element. The effects of the disturbances produced by several kinds of fittings have been extensively studied in connection with orifice meters and, to a lesser extent, with Venturi tubes and flow nozzles. In general, when disturbed flow is produced by the presence of fittings, the value of the discharge coefficient is more likely to be higher than when determined under normal flow conditions. The magnitude of the effect will depend mostly upon the type of fittings, the distance from the primary element, and the diameter ratio, β. The minimum installation requirements that should be fulfilled to reduce the effects produced by such fittings to the minimum are given in Part II of this edition.

The condition of the channel surface, that is, its relative roughness, will also affect the flow of a fluid through a differential pressure meter. In general, a higher discharge coefficient appears to be obtained with a rough-surface pipe than with a smooth-surface pipe of the same size. To date, no index for relative roughness, especially with respect to the interior surface of a pipe, has received general recognition. Thus, at present there is no reference basis by which to correlate such data as are now available on this factor. On the other hand, if a given surface grain size, that is, a given degree of absolute roughness, is assumed, the relative roughness will be greater with a small pipe than with a large one. Therefore, a reasonable expectation would be that a higher discharge coefficient will be obtained with square-edged orifices with the smaller pipe, whereas, with Venturi tubes, the coefficient will be slightly lower. This is exactly what has been observed when different sizes of pipe of ordinary commercial smoothness have been used. Thus, the size of the pipe is one of the factors influencing the values of discharge coefficients, particularly of orifices. With nozzles, the direction of pipe size effect may vary, depending to some extent upon the nozzle-approach curvature.

Note: The evaluation of a surface in terms of microinches is becoming a common practice. However, this does not give a measure of the "waviness," either circumferential or longitudinal, of the interior surface of a pipe. Also, some surface instruments measure the peak to hollow height, automatically integrate the area on either side of a mean center line, and divide this by the length of travel to give a "center-line average" (CLA) value of the surface.

I-5-28 Turning to the influences of primary-element construction, the discharge coefficient of a Venturi tube or a nozzle having a smooth finish, of the inlet cone or curved entrance, has often been observed to be higher than that of a tube or nozzle of the same size but having a rough finish.

With orifices, the jet contraction and the value of the discharge coefficient are influenced in a marked degree by the condition of the upstream edge, or corner, of the orifice. Dulling or rounding of this edge decreases the jet contraction; and, if the rounding is continued, contraction will be entirely suppressed as in the case of a nozzle. Dulling or a very slight rounding of a sharp square corner produces relatively much more effect than a small change in the curvature of an orifice of well-rounded approach.

The width of the cylindrical face or edge of the orifice, as measured normal to the plane of the inlet face of the orifice plate, has also to be considered. If this edge width is relatively large (e.g., one-third of the orifice diameter), the amount of jet contraction is decreased, if not suppressed altogether. Therefore, to make possible the comparison of data and to facilitate reproducing given conditions, the orifice edge width must be kept small in proportion to the other dimensions of the orifice plate.

The position of the center of the orifice with respect to the axis of the pipe, that is, whether the orifice is placed concentrically or eccentrically in the pipe, will affect the pressure gradient, particularly on the outlet side. If the orifice is displaced away from the pressure taps, the differential pressure may be a few per cent different than when centered. On the other hand, as the orifice is displaced toward the pressure taps, the differential pressure will become increasingly erratic. Therefore, the orifice must be placed concentric with the pipe, particularly when the pressure taps are within one or two pipe diam of the orifice plate. This, of course, does not apply to eccentric and segmental orifices.

The variation of the static pressure or pressure gradient on the inlet and outlet sides of the orifices was mentioned in Par. I-5-25 and depicted in Fig. I-5-4. Evidently, the value of the differential pressure, $(p_1 - p_2)$ or h, which will be observed, will depend upon the positions of the pressure taps. Consequently, the discharge coefficient value determined or used will also be influenced by the position of the pressure taps. This applies to flow nozzles also, but to a lesser degree.

I-5-29 Correlation of Discharge Coefficients: Flow Nozzles and Orifices. In Pars. I-5-20 – I-5-22, it was indicated that the properties of the flowing fluid may influence the character of the flow through a differential pressure meter and, thereby, the coefficient. This fact may be represented by the relation

$$C = \psi \left(\frac{d V \rho}{\mu}, \frac{V}{V_s}, \beta \right)$$

$$C = \psi \ (R_d, M, \beta) \qquad \text{(I-5-61)}$$

Or

$$K = \psi' \ (R_d, M, \beta)$$

In the paragraphs above, most of the constructional features that will affect the character of the flow and, therefore, the coefficients are described briefly. All of them can be described in terms of a ratio of either the pipe diameter, D, or the throat diameter, d. For example, if the actual roughness, i.e., the peak to hollow height of the inner surface of a pipe, is ζ, then the relative roughness is ζ/D. Again, let v be the radius of curvature of the inlet edge of an orifice, and the relative degree of curvature would be v/d. But d may be specified in terms of $\beta = d/D$; hence, v/d may be replaced by v/D. For an absolutely square edge, $v = 0$.

Thus, the general relation between the discharge or flow coefficients and the fluid characteristics plus the constructional features of the primary element may be represented by

$$C = \psi_2 \ (R_d, M, \beta, D)$$
$$\qquad \text{(I-5-62)}$$
$$K = \psi_2' \ (R_d, M, \beta, D)$$

With incompressible fluids, the Mach number, M, has so little if any effect upon C or K that it can be and usually is omitted from the relation. On the other hand, the diameter ratio, $\beta \ (= d/D)$, must remain an independent variable, because limiting it to a single value, which would be implied if represented by D only, would destroy the flexibility of these meters, especially the orifice. The evaluation of the actual relations represented by equations (I-5-62) must be derived from the results of actual tests.

I-5-30 Application of Coefficients for Incompressible Fluids to Compressible Fluids: Expansion Factors. As stated previously, with compressible fluids there is a change of the fluid density accompanying the pressure change from p_1 to p_2. In the general relationship, represented by equation (I-5-62), this was represented by the Mach number, M. However, in considering the effects of fluid compressibility when both the initial and jet velocities are less than the acoustic velocity, V_s, it is usually convenient to replace M with the acoustic ratio, x/γ, as discussed in Par. I-3-42, resulting in equation (I-3-54). Thus the general relation may be written

$$C = \psi_3 (R_d, \ x/\gamma, \ \beta, \ D)$$

$$K = \psi' (R_d, \ x/\gamma, \ \beta, \ D)$$

(I-5-63)

In Venturi tubes and flow nozzles, the expansion which accompanies the change in pressure takes place in an axial direction *only*, due to the confining walls of these differential producers. The adiabatic expansion factor, Y, equation (I-5-26), compensates for this unidirectional expansion. With the thin-plate orifice, there are no confining walls, and the expansion takes place *both* radially and axially. To take account of this multidirectional expansion, an empirical equation for the expansion factor is derived from tests.

I-5-31 Expansion Factor, Y, for Square-Edged Orifices: Dependence upon the Acoustic Ratio. Assume that a particular orifice can be tested with air

($\gamma = 1.4$) over a fairly wide range of differential pressures so that $(p_1 - p_2)/p_1 = x$ will range from about 0.01 to 0.45 or over. Also, the static pressure is to be taken on the upstream or inlet side of the orifice, so that the inlet density, ρ_1, can be determined directly. For defining the flow coefficient, K, based on the inlet conditions, the hydraulic equation in the following form will be used

$$m = \frac{\pi d^2}{4} F_a \ (nK_1) \sqrt{2g_c \ \rho_1 \ (p_1 - p_2)}$$

(I-5-64)

The values of K_1 obtained from such a series of tests will be found to decrease as the ratio, x, increases. Moreover (except for large-diameter ratios), this relation is very nearly linear, so nearly so that the plotted values of K_1 to x can be represented by a straight line, such as line I in Fig. I-5-6.

Note: In equation (I-5-64) n is a numerical constant that takes account of the units in which the several factors are measured.

Now, suppose that the series of tests is repeated, using the same orifice but a different gas, for which $\gamma = 1.28$. As before, the plotted values of K_1 to x can be represented by a straight line. However, this line will be steeper than the first, as illustrated by line II

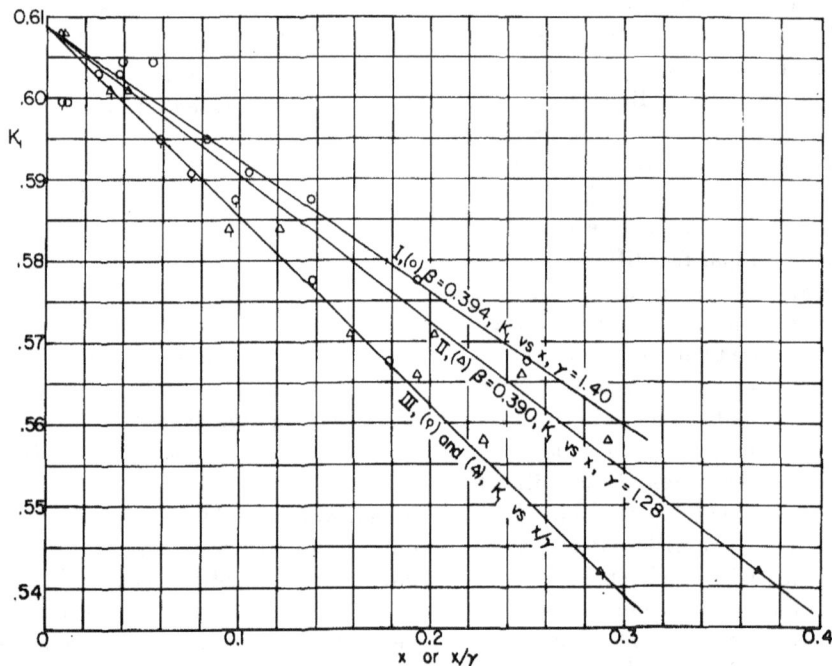

FIG. I-5-6 COEFFICIENTS FOR HYDRAULIC EQUATION OBTAINED FROM TESTS WITH GASES, PLOTTED AGAINST DIFFERENTIAL PRESSURE RATIO AND ACOUSTIC RATIO (K_1 IS THE FLOW COEFFICIENT, INCLUDING THE EXPANSION FACTOR, Y_1.)

in Fig. I-5-6. The slopes of these two lines are seen to be very nearly inversely proportional to the ratios of the specific heat. Hence, if values of x/γ instead of x are used as abscissas, the results from both tests could be expected to be represented by a single line, such as is shown to be the case by line III in Fig. I-5-6.

If K_0 denotes the value of K_1 when $x = 0$ (as obtained by extending line III to the 0-abscissa line) and ϵ (epsilon) is the slope of the K_1-to-x/γ line, then the value of K_1 corresponding to any particular value of x will be given by

$$K_1 = K_0 - \epsilon \frac{x}{\gamma}$$

$$= K_0 \left(1 - \frac{\epsilon}{K_0} \frac{x}{\gamma}\right) \qquad (I\text{-}5\text{-}65)$$

$$= K_0 Y_1$$

from which

$$Y_1 = \left(1 - \frac{\epsilon}{K_0} \cdot \frac{x}{\gamma}\right) \qquad (I\text{-}5\text{-}66)$$

Y_1 is termed the "net expansion factor" since it is introduced to take account of the effects of expansion as an expansible fluid flows through an orifice, and the "hydraulic" equation is used for computing the rate of flow [1, 7]. As indicated, the value of Y_1 depends upon the acoustic ratio and the ratio of ϵ/K_0. From many experiments, the slope factor, ϵ, has been found to be practically independent of all orifice shape factors except β.

K_0 has been defined as the limiting value of K_1 as $x \to 0$; and, sometimes, it is referred to as the water flow coefficient, K_w. In Par. I-5-29, this value of K_0 or K_w was shown to be a function of D, R_d and β. Hence, the ratio, ϵ/K_0, apparently should also be a function of D, R_d and β. However, the effect of D and R_d upon the ratio, ϵ/K_0, is generally so small that it may be neglected and ϵ/K_0 considered as a function of β only.

I-5-32 For gases, the product of the flow coefficient and the expansion factor for a given orifice meter shape (i.e., "shape" = a constant) is seen to be dependent upon the Reynolds number and the acoustic ratio. The effects of these two ratios are doubtlessly interdependent to a slight extent but not enough to be of any practical significance. A possible reason for the inability to separate the effects of the two ratios is that the Reynolds numbers of

most, if not all, of the tests from which the expansion factor was determined were fairly high. Thus, any changes of K due to the Reynolds-number effect would be small enough to be entirely masked by the much greater effect of expansion [7].

If $K = K_0 = K_w$, a relation which is used often without stating it, the complete hydraulic equation for use with orifice meters when metering gases is obtained. This equation is

$$m = 0.52502\, K\, Y_1\, d^2 F_a \sqrt{\rho_1 \Delta p} \qquad (I\text{-}5\text{-}67)$$

$$m = 0.52502\, \frac{C}{\sqrt{1-\beta^4}}\, Y_1\, d^2 F_a \sqrt{\rho_1 \Delta p}$$

Repeating what has been stated above, K is a function of D, R_d and β, while Y_1 is a function of x/γ and β. When the procedure as outlined, is followed and the inlet pressure and temperature are used to evaluate the density, ρ_1, (or ρ_1 is obtained directly with a densitometer), the corresponding expansion factor Y_1, is given by

$$Y_1 = 1 - (0.410 + 0.350\beta^4)x/\gamma \qquad (I\text{-}5\text{-}68)$$

This equation applies when the inlet pressure tap is between 0 and D, as measured from the inlet face of the orifice plate, and the outlet pressure tap is at the downstream corner or within the distance shown by the "mean" line of Fig. I-5-5, as measured from the inlet face of the orifice plate, also. In other words, it applies to corner, flange, vena contracta and D and $1/2\, D$ taps. Furthermore, this equation applies *only* with jet velocities *below* the velocity of sound at the conditions in the jet, that is, where $V_2 < V_s$ [1].

In equation (I-5-65) and the development of equation (I-5-68), the coefficient of discharge, C, could have been used instead of the flow coefficient, K. However, if this had been done, the numerical factors in equation (I-5-68) would be different. As it is, expansion factors computed with equation (I-5-68) *must* be used with the flow coefficient, K, or the equivalent, $C/\sqrt{1-\beta^4}$.

Note: Another equation for the expansion factor is

$$Y = 1 - (0.3707 + 0.3184\beta^4)(1 - r^{1/\gamma})^{0.925} \qquad (I\text{-}5\text{-}69)$$

This equation is given in ISO Document No. R 541 (1966), "Measurement of Fluid Flow," and is intended, primarily, for use with corner taps [8].

I-5-33 In many cases, especially where flange taps are used, the static pressure is taken from the downstream or outlet pressure tap. If the fluid is a gas, this means that the density, ρ_2, is determined from p_2 and either T_1 or T_2, whichever is measured (usually there is very little difference between T_1 and T_2). The expansion factor, Y_2, which corresponds with the use of ρ_2 (i.e., p_2), can be evaluated from Y_1 in the following manner. Using the relation given by equation (I-5-27), equation (I-5-67) may be written

$$m = nKY_1 \sqrt{\Delta p \, p_1} = nKY_2 \sqrt{\Delta p \, p_2} \qquad \text{(I-5-70)}$$

Here, as before

$$\Delta p = p_1 - p_2$$

and also, by definition,

$$x = \Delta p/p_1 = 1 - r$$

and

$$r = p_2/p_1$$

From equation (I-5-70),

$$Y_2 = Y_1 \sqrt{\frac{p_1}{p_2}} = Y_1 \sqrt{\frac{1}{r}} = Y_1 \sqrt{\frac{1}{1-x}} \qquad \text{(I-5-71)}$$

Let $x_2 = \Delta p/p_2$. Then, $\Delta p = x_2 p_2 = x p_1$, or

$$x = x_2 \frac{p_2}{p_1} = x_2 \frac{p_2}{p_2 + \Delta p} = \frac{x_2}{1 + x_2} \qquad \text{(I-5-72)}$$

Applying this to equation (I-5-71) gives

$$Y_2 = Y_1 \sqrt{1 + x_2} \qquad \text{(I-5-73)}$$

Using the value of Y_1 from equation (I-5-68) and the value of x from equation (I-5-72) gives

$$Y_2 = \left(\sqrt{1 + x_2}\right) - (0.41 + 0.35\beta^4) \frac{x_2}{\gamma} \frac{1}{\sqrt{1 + x_2}} \qquad \text{(I-5-74)}$$

I-5-34 **Mass Flow by Modifications to the Orifice Meter.** It has been shown both analytically and experimentally that by using the principle of the "magnus effect" a pressure difference directly proportional to mass flow rate may be obtained. To do this, a cylinder is rotated in the flow conduit, and the circumferential velocity of the cylinder causes the fluid velocity in one gap between the cylinder and conduit wall to increase, while in the opposite

gap the fluid velocity is decreased. The difference in the static pressures at the two gaps is directly proportional to the total mass rate of flow irrespective of fluid density [9].

Again, it has been shown both analytically and experimentally that by placing two matched orifices in series a pressure difference can be obtained that is directly proportional to the mass rate of flow. To do this, a constant volumetric flow is recirculated through the system in such a way as to be subtracted from the flow through one of the orifices and added to the flow through the other. The difference between the separate differential pressures is proportional directly to the net mass rate of flow. Several modifications of this procedure are possible [10].

It is to be noted that, with both of the procedures outlined above, auxiliary power is required. In the first procedure, power is required to rotate the cylinder; in the second, a power-driven pump is required to produce the recirculating flow. The procedures in both cases are proprietary.

I-5-35 **Definite Relationships between C, R_d, β, and D: Venturi Tubes.** The discharge coefficients for the Herschel type Venturi tube, Fig. I-5-1, recommended by the Technical Committee on Measurement of Fluid Flow in Closed Conduits, of the International Organization for Standardization [11], are:

1. For a Venturi tube with a rough-cast convergent inlet section and

$$4 \text{ in.} \leq D \leq 32 \text{ in.}$$
$$0.3 \leq \beta \leq 0.75$$
$$2.10^5 \leq R_D \leq 2.10^6$$
$$C = 0.984 \pm 0.70 \text{ per cent.}$$

2. For a Venturi tube with a machined convergent inlet section and

$$2 \text{ in.} \leq D \leq 10 \text{ in.}$$
$$0.4 \leq \beta \leq 0.75$$
$$2.10^5 \leq R_D \leq 1.10^6$$
$$C = 0.995 \pm 1.0 \text{ per cent.}$$

3. For a Venturi tube with a rough welded sheet-iron convergent inlet section and

$$8 \text{ in.} \leq D \leq 48 \text{ in.}$$
$$0.4 \leq \beta \leq 0.7$$
$$2.10^5 \leq R_D \leq 2.10^6$$
$$C = 0.985 \pm 1.50 \text{ per cent.}$$

Note: The tolerance values following the values of C are twice the standard deviation as given in the ISO Recommendation. This is in agreement with the method of evaluating tolerances used in this report.

I-5-36 **Flow Nozzle Coefficients.** From a review of several thousand tests of flow nozzles of the ASME long-radius type, in 2-in. and larger pipes, the following equation has been derived for use with pipe-wall pressure taps at $1\,D$ and $1/2\,D$, and $0.30 \gtrless \beta \gtrless 0.825$ and $10^4 \gtrless R_d \gtrless 10^6$

$$C = 0.99622 + 0.00059\,D$$

$$- (6.36 + 0.13D - 0.24\,\beta^2)\frac{1}{\sqrt{R_d}} \quad \text{(I-5-75)}$$

A table of flow nozzle coefficients computed from the above equation is given in Part II.

I-5-37 Coefficients for Thin-Plate Square-Edged Orifices. The equations given below, in Chapter II-III, and the values given in Tables II-III-2, II-III-3 and II-III-4 are based on determinations made at Ohio State University and reported by the Joint AGA-ASME Committee on Orifice Coefficients [12, 13]. Although there have been many determinations made in different laboratories since those at OSU, the tests at any one laboratory have covered only a limited set of conditions. Attempts to correlate these later determinations as a group or with the OSU values have not resulted satisfactorily. Therefore, any improvement to the equations and tabulated values given in this report must await new determinations covering a sufficiently wide and comprehensive range of conditions [14, 44].

Since it may aid in making interpolations as well as providing a basis for tables for pipe sizes other than those given in Chapter II-III, the equations are given in detail. These equations apply particularly to pipes 2 in. and larger, to values of β between 0.2 and 0.75 and to values of R_d above 10^4. The letter symbols are as defined in Par. I-5-3 above, with special values as follows:

K = Flow coefficient corresponding to any specific set of values of D, β, and R_d (or R_D)

K_0 = The limiting value of K for any specific values of D and β when R_d (or R_D) becomes infinitely large

C = $K/E = K\sqrt{1-\beta^4}$ and $R_D = \beta R_d$

I-5-38 For flange pressure taps, K_e = the particular value of K for any specific values of D and β, when $R_d = (10^6 d)/15$. $K = K_0\,(1 + A/R_d)$, and $K_0 = K_e\,[(10^6 d)/(10^6 d + 15A)]$.

$$K_e = 0.5993 + \frac{0.007}{D} + \left(0.364 + \frac{0.076}{\sqrt{D}}\right)\beta^4$$

$$+\, 0.4\left(1.6 - \frac{1}{D}\right)^5\left[\left(0.07 + \frac{0.5}{D}\right) - \beta\right]^{5/2}$$

$$-\left(0.009 + \frac{0.034}{D}\right)\left(0.5 - \beta\right)^{3/2}$$

$$+\left(\frac{65}{D^3} + 3\right)\left(\beta - 0.7\right)^{5/2} \quad \text{(I-5-76)}$$

and

$$A = d\left(830 - 5000\beta\right.$$

$$+\, 9000\beta^2 - 4200\beta^3 + \left.\frac{530}{\sqrt{D}}\right) \quad \text{(I-5-77)}$$

Note: In equation (I-5-76), each of the last three terms, for some value of β, reduces to the form $x\sqrt{-1}$, i.e., to an "imaginary" number. In such cases the term is to be dropped.

I-5-39 For $1\,D$ and $1/2\,D$ taps and vena contracta taps, $K = K_0 + b\lambda$, and $\lambda = 1000/\sqrt{R_D} = 1000/\sqrt{\beta R_d}$. For the $1\,D$ and $1/2\,D$ taps,

$$K_0 = (0.6014 - 0.01352D^{-1/4})$$

$$+\, (0.3760 + 0.07257D^{-1/4})$$

$$\left(\frac{0.00025}{D^2\beta^2 + 0.0025D} + \beta^4 + 1.5\beta^{16}\right) \quad \text{(I-5-78)}$$

and

$$b = \left(0.0002 + \frac{0.0011}{D}\right) + \left(0.0038 + \frac{0.0004}{D}\right)$$

$$[\beta^2 + (16.5 + 5D)\,\beta^{16}] \quad \text{(I-5-79)}$$

For vena contracta taps,

$$K_0 = 0.5922 + 0.4252$$

$$\left(\frac{0.0006}{D^2\beta^2 + 0.01D} + \beta^4 + 1.25\beta^{16}\right) \quad \text{(I-5-80)}$$

and

$$b = 0.00025 + 0.002325$$

$$(\beta + 1.75\beta^4 + 10\beta^{12} + 2D\beta^{16}) \quad \text{(I-5-81)}$$

I-5-40 Sonic-Flow Primary Elements. In 1839, from theoretical studies of Bernoulli's and Venturi's works, Saint Venant and Wantzel developed a general equation of the discharge of fluids from aperatures by which the existence of a sonic-flow limit could be inferred. The phenomenon that the mass rate of flow of a gas through a nozzle reaches a maximum that is directly proportional to the inlet pressure was observed by Weisbach in 1866 and again by Fleigner in 1874. In recent years, the sonic-flow nozzle has been used as a reference meter, as a transfer standard, as a control for regulating the flow of a gas, and as a propulsion engine.

I-5-41 Maximum Theoretical Flow Rate of an Ideal Gas. From the equation of continuity, the theoretical mass flow rate per unit area is

$$\frac{m_T}{a} = \rho_2 V_2 \tag{I-5-82}$$

Using the relation $V_2 = M_2 V_s$ and the equation for an ideal gas in the form,

$$\rho_2 = \frac{p_2 (MW)}{R T_2} \tag{I-5-83}$$

equation (I-5-82) becomes

$$\frac{m_T}{a} = \frac{p_2 (MW) M_2 V_s}{R T_2} \tag{I-5-84}$$

For an ideal gas the acoustic speed is

$$V_s = \sqrt{\gamma g_c R T_2 / (MW)} \tag{I-5-85}$$

Inserting this value into equation (I-5-84) and multiplying all terms by $\sqrt{T_{1t}} \big/ p_{1t}$ develops the flow function, θ_i, for an ideal gas

$$\theta_i = \frac{m_T \sqrt{T_{1t}}}{a \, p_{1t}} = \frac{p_2}{p_{1t}} \sqrt{\frac{T_{1t}}{T_2}} M_2 \sqrt{\frac{g_c (MW) \gamma}{R}} \tag{I-5-86}$$

In order to simplify equation (I-5-86), the ratios of p_2 / p_{1t} and T_{1t} / T_2 are needed. For an ideal gas the relation between the specific heats and the gas constant is

$$c_p - c_v = R/J \ (MW) \tag{I-5-87}$$

or

$$c_p = [\gamma/(\gamma - 1)] \ [R/J \ (MW)] \tag{I-5-88}$$

The first law of thermodynamics may be written

$$c_p T_{1t} = c_p T_2 + V_2^2 / 2 g_c J \tag{I-5-89}$$

Combining equations (I-5-88) and (I-5-89) and since $V_2 = M_2 V_s$,

$$\frac{T_{1t}}{T_2} = 1 + \frac{V_2^2}{2 g_c c_p J T_2} = 1 + \frac{(\gamma - 1) V_2^2 (MW)}{2 g_c \gamma R T_2} \tag{I-5-90}$$

Inserting the acoustic speed from equation (I-5-85),

$$\frac{T_{1t}}{T_2} = 1 + \frac{\gamma - 1}{2} M_2^2 \tag{I-5-91}$$

According to the isentropic process equation,

$$\frac{T_{1t}}{T_2} = \left(\frac{p_{1t}}{p_2} \right)^{\frac{\gamma - 1}{\gamma}} \tag{I-5-92}$$

from which the ratio of p_{1t}/p_2 is found to be

$$\frac{p_{1t}}{p_2} = \left(1 + \frac{\gamma - 1}{2} M_2^2 \right)^{\frac{\gamma}{\gamma - 1}} \tag{I-5-93}$$

Inserting these relations into equation (I-5-86) defines the flow function for an ideal gas entirely in terms of the throat Mach number and the gas properties.

$$\theta_i = \frac{m_T \sqrt{T_{1t}}}{a \, p_{1t}}$$

$$= M_2 \sqrt{ \left(1 + \frac{\gamma - 1}{2} M_2^2 \right)^{-\frac{\gamma + 1}{\gamma - 1}} } \sqrt{\frac{\gamma g_c (MW)}{R}} \tag{I-5-94}$$

The maximum mass rate of flow can be found by setting the logarithmic derivative of the flow function, equation (I-5-94), to zero.

$$\frac{d\theta_i}{\phi_i} = 0 = \frac{dM_2}{M_2} - \frac{\frac{(\gamma + 1)}{2} M_2 dM_2}{1 + \left(\frac{\gamma - 1}{2} \right) M_2^2} \tag{I-5-95}$$

$$\frac{dM_2}{M_2}\left[1 + \frac{\gamma - 1}{2}M_2^2 - \frac{(\gamma + 1)}{2}M_2^2\right] = 0 \qquad (I-5-96)$$

Setting the bracketed term to zero,

$$1 - \frac{1}{2}M_2^2 - \frac{1}{2}M_2^2 = 0 \qquad (I-5-97)$$

$$M_2^2 = 1 \quad \text{and} \quad M_2 = 1 \qquad (I-5-98)$$

Thus, it is seen that the theoretical maximum mass rate of flow per unit area exists at a throat Mach number of unity. This is called "choked" flow and, frequently, also "critical" flow.

To show that the sonic flow occurs at the throat, the logarithmic differential of the continuity equation,

$$\frac{d\rho}{\rho} + \frac{dV}{V} - \frac{dA}{A} = 0 \qquad (I-5-99)$$

is combined with the compressible-flow momentum equation

$$dp = -\frac{\rho}{g_c}VdV \qquad (I-5-100)$$

Then,

$$\frac{da}{a} = g_c\frac{dp}{\rho V^2} - \frac{d\rho}{\rho} = g_c\frac{dp}{\rho V^2}\left(1 - \frac{V^2 d\rho}{g_c dp}\right) \quad (I-5-101)$$

In isentropic flow the term, $g_c(dp/d\rho)$, is recognized as the square of the acoustic speed.

$$\frac{da}{a} = g_c\frac{dp}{\rho V^2}(1 - M^2) \qquad (I-5-102)$$

A Mach number of unity occurs at $da = 0$, which means at the minimum conduit area. Additionally, this shows why sonic flow is not observed with a thin, square-edged orifice. With such an orifice, the function, da, is discontinuous at the "throat," and it does not smoothly attain the unique value of zero.

Since the maximum flow occurs in the throat at a Mach number of unity, the sonic flow function, θ_i^*, can be obtained from equation (I-5-94) by setting M_2 equal to unity. Thus,

$$\theta_i^* = \frac{m_T}{a^*}\frac{\sqrt{T_{1t}}}{p_{1t}} \qquad (I-5-103)$$

$$= \sqrt{\gamma\left(\frac{1 + \gamma}{2}\right)^{-\left(\frac{\gamma + 1}{\gamma - 1}\right)}}\sqrt{\frac{g_c(MW)}{R}}$$

Let

$$F_i = \sqrt{\gamma\left(\frac{1 + \gamma}{2}\right)^{-\left(\frac{\gamma + 1}{\gamma - 1}\right)}} \qquad (I-5-104)$$

F_i is called the ideal isentropic expansion function. Then the sonic flow function is

$$\theta_i^* = F_i\sqrt{g_c\frac{(MW)}{R}} \qquad (I-5-105)$$

I-5-42 The choking pressure ratio can be determined from equation (I-5-93) by setting M_2 equal to unity and inverting:

$$\frac{p_2}{p_{1t}} = \left(\frac{2}{\gamma + 1}\right)^{\frac{\gamma}{\gamma - 1}} \qquad (I-5-106)$$

When β approaches zero, the static pressure, p_1, approaches the total or stagnation pressure, p_{1t}, of the flowing fluid. For any value of the ratio, β, this total pressure can be measured with an impact or simple Pitot tube. But often, in the interest of minimizing upstream flow disturbances, the static pressure is measured with a pipe-wall pressure tap. The relation between p_1 and p_{1t} for any value of β is

$$\left(\frac{p_1}{p_{1t}}\right)^{2/\gamma} - \left(\frac{p_1}{p_{1t}}\right)^{\frac{\gamma + 1}{\gamma}}$$

$$= \beta^4\frac{\gamma - 1}{2}\left(\frac{2}{\gamma + 1}\right)^{\frac{\gamma + 1}{\gamma - 1}} \qquad (I-5-107)$$

For β less than 0.5 the following simpler equation is sufficiently accurate:

$$\frac{p_1}{p_{1t}} \approx 1 - \beta^4 \frac{\gamma}{2}\left(\frac{2}{\gamma+1}\right)^{\frac{\gamma+1}{\gamma-1}} \qquad (I-5-108)$$

Note: The ratio between the outlet and inlet static pressures, $r_c = p_2/p_1$, at throat sonic velocity of an ideal gas, may be derived from equation (I-5-23) written in the form

$$m_T^2\left(\frac{1}{a^2}\right)\left(\frac{1}{2g_c\,p_1\,\rho_1}\right)\left(\frac{\gamma-1}{\gamma}\right)$$

$$=\left(r^{2/\gamma} - r^{\frac{\gamma+1}{\gamma}}\right)(1-\beta^4\,r^{2/\gamma}) \qquad (I-5-109)$$

The derivative of equation (I-5-109), dm_T/dr, equated to zero reduces to

$$r_c^{\frac{1-\gamma}{\gamma}} + \left(\frac{\gamma-1}{2}\right)\beta^4\,r_c^{2/\gamma} = \frac{\gamma+1}{2} \qquad (I-5-110)$$

Values of r_c for several values of β and γ are given in Table I-5-1.

I-5-43 As discussed in Par. I-3-17, the temperature indicated by a thermometer held in a moving stream is between the stagnation and static values. The use of thermometer wells designed to measure the stagnation temperature is recommended [15, 16, 17]. However, in isentropic flow, the stagnation temperature is conserved, even across the shock plane, so that a thermometer may be placed downstream and thus cannot affect the inlet flow.

I-5-44 Determination of the Maximum Ideal Flow Rate of a Real Gas. In actual use the gas discharged from a sonic flow primary element is a real gas. Real gases have two major deviations from the behavior of

an ideal gas. First, they do not follow the ideal equation of state, $pv/T = $ constant. Instead the real gas equation of state usually is written

$$\frac{p\,(MW)}{\rho R\,T} = Z\,(p,T) \neq 1 \qquad (I-5-111)$$

As indicated, Z is actually a function of pressure and temperature and in general varies between 0.2 and 4.2 [21]. Z is within 20 per cent of unity in the reduced pressure range from 6 to 9 and for reduced temperatures greater than unity. Also, Z is between 1.0 and 0.8 for reduced pressures between 0.0 and 0.5 and reduced temperatures greater then unity. The reduced pressure is the ratio of the actual pressure of the fluid to its critical pressure. Likewise, the reduced temperature is the ratio of the actual temperature of the fluid to its critical temperature. At the critical temperature of a fluid, the density of the gaseous and liquid phases are identical. Also, at temperatures above the critical no amount of pressure will cause a gas to condense. Second, the isentropic exponent of a real gas, Γ, is not constant and is not equal to the ratio of specific heats, c_p/c_v. Γ is also a function of stagnation pressure and temperature; but the effect of pressure is slight, and its value is given, usually, as a function of temperature alone.

Note: The effect of pressure on Γ is given by the relation

$$\frac{c_p/c_v}{\Gamma^*} = \frac{1-2cp^2}{Z} \approx \frac{1}{Z} \quad \text{when } p < 450 \text{ psia} \qquad (I-5-112)$$

where c is the second virial coefficient.

These differences must be included from the beginning in the derivation of a sonic flow function, ϕ^*, of a real gas. It is assumed that the gas flows isentropically and that sonic flow exists at the throat.

Table I-5-1 Values of r_c for Different Values of β and γ, as Determined by Equation (I-5-110)

$\beta \backslash \gamma$	1.10	1.15	1.20	1.25	1.30	1.35	1.40	1.45	1.667
0.00	0.5847	0.5744	0.5645	0.5550	0.5458	0.5369	0.5283	0.5200	0.4872
0.20	0.5849	0.5746	0.5647	0.5552	0.5460	0.5371	0.5285	0.5202	0.4874
0.40	0.5877	0.5775	0.5675	0.5580	0.5489	0.5401	0.5315	0.5232	0.4905
0.50	0.5923	0.5820	0.5722	0.5627	0.5537	0.5450	0.5364	0.5280	0.4956
0.60	0.6006	0.5905	0.5809	0.5715	0.5625	0.5538	0.5454	0.5372	0.5050
0.65	0.6072	0.5972	0.5877	0.5784	0.5694	0.5609	0.5526	0.5444	0.5124
0.70	0.6159	0.6061	0.5968	0.5875	0.5788	0.5703	0.5620	0.5541	0.5223
0.75	0.6280	0.6183	0.6092	0.6002	0.5914	0.5832	0.5750	0.5673	0.5358
0.80	0.6440	0.6347	0.6258	0.6171	0.6084	0.6005	0.5925	0.5849	0.5542
0.85	0.6671	0.6580	0.6496	0.6412	0.6332	0.6251	0.6177	0.6104	0.5809

Using * to denote the gas properties and conditions at sonic flow, then as for an ideal gas

$$\frac{p_1}{p_2} = \left(\frac{\rho_1}{\rho_2}\right)^{\Gamma} \qquad \text{(I-5-113)}$$

For pressure waves of small intensity, the acoustic speed is defined by

$$V_s = \sqrt{\left(\frac{\partial p}{\partial \rho}\right)_s} = \sqrt{\Gamma^* g_c \frac{Z R T}{(MW)}}$$
$$= \sqrt{\Gamma^* g_c \frac{p}{\rho}} \qquad \text{(I-5-114)}$$

where Γ^* is the local value of the isentropic exponent at the speed of sound and not the mean value, Γ, required by equation (I-5-113). In order to proceed to the flow function, the choking pressure ratio must be found. Equation (I-5-113) is incorporated into the momentum equation, (I-5-100), and p is substituted for p_1 and is allowed to vary. This means that equation (I-5-113) must define the path of integration,

$$\int_0^V \frac{V dV}{g_c} = - \int_{p_{1t}}^{p_2} \frac{dp}{\rho} = \frac{p_2^{1/\Gamma}}{\rho_2} \int_{p_2}^{p_{1t}} p^{-1/\Gamma} \, dp \qquad \text{(I-5-115)}$$

where p_2 is taken as the fixed lower limit, and

$$\frac{V^2}{2g_c} = \frac{p_2^{1/\Gamma}}{\rho_2}\left(\frac{\Gamma}{\Gamma-1}\right)\left(p^{\frac{\Gamma-1}{\Gamma}}\right)\bigg|_{p_2}^{p_{1t}} \qquad \text{(I-5-116)}$$

Inserting the limits

$$\frac{V^2}{2g_c} = \frac{p_2^{1/\Gamma}}{\rho_2}\left(\frac{\Gamma}{\Gamma-1}\right)\left(p_{1t}^{\frac{\Gamma-1}{\Gamma}} - p_2^{\frac{\Gamma-1}{\Gamma}}\right) \qquad \text{(I-5-117)}$$

Multiplying by $(p_2/p_2)^{\frac{\Gamma-1}{\Gamma}}$ and setting $V = V_s$ from equation (I-5-114),

$$\frac{p_2}{\rho_2}\left(\frac{\Gamma}{\Gamma-1}\right)\left[\left(\frac{p_{1t}}{p_2}\right)^{\frac{\Gamma-1}{\Gamma}} - 1\right] = \frac{\Gamma^* g_c p_2}{2 g_c \rho_2} \qquad \text{(I-5-118)}$$

from which the choking pressure ratio for the flow of a real gas is

$$\frac{p_2}{p_{1t}} = \left[\frac{\Gamma^*}{2}\left(\frac{\Gamma-1}{\Gamma}\right) + 1\right]^{-\frac{\Gamma}{\Gamma-1}} \qquad \text{(I-5-119)}$$

It is to be noted that this reduces to the pressure ratio for an ideal gas when $\Gamma^* = \Gamma = \gamma$.

As before, the maximum ideal rate of flow per unit area is determined by the continuity equation at sonic speed,

$$\frac{m_T}{a^*} = \rho_2 V_s = \sqrt{\Gamma^* g_c p_2 \rho_2} \qquad \text{(I-5-120)}$$

Inserting the stagnation pressure from equation (I-5-119) and the density from equation (I-5-113), which is

$$\rho_2 = \frac{p_{1t}(MW)}{Z R T_{1t}}\left(\frac{p_2}{p_{1t}}\right)^{1/\Gamma} \qquad \text{(I-5-121)}$$

equation (I-5-120) becomes

$$\frac{m_T}{a^*} = \sqrt{\frac{\Gamma^* g_c p_{1t}^2 (MW)}{Z R T_{1t}}\left[1 + \frac{\Gamma^*}{2}\left(\frac{\Gamma-1}{\Gamma}\right)\right]^{-\frac{\Gamma+1}{\Gamma-1}}} \qquad \text{(I-5-122)}$$

Multiplying all terms by $\sqrt{T_{1t}}/p_{1t}$ completes the development of the sonic-flow function of a real gas [18, 19, 20],

$$\phi^* = \frac{m_T}{a^*}\frac{\sqrt{T_{1t}}}{p_{1t}} = \sqrt{\Gamma^*\left[1 + \frac{\Gamma^*}{2}\left(\frac{\Gamma-1}{\Gamma}\right)\right]^{-\frac{\Gamma+1}{\Gamma-1}}}$$
$$\sqrt{\frac{g_c (MW)}{Z R}} \qquad \text{(I-5-123)}$$

Analogously to the ideal-gas derivation, the function, F, is called the real-gas isentropic expansion function, that is,

$$F = \sqrt{\Gamma^*\left[1 + \frac{\Gamma}{2}\left(\frac{\Gamma-1}{\Gamma}\right)\right]^{-\frac{\Gamma+1}{\Gamma-1}}} \qquad \text{(I-5-124)}$$

and the sonic-flow function may be written

$$\phi^* = F \sqrt{\frac{g_c (MW)}{Z R}} \qquad \text{(I-5-125)}$$

I-5-45 For certain gases, such as steam, ammonia and some of the refrigerants, the sonic-flow function varies much more than for other real gases. Also, the equations of state for most of these particular gases or "vapors" are tabulated in terms of pressure, temperature and specific volume. The existence of these tables makes it more convenient to calculate the mass rate of flow using the pressure and specific volume. Substituting the relation,

$$P_{1t} v_{1t} = \frac{Z R T_{1t}}{(MW)} \qquad (I\text{-}5\text{-}126)$$

into equation (I-5-22) and using equation (I-5-124),

$$\frac{m_T}{a^*} = F \sqrt{g_c} \sqrt{\frac{P_{1t}}{v_{1t}}} \qquad (I\text{-}5\text{-}127)$$

For the gases tabulated in Part II, the sonic flow functions and the isentropic expansion functions are presented as dimensionless ratios of the real gas function to the ideal-gas functions, (ϕ^*/ϕ_i^*) and (F/F_i). These tables show the departure of the real-gas sonic-flow from that of an ideal gas of like molecular structure and equal molecular weight, whose isentropic exponent is constant and equal to the theoretical ratio of specific heats, c_p/c_v. Applying the calibration coefficient relation,

$$Ca = a^* \qquad (I\text{-}5\text{-}128)$$

completes the development of the working sonic flow equations.

From equation (I-5-123)

$$m = a C \left(\frac{\phi^*}{\phi_i^*}\right) \phi_i^* \left(\frac{P_{1t}}{\sqrt{T}_{1t}}\right) \qquad (I\text{-}5\text{-}129)$$

The quantity, ϕ_i^*, is included in the tables of each gas and can be computed from equation (I-5-94).

From equation (I-5-127)

$$m = a C \sqrt{g_c} \left(\frac{F}{F_i}\right) F_i \sqrt{\frac{P_{1t}}{v_{1t}}} \qquad (I\text{-}5\text{-}130)$$

The quantity, F_i, is included in each table of F/F_i factors and can be computed from equation (I-5-104).

Note: In equation (I-5-129), if a is in sq. ft., p must be in psfa; and, if a is in sq in., p must be in psia. Likewise, in equation (I-5-130), as written a is ft^2 and p is psfa; but, if a is in.^{2} and p is psia, then the right side must be multiplied by 1/12.

I-5-46 The effect on ϕ^* of varying Γ^*, Γ and Z may be shown by the following procedure. Let

$$\Gamma^* = \Gamma + \Delta\Gamma = \Gamma\left(1 + \frac{\Delta\Gamma}{\Gamma}\right) = \Gamma(1+\epsilon) \qquad (I\text{-}5\text{-}131)$$

where ϵ represents $(\Gamma^* - \Gamma)/\Gamma$. Applying this to the right side of equation (I-5-116) gives approximately,

$$\sqrt{\frac{g_c(MW)\Gamma}{R Z \left(1 + \frac{\Gamma-1}{2}\right)^{\frac{\Gamma+1}{\Gamma-1}}} \left(\frac{1+\epsilon}{1 + \frac{\Gamma+1}{2}\epsilon}\right)}$$

$$\cong \sqrt{\frac{\Gamma}{\left(\frac{\Gamma+1}{2}\right)^{\frac{\Gamma+1}{\Gamma-1}}}} \left[1 - \left(\frac{\Gamma-1}{4}\right)\epsilon\right]$$

$$\sqrt{\frac{g_c(MW)}{Z R}} \qquad (I\text{-}5\text{-}132)$$

By a plot of the value of $\sqrt{\Gamma/[(\Gamma+1)/2]^{(\Gamma+1)/(\Gamma-1)}}$ versus Γ, the value of this radical between $\Gamma = 1.4$ and $\Gamma = 1.5$ is found to be very close to the straight line, $0.68[1 + 0.24(\Gamma - 1.4)]$, and the plotted points are only about 0.1 per cent below this line at $\Gamma = 1.3$ and 1.6. Using this value for the radical gives

$$\phi^* = 0.68 [1 + 0.24 (\Gamma - 1.4)] \left[1 - \left(\frac{\Gamma-1}{4}\right)\right]$$

$$\sqrt{\frac{g_c(MW)}{Z R}} \qquad (I\text{-}5\text{-}133)$$

Using $\sqrt{Z} = 1 + (1 - Z)/2$ as an approximation and neglecting the cross products of the second term in each bracket yields

$$1.46 \sqrt{\frac{R}{g_c(MW)}} \phi^* = 1 + 0.50(1 - Z)$$

$$+ 0.24 (\Gamma - 1.4) - \left(\frac{\Gamma-1}{4}\right)\epsilon \qquad (I\text{-}5\text{-}134)$$

Equal incremental changes in $(1 - Z)$, $(\Gamma - 1.4)$ and ϵ affect the value of ϕ^* in a decreasing order of importance of about 1/2, 1/4 and 1/8 times each change, respectively. Although the value of Z may be determined from tables for many real gases, the mean value of Γ may not be tabulated, so that its value would have to be determined. The local value of Γ can be determined from the acoustic speed data,

$$\Gamma^* = V_s^2/g_c Z R T = \frac{(c_p/c_v) Z}{1 - 2cp^2} \qquad (I\text{-}5\text{-}135)$$

where c is the second virial coefficient (about $10^{-6}/$ atm²).

I-5-47 Determination of the Maximum Flow Rate of a Real Gas. The sonic-flow functions for a real gas such as steam, most refrigerants and many other commercial gases can be determined by the following theoretical method, which is sometimes called the "enthalpy" method. It is the most practical procedure to use for those gases of which the properties are tabulated. The downstream rate of flow is determined from the continuity equation and from an isentropic conservation of energy flow.

$$dw + J\,dq = dZ + \frac{V\,dV}{g_c} + J\,dH \qquad \text{(I-5-136)}$$

where w = work done by the fluid. For no work done, i.e., $dw = 0$, no heat added, and horizontal flow

$$\int_0^{V_2} \frac{V\,dV}{g_c} = -J\int_{1t}^{2} dH \qquad \text{(I-5-137)}$$

or

$$\frac{V_2^2}{2g_c} = J(H_{1t} - H_2) \qquad \text{(I-5-138)}$$

The equation of continuity for use with tables is

$$\frac{m_T}{a} = \frac{V_2}{v_2} = \frac{1}{v_2}\sqrt{2g_c\,J(H_{1t} - H_2)} \qquad \text{(I-5-139)}$$

The procedure consists of first fixing the inlet stagnation conditions and then using in turn several values of the downstream (i.e., throat) pressure and iteratively solving for the mass rate of flow until a maximum is indicated. Referring to Fig. I-5-7(a), the inlet stagnation conditions are known or assumed which fixes the inlet enthalpy. Then,

1. A value of the downstream pressure, p_2, is assumed and thereby a value of H_2, with which a value of the downstream velocity, V_2, is computed by equation (I-5-138).

2. Using equations (I-5-113) (if v_2 is not tabulated), (I-5-139), (I-5-127) and (I-5-125) in that order, values of ρ_2, m_T/a, F and ϕ^* are determined and the value of ϕ^* plotted as in Fig. I-5-7(b).

3. Steps 1 and 2 are repeated until the maximum value of ϕ^* is established.

For these computations it has been found advantageous to use the temperature the gas would have if it

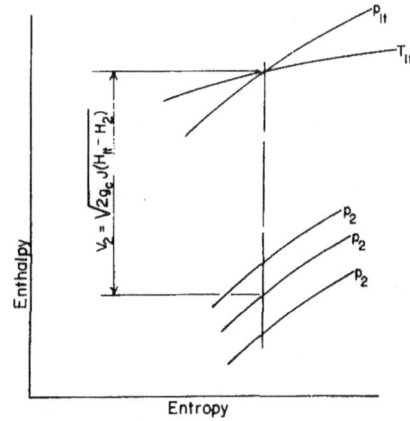

(a)

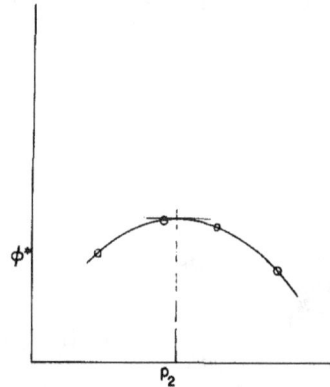

(b)

FIG. I-5-7 FLOW-MAXIMIZATION PROCEDURE

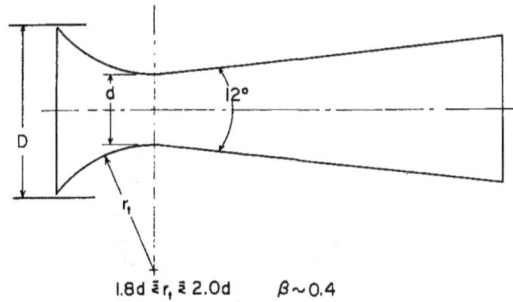

$1.8d \lesseqgtr r_1 \lesseqgtr 2.0d \qquad \beta \sim 0.4$

FIG. I-5-8 PROPORTIONS OF INTERIOR SURFACE OF CIRCULAR-ARC VENTURI

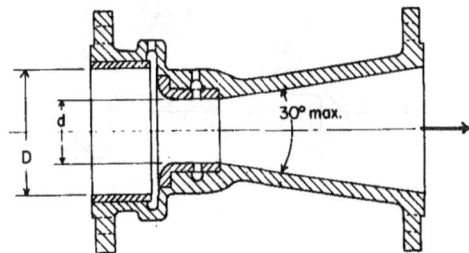

FIG. I-5-9 NOZZLE-VENTURI TUBE

were throttled isenthalpically, i.e., H = constant, to a very low pressure.

This method may be programmed for digital computer use where such equipment is used or available.

In Part II the functions are presented in the same format as in the preceding derivations for real gases. The use of the "working" equations, (I-5-127) and (I-5-129), is not affected.

I-5-48 Figure I-5-8 shows the proportions of the interior surface of a nozzle-Venturi, or circular-arc Venturi, used at sonic throat-velocity conditions in tests of jet-type propulsion engines. The overall pressure loss of these nozzle-Venturis may be as low as 15 percent of the pressure drop required to produce sonic velocity at the throat. The coefficient of discharge is between 0.990 and 0.995 [22, 23, 24, 25].

I-5-49 Modifications of the Three Basic Primary Elements: The Nozzle-Venturi. The nozzle-Venturi (Fig. I-5-9) was developed to provide a primary element of shorter overall length than the conventional Venturi tube but without a too greatly increased pressure loss. Although it has been used to a slight extent in this country, it has been used abroad enough to justify including it in the recommendation of the ISO Committee on Flow Measurement [11]. An abstract of these recommendations on the proportions and discharge coefficient is given in Part II.

I-5-50 Other Modifications of Venturi Tubes and Flow Nozzles. Several forms of flow tubes have been developed which combine features of the Venturi tube and the flow nozzle, as shown in Fig. I-5-10. The primary objective in the design of these special

(a)

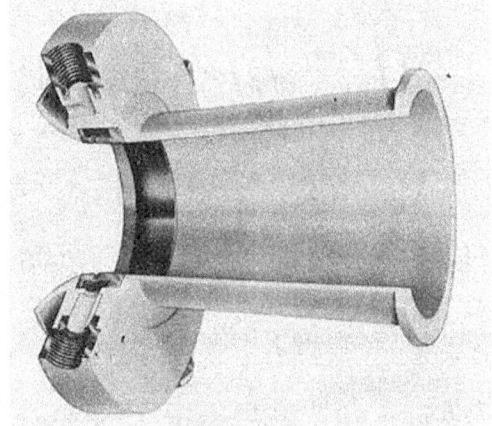

(b)

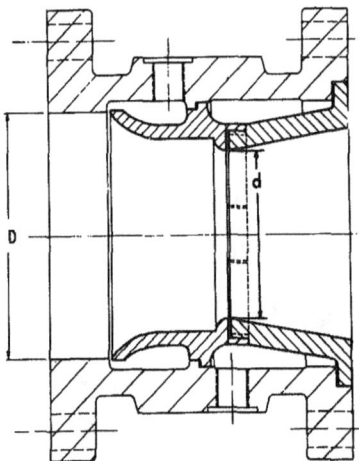

(c)

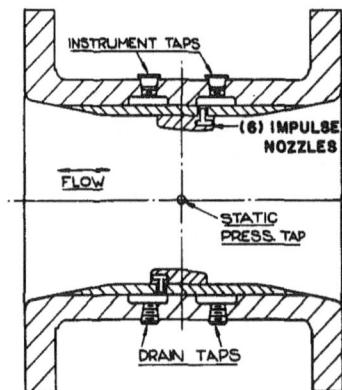

(d)

FIG. I-5-10 MODIFICATIONS OF FLOW NOZZLES AND VENTURI TUBES
(PROPRIETARY)

forms of flow tubes was to obtain a high differential pressure with as low an overall pressure loss as possible. To a notable extent these objectives were achieved, due in part to boundary layer effects. Because these special forms are proprietary, the committee presents no specific data on individual tubes [26–29].

I-5-51 Quadrant-Edge and Conical-Edge Orifices. The inlet edge of the quadrant-edge orifice is rounded, and usually the radius of the rounding is equal to the plate thickness (Fig. I-5-11). The angle of the entrance cone of the conical-edge orifice may range from 40 to 80 deg as measured from the face of the plate. These forms of orifices have the characteristic of having an almost flat, or constant, value of the discharge coefficient over a relatively low range of flow rates, as may be represented by values of R_d below about 50,000. For this reason these orifices have been the subject of test programs in several laboratories. Some of these programs have shown that the installation conditions and the velocity profile of the approaching fluid affect both the value and range of "flatness" of the coefficient. The ISO Technical Committee on Flow Measurement is asking that further studies be made on these orifices, especially the conical-inlet orifice [30, 31].

I-5-52 Eccentric and Segmental Orifices. A circular eccentric orifice is constructed and installed so that its center is not on the axis of the pipe. Generally, the eccentricity is such that one side of the

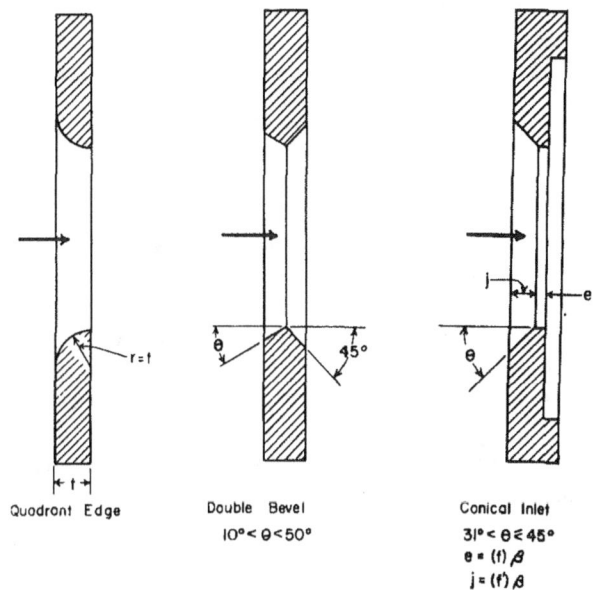

FIG. I-5-11 SPECIAL ORIFICE FORMS FOR MEASUREMENT OF LOW RATES OF FLOW

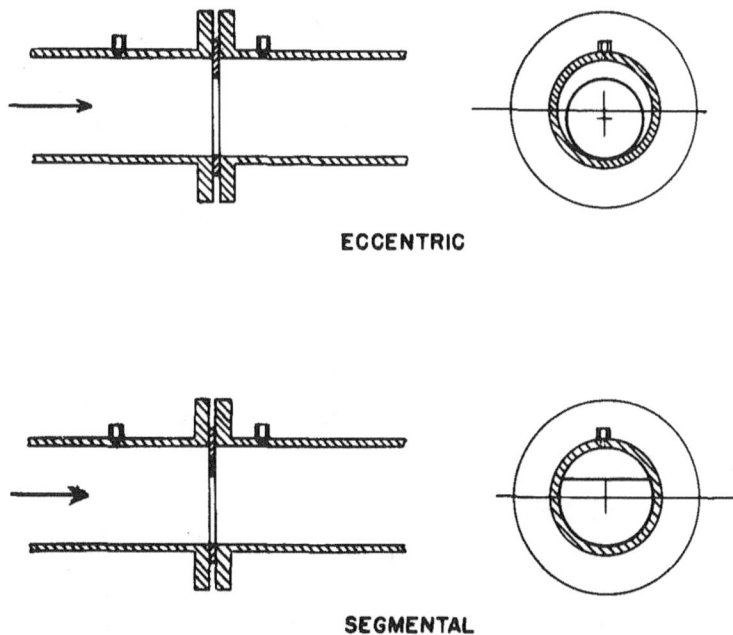

FIG. I-5-12 ECCENTRIC AND SEGMENTAL ORIFICES

circular opening is substantially flush with the inside wall of the pipe. The point at which the eccentric opening is nearest to, or flush with, the wall of the pipe is diametrically opposite the pressure taps, thus placing the maximum depth of dam height on the side of the pressure taps, as shown in Fig. I-5-12.

Note: Sometimes with eccentric orifices, the pressure taps are located 90 deg around the pipe from the top when, because of the pipe location, the taps must be on the side of the pipe while the orifice hole must be at the bottom to provide maximum flushing.

With the segmental orifice, the shape of the restricting area is a segment of a circle of approximately the same diameter as that of the pipe. Thus, the orifice opening is irregular in shape. Sometimes, with very large pipes, only that part of the orifice plate forming the segment is used. The segmental orifice is installed with the segment on the same side of the pipe as the pressure taps.

Note: With segmental orifices the primary dimensional relation is the area ratio, a/A. The diameter ratio has the relative significance given by the relation, $\beta = \sqrt{a/A}$.

However, it may be convenient, particularly for comparative purposes, to use this equivalent value of β as a parameter in the presentation of data on segmental orifices.

One reason for using an eccentric or segmental orifice is that the plane of the vena contracta is displaced farther downstream from where it would be with a concentric circular orifice of the same area ratio. Another reason for using them is that, by placing those portions of the openings, which are flush with the inside surface of the pipe, at the bottom (of horizontal lines), complete drainage of any extraneous matter will be obtained. This minimizes the danger of such matter affecting the accuracy of the meter.

Curves of the discharge coefficients and expansion factors for eccentric and segmental orifices are given in Part II [32, 33].

I-5-53 Overall Pressure Loss. The relative overall pressure losses associated with the several general types of differential pressure primary elements so far described and discussed are shown in Fig. I-5-13. It will be seen that the thin-plate square-edged orifice

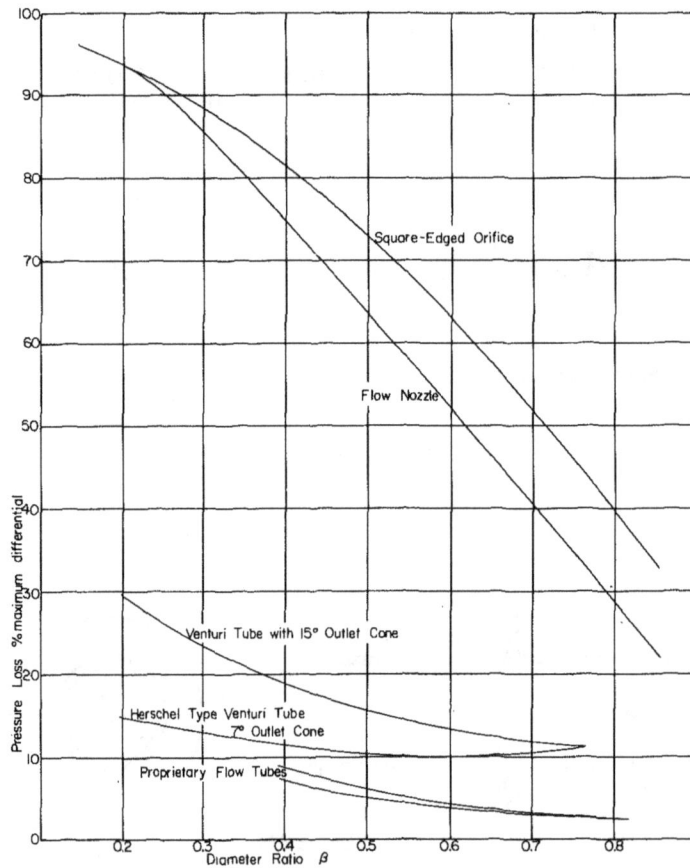

FIG. I-5-13 OVERALL PRESSURE LOSS
THROUGH SEVERAL PRIMARY
ELEMENTS

has the highest loss, while some of the special forms of Fig. I-5-10 have the lowest. However, the extreme simplicity, reproducibility and adaptability of the orifice are largely responsible for its being the most widely used differential pressure element.

I-54 Centrifugal Meters: The Elbow Meter. When a fluid flows along a curved channel, it is subjected to angular acceleration, and the basic relation between acceleration, force and mass applies. The force in this case is evidenced by the difference between the pressures which are observed at the outside and inside of the curve, especially in closed channels. Thus, a very simple form of a centrifugal meter is a common pipe elbow, with pressure taps in the outer and inner surfaces in the plane determined by the curved center line of the elbow (Fig. I-5-14). Although for most of the tests of elbow meters that have been reported the pressure taps were located in a radial plane 45 deg from the elbow inlet, other locations have been used, notably 22 ½ deg from the inlet. An advantage of the 45-deg location is that flow in either direction can be measured. Special forms of elbows have been tried but gained little attention.

I-5-55 To develop a rational flow equation for an elbow meter, it is helpful to assume that the cross section of the elbow is rectangular. That the final equation can be applied to any cross-sectional shape is shown later. Other simplifying assumptions are the following:

1. The elbow is in a horizontal plane.
2. A uniform distribution of velocity exists over the cross section of the elbow.
3. There is no loss of either pressure or velocity from inlet face to outlet face.
4. There is no effect of fluid viscosity.
5. A uniform distribution of pressure exists over the inside and outside walls of the elbow.
6. There is a steady state of flow with respect to time.

The symbols and units used in the development of an equation are:

A = Radial cross-sectional area of elbow   ft²
D = Radial width of a rectangular elbow, also diameter of a circular cross section   ft
F = Force, due to fluid momentum   lb$_f$
p = Pressure   lb$_f$/ft²
R = Radius of elbow center line   ft
S = Resultant sidewall force   lb$_f$
V = Fluid velocity   ft/sec
η (eta) = Height of duct section   ft
ρ (rho) = Fluid density   slug/ft³

Subscript i = Inside
o = Outside
L = Pipe line (for pressure and velocity at inlet and outlet faces)
x = x-direction vector component
y = y-direction vector component
z = z-direction vector component perpendicular to the x-y plane (not shown in Fig. I-5-15).

The momentum flux equation as applied to an elbow states that the net force on the surface of the elbow equals the net flux of momentum through the surfaces of the elbow plus the rate of change of momentum inside the elbow, that is,

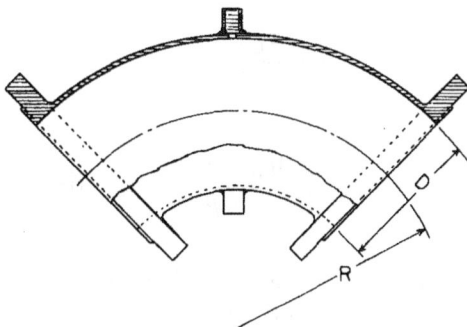

FIG. I-5-14 ELBOW METER

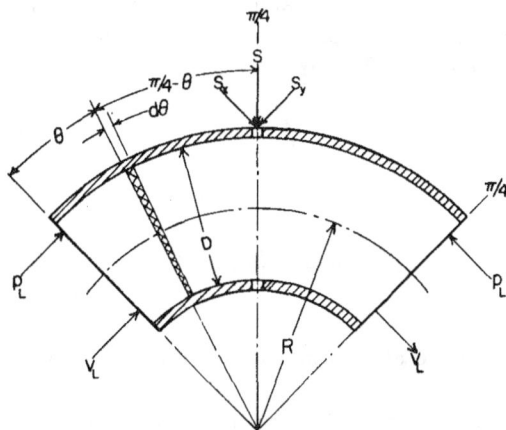

FIG. I-5-15 ELEMENTS OF FLUID FLOW THROUGH ELBOW METER

$$F + \int (-g\,\rho\,z)\,d(\text{vol})$$

$$= \int V(\rho\,V \cdot dA) + \frac{\partial}{\partial t}\int V\,\rho \cdot d(\text{vol}) \quad (I\text{-}5\text{-}140)$$

By the first assumption above, the second term disappears; and the last term disappears by the last assumption, so that equation (I-5-140) reduces to

$$F = \int V(\rho \cdot dA) \qquad (I\text{-}5\text{-}141)$$

Taking force components in the x- and y-direction

$$F_x = \int V_x(\rho\,V \cdot dA) \text{ and } F_y = \int V_y(\rho\,V \cdot dA) \quad (I\text{-}5\text{-}142)$$

The force, F, is composed of two parts, the pressure force on the flow areas of the elbow and the resultant force on the sidewalls of the elbow, S, for which the equation will be solved.

$$F_x = S_x - p_L A = \int V_{Lx}(\rho\,V_L \cdot dA) = \rho A V_L^2 \quad (I\text{-}5\text{-}143)$$

$$F_y = -S_y + p_L A = \int V_{Ly}(\rho\,V_L \cdot dA) = \rho A V_L^2 \quad (I\text{-}5\text{-}144)$$

$$S_x = \rho A V_L^2 + p_L A \text{ and } S_y = -\rho A V_L^2 - p_L A \qquad (I\text{-}5\text{-}145)$$

$$S = \sqrt{S_x^2 + S_y^2} = \sqrt{2}\,\sqrt{(\rho A\,V_L^2 + p_L A)^2} \qquad (I\text{-}5\text{-}146)$$

This is the net force on the sidewalls which is responsible for turning the flow. Equations (I-5-143) and (I-5-144) are easily generalized to include elbows of more or less than 90 deg. The number, S, in equation (I-5-146) is the value of a three-dimensional vector integral of the pressure over the sidewall surface of the elbow. To formulate this integral,

$$S = \int p\,dA \qquad (I\text{-}5\text{-}147)$$

where dA is a vector quantity, and

$$dA_i = (R - D/2)\,\eta\,d\theta \text{ and } dA_o = (R + D/2)\,\eta\,d\theta \qquad (I\text{-}5\text{-}148)$$

where $d\theta$ is the elemental angular movement of the fluid (Fig. I-5-15).

A uniform pressure, p_o, on the outside and p_i on the inside multiplying the differential area vector forms a differential force which has two components: one parallel to S, the other normal to it. These normal components taken about the axis of symmetry, $\pi/4$ (Fig. I-5-15), are symmetrically opposing. In addition, the pressure distribution on the top and bottom of the duct may be assumed to be symmetrically opposing. Thus, the total value of the force, S, over the sidewalls is

$$S = \int_o^{\pi/2} p_o\eta(R + D/2)\cos(\pi/4 - \theta)\,d\theta$$

$$- \int_o^{\pi/2} p_i\eta(R - D/2)\cos(\pi/4 - \theta)\,d\theta \qquad (I\text{-}5\text{-}149)$$

Now,

$$\cos(\pi/4 - \theta) = \cos(\pi/4)\cos\theta + \sin(\pi/4)\sin\theta$$

and

$$\cos(\pi/4) = \sin(\pi/4) = 1/\sqrt{2}$$

Also,

$$\int_o^{\pi/2} \cos\theta\,d\theta = \int_o^{\pi/2} \sin\theta\,d\theta = 1$$

so that

$$S = p_o\eta(R + D/2)2/\sqrt{2}$$

$$- p_i\eta(R - D/2)2/\sqrt{2} \qquad (I\text{-}5\text{-}150)$$

or

$$S = (p_o - p_i)\eta R\sqrt{2} + \frac{(p_o + p_i)}{2}\,\eta D\sqrt{2} \qquad (I\text{-}5\text{-}151)$$

Combining equations (I-5-151) and (I-5-146)

$$(p_o - p_i)\eta R\sqrt{2} + \frac{(p_o + p_i)}{2}\,\eta D\sqrt{2}$$

$$= \sqrt{2}\,\sqrt{(\rho\eta D V_L^2 + p_L D)^2} \qquad (I\text{-}5\text{-}152)$$

from which

$$(p_o - p_i) = \frac{\rho\eta D V_L^2}{\eta R} + \frac{p_L \eta D}{\eta R}$$

$$- \frac{(p_o + p_i)\eta D}{2\eta R} \qquad (I\text{-}5\text{-}153)$$

By equation (I-5-153) it appears that the height of a rectangular duct, η, is self-eliminating. On the basis of the assumption of no pressure loss between inlet and outlet, the approximate relation,

$(p_o + p_i)/2 \cong p_L$, may be applied to equation (I-5-153) to give

$$p_o - p_i = \Delta p = \frac{D}{R} \rho V_L^2 \qquad (I\text{-}5\text{-}154)$$

I-5-56 Equation (I-5-154) is an approximate theoretical equation for the measurement of fluid flow with an elbow and is applicable to any cross section that is symmetric with respect to the center-line plane. The theoretical rate of flow is

$$q_T = A V_L = A \sqrt{\frac{R}{D} \frac{\Delta p}{\rho}} \qquad (I\text{-}5\text{-}155)$$

Introducing a flow coefficient, K, determined by a calibration, and using the customary units of measurement (namely, A, in.^{2}; D and R, in.; p_L and Δp, psi; and ρ, lb$_m$/ft^3) gives as the equation for computing the rate of flow

$$q = 0.47268 \, A \, K \sqrt{\frac{R}{D} \frac{\Delta p}{\rho}} \qquad (I\text{-}5\text{-}156)$$

cfs at the fluid density, ρ

or

$$q = 0.37125 \, K \, D \, \sqrt{D \, R \, \Delta p/\rho} \quad \text{cfs} \qquad (I\text{-}5\text{-}157)$$

and

$$m = 0.37125 \, K \, D \, \sqrt{D \, R \, \rho \, \Delta p} \quad \text{lb}_m/\text{sec} \quad (I\text{-}5\text{-}158)$$

I-5-57 Unlike the three principal differential pressure meters, elbow meters have not been the object of tests and structual recommendations by a committee of the ASME or other technical societies. However, a review of published experimental data on elbow meters indicated that the relative roughness of the elbow surface had no significant effect on flow measurements. Also, for 90-deg elbows with pressure taps at 45 deg and tap hole diameters, δ, as recommended for orifice meters, the value of K may be computed by

$$K = 1 - \frac{6.5}{\sqrt{R_D}} \qquad (I\text{-}5\text{-}159)$$

when $10^4 \leq R_D \leq 10^6$, and $R/D \geq 1.25$. Flows computed with this value of K and uncalibrated elbows will be subject to a tolerance (uncertainty) of about $\pm$ 4 per cent. With a calibrated elbow, the tolerance should be comparable to that for other types of differential pressure meters. With either calibrated or uncalibrated

elbows meters, a high degree of repeatability is attainable [34, 35].

I-5-58 The use of the scroll case of a turbine or pump and the guide-vane speed ring of a turbine may be classed as forms of centrifugal meters. In order to use either of these as a means of flow measurement, a calibration must be made using some other method of measurement, such as a tracer method. (See Chapter I-I-9 and [36].)

I-5-59 Linear-Resistance Meters. With these meters, the outstanding characteristic is the linear relationship between the pressure drop and the flow rate from zero flow up to some maximum rate. Because of this linear relationship, these meters have the same flow coefficient over this range; and, within this range, the determination of this coefficient at one rate is sufficient. Above this linear maximum rate, the pressure drop begins to increase very gradually at a faster rate than the flow. Hence, when calibrating one of these meters, enough different rates should be covered to determine the upper limit of linearity. If the meter is to be used above this point, the relation between rate of flow and pressure drop should be determined.

The two most common forms of primary elements are capillary tubes and porous plugs, as illustrated in Fig. I-5-16. With the capillary tube, the ratio of length to bore is the principal factor determining the relationship between rate of flow and pressure drop. Some writers recommend that the length should be 150 or more times the bore. A second factor is the character of the entrance and exit ends of the tube, i.e., whether these are square or smooth and well tapered [37]. Instead of a single tube, a bundle of capillary passages may be used to increase the flow capacity. Also, it is not necessary for the passage to be of circular cross section or even of a uniform shape [38].

The porous-plug form can be made by fastening a plug of suitable porous material within a section of pipe or tubing and providing connections for measuring the pressure drop across the plug. Some of the materials which may be satisfactory for the plug are steel wool, cotton waste, sintered alumina, glass wool, and layers of fine screening so placed that the wires of adjacent layers are not parallel. The first two are suited to use with dry gases; the sintered alumina could be used at elevated temperatures; and the last two could be used with wet gases or even liquids. With any given plug material, the principal factors influencing the rate-pressure drop relation are the tightness with which the material is packed and the length of the plug [39].

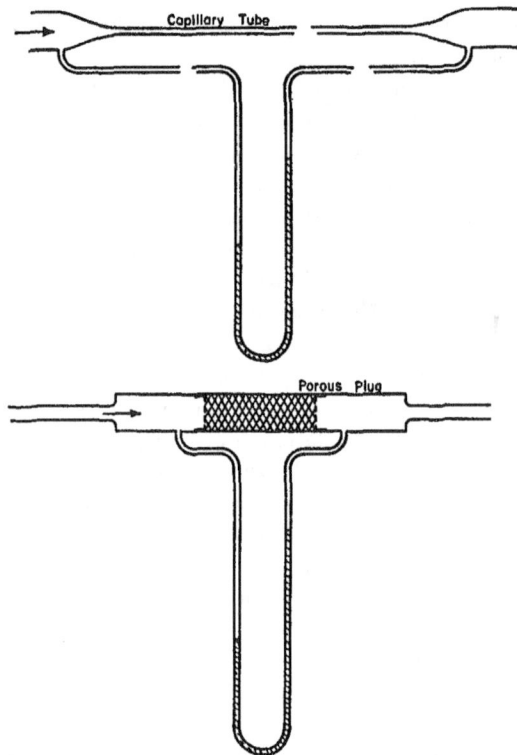

FIG. I-5-16 TWO FORMS OF LINEAR-RE-
SISTANCE FLOW METERS

I-5-60 In order to have a linear relationship between the rate of flow and the pressure drop across
the metering unit, it is necessary that the flow rate
is low enough for the pressure drop to be a measure
of the viscous drag. The transition from laminar
flow (or viscous drag flow) to turbulent flow usually
takes place at a Reynolds number between 2000 and
2500. Therefore, in order to meet the above requirement the Reynolds number applying to the metering
passage should be less than 2000 (and below 1500 if
possible). The basic relation for the flow through
capillary tubes, commonly referred to as Poiseuille's
equation, is [40]:

$$q = \frac{\pi}{8} C \frac{D^4}{16} \frac{1}{L\mu} (p_1 - p_2) \qquad \text{(I-5-160)}$$

If normal units were used, all linear dimensions
would be in ft, ft² and ft³. However, the common
practice in fluid metering is to have q in ft³/sec,
while D, L and $(p_1 - p_2)$ are measured in in. and
psi. To take account of this, we must write

$$q = \frac{\pi}{8} \frac{1}{16} C \frac{D^4}{(12)^4} \frac{12}{L} \frac{144 (p_1 - p_2)}{\mu}$$
$$\qquad \text{(I-5-161)}$$
$$= \frac{\pi C D^4}{1536 \ L} \left(\frac{p_1 - p_2}{\mu} \right)$$

The coefficient, C, must be determined by a calibration. The other quantities are:

D = Diameter of the capillary tube, in.

L = Length of tube between points of pressure
   measurements, in.

$(p_1 - p_2)$ = pressure drop over the tube length, L,
   psi.

q = Volume rate of flow at the mean temperature
   and the mean pressure, cfs

μ = Viscosity of the fluid, lb_m/ft-sec

For use with a bundle of tubes, including the irregular areas between them, a multiplicity of thin,
slit-like passages, or even a porous plug, the diameter term may be replaced with $D^4 = (16A)^2/\pi^2$,
where A = the total cross-sectional area of the passages through which the fluid flows, in². Thus, including the coefficient, C, the flow equation becomes

$$q = \frac{C \ A^2}{96 \ \pi \ L} \left(\frac{p_1 - p_2}{\mu} \right) \qquad \text{(I-5-162)}$$

Since a direct determination of A may be difficult, or
impossible, the product CA^2, or even CA^2/L, may be
determined by the calibration [39].

In the general use of linear-resistance meters,
especially capillary tubes, the effects of inlet and
exit losses, and in the case of a coiled capillary,
the effects of "slip" and curvilinear flow are usually
included in the calibration coefficient, C. Also,
when the flow rate increases to the point where laminar flow begins to break into turbulent flow, these equations will cease to hold [41, 42].

I-5-61 Frictional-Resistance Meters: Pipe Section.
When the velocity of the fluid in a pipe line is very
high, as in the steam and water lines of some modern power plants, the frictional pressure drop becomes significant. Hence, the difference in the
static pressures between two sidewall pressure taps
located an appreciable distance apart in a straight
section of pipe may be measured readily with an appropriate differential pressure gage. Thus, when
adequately calibrated, such a section of pipe with
its pressure taps may be a satisfactory primary
element for some purposes. Of course, each such
section of pipe must be individually calibrated for
the range of flows and the fluids to be metered.

If the object of the measurement is simply to
monitor the relative steadiness of the flow, with

little or no interest in the actual rate, it may suffice to use the pipe-line flow equation with an appropriate value of the friction factor. For this purpose, the equation for flow in a pipe may be written

$$m = 0.1515\, D^{5/2}\, \sqrt{\frac{\rho\, \Delta p}{f\, L}} \qquad \text{lb}_m/\text{sec} \qquad (\text{I-5-163})$$

in which D is in inches, Δp, in psi, and L, in ft, is the distance between the pressure taps. The friction factor, f, is a function of the relative roughness, ϵ/D and R_D. For steel pipe 4 in. i.d. and over and $R_D \geq 5(10^5)$, an average value of $f = 0.015$ may be assumed, and [43]

$$m = 1.237\, D^{5/2}\, \sqrt{\frac{\rho\, \Delta p}{L}} \qquad (\text{I-5-164})$$

References

[1] "Notes on the Orifice Meter: The Expansion Factor for Gases," E. Buckingham; *Journal of Research*, National Bureau of Standards, vol. 9, July 1932, RP. 459, p. 61.

[2] "Einfluss der Expansion auf die Kontraktion hinter Stauranden (Influence of Expansion Upon the Contraction After an Orifice)," G. Ruppel; *Technische Mechanik und Thermodynamik*, vol. 1, Apri. 1930, p. 151 and Sept. 1930, p. 338.

[3] Discharge Coefficients of Square-Edged Orifices for Measuring the Flow of Air," H. S. Bean, E. Buckingham and P. S. Murphy; *Journal of Research*, National Bureau of Standards, vol. 2, Mar. 1929, RP.49, p. 561.

[4] "Coefficients of Discharge of Sharp-Edged Concentric Orifices in Commercial 2-In., 3-In., and 4-In. Pipes at Low Reynolds Numbers Using Flange Taps," E. E. Ambrosius and L. K. Spink; *Trans. ASME*, vol. 69, Nov. 1947, p. 805.

[5] "Orifice Discharge Coefficients for Viscous Liquids," G. L. Tuve and R. E. Sprenkle; *Instruments*, vol. 6, Nov. 1933, p. 201.

[6] "Experiments on the Metering of Large Volumes of Air," H. S. Bean, M. E. Benesh and E. Buckingham; *Journal of Research*, National Bureau of Standard, vol. 7, July 1931, RP 335, p. 93.

[7] "Values of Discharge Coefficients of Square-Edged Orifices, Comparison of Results Obtained by Tests Using Gases with Those Obtained by Tests Using Water," H. S. Bean; *American Gas Association Monthly*, vol. 17, July 1935, p. 259.

[8] "Uber die Expansionszahl bei der Durchflussmessung mit Normblenden (About the Expansion Factor for the Measurement of Flow with Standard Orifices)," Guy Thibessard; *Sonderdruck aus Brennstoff-Warme-Kraft*, Bd. 12, 1960, Nr. 3 S. 97/101.

[9] "The Mass Flow Meter: A Method of Measuring Flow," D. Brand and L. A. Ginsel; *Instruments*, Mar. 1951, p. 331.

[10] "An Orifice Meter that Measures True Mass Flow," B. Fishman; 17th Annual Instrumentation-Automation Conference, Oct. 1962, paper 16.4.

[11] ISO Recommendation R 541, *Measurement of Fluid Flow by Means of Orifice Plates and Nozzles*, and ISO Recommendation 781, *Measurement of Fluid Flow by Means of Venturi Tubes*.

[12] "The Flow of Water Through Orifices," S. R. Beitler; Ohio State University Engineering Experiment Station Bulletin No. 89, May 3, 1935.

[13] "History of Orifice Meters and the Calibration, Construction, and Operation of Orifices for Metering;" Report of the Joint AGA–ASME Committee on Orifice Coefficients, 1935; Reprint by ASME, Dec. 1936.

[14] "A Statistical Approach to the Prediction of Discharge Coefficients of Concentric Orifice Plates," Rodger B. Dowdell and Yu-Lin Chen; ASME Paper, 1969.

[15] "Thermodynamics of Compressible Fluid Flow," A. H. Shapiro; The Roland Press, vol. 1, 1953, p. 79–83.

[16] "Measurement of Temperatures in High-Velocity Steam," J. W. Murdock and E. F. Fiock; *Trans. ASME*, vol. 72, Nov. 1950, p. 1155.

[17] "Measurement of Temperatures in High-Velocity Gas Streams," W. J. King; *Trans. ASME*, vol. 65, July 1943, p. 421.

[18] "Computation of the Critical Flow Function, Pressure Ratio and Temperature Ratio for Real Air," R. M. Reimer; *Trans. ASME, Journal of Basic Engineering*, vol. 86–2, June 1964, p. 169.

[19] "The Critical Flow Function for Superheated Steam," J. W. Murdock and J. M. Bauman; *Trans. ASME, Journal of Basic Engineering*, Sept. 1964, p. 507.

[20] "Review of Critical Flowmeters for Gas Flow Measurements," B. T. Arnberg, *Trans. ASME, Journal of Basic Engineering*, vol. 84–4, Dec. 1964, p. 447.

[21] "Tables of Thermal Properties of Gases," National Bureau of Standards, Circular 564, 1955.

[22] "A Theoretical Method of Determining Discharge Coefficients for Venturis Operating at Critical Flow Conditions," R. E. Smith and R. J. Matz; *Trans. ASME, Journal of Basic Engineering*, vol. 84–4, Dec. 1962, p. 434.

[23] "Accurate Measurement of Airflow Rate During Tests of Air-Breathing Propulsion Systems," R. E. Smith, Jr. and H. E. Wolff; Arnold Engineering Development Center, Tr.–68–22.

[24] "The Calculation of the Discharge Coefficient of Profiled Choked Nozzles and the Optimum Profile for Absolute Air Flow Measurements," B. S. Stratford; *Aeronautical Journal*, Royal Aeronautical Society (London), vol. 68, Apr. 1964, p. 237.

[25] "A Standard Choked Nozzle for Absolute Calibration of Air Flowmeters," D. W. Sparkes; *Aeronautical Journal*, Royal Aeronautical Society (London), vol. 72, Apr. 1968, p. 335.

[26] "The Dahl Flow Tube," I. O. Miner; *Trans. ASME*, vol. 78, Apr. 1956, p. 475.

[27] "Design and Calibration of the Lo-Loss Tube," L. J. Hooper; *Trans. ASME, Journal of Basic Engineering*, vol. 84–4, Dec. 1962, p. 461.

[28] "Calibrations of Six Beth-Flow Meters at Alden Hydraulic Laboratory, Worcester Polytechnic Institute," L. J. Hooper; *Trans. ASME*, vol. 72–8, Nov. 1950, p. 1099.

[29] "The Twin Throat Venturi: A New Fluid-Flow Measuring Device," A. A. Kalinske; *Trans. ASME, Journal of Basic Engineering*, vol. 82–3, Sept. 1960, p. 710.

[30] "The Quadrant-Edge Orifice: A Fluid Meter for Low Reynolds Numbers," M. Bogama and P. L. Monkmeyer; *Trans. ASME, Journal of Basic Engineering,* vol. 82—3, Sept. 1960, p. 729.

[31] "Quadrant-Edge Orifice Performance, Effect of Upstream Velocity Distribution," M. Bogama, B. Spring and M. V. Ramamoorthy; *Trans. ASME, Journal of Basic Engineering,* vol. 84—4, Dec. 1962, p. 415.

[32] "Calibration of Eccentric and Segmental Orifices in 4-in. and 6-in. Pipe Lines," S. R. Beitler and D. J. Masson; *Trans. ASME,* vol. 71, Oct. 1949, p. 751.

[33] "Experimental Study of the Effects of Orifice Plate Eccentricity on Flow Coefficients," R. W. Miller and O. Kneisel; *Trans. ASME, Journal of Basic Engineering,* vol. 91, Mar. 1, 1969, p. 121.

[34] Research Report RR 1790, "Pipe Bends as Flow Measuring Devices," J. A. Landstra and T. J. Vermeuten; Bataafse Internationale Petroleum Maatschappij, N. V.

[35] "Performance Characteristics of Elbow Flowmeters," J. W. Murdock, C. J. Foltz and C. Gregory; *Trans. ASME, Journal of Basic Engineering,* vol. 86—3, Sept. 1964, p. 498.

[36] "Improved Type of Flow Meter for Hydraulic Turbines," I. A. Winter; *Proceedings ASCE,* vol. 59, Apr. 1933, p. 565.

[37] "Gas Flow Meters for Measuring Small Rates of Flow," A. F. Benton; *Journal of Industrial and Engineering Chemistry,* July 1, 1919.

[38] "Liquid Metering with Capillary Flow Elements," C. E. Greeff; 19th Annual ISA Conference and Exhibit, Oct. 1964, Reprint No. 5.2-2.64.

[39] "Permeability of Glass Wall and Other Highly Porous Media," A. S. Iberal; *Journal of Research,* National Bureau of Standards, vol. 45, Nov. 1950, RP 2150, p. 398.

[40] "Recueil des Savants Etrangers, (Collection of Works of Foreign Scientists)," J. L. M. Poiseuille; pour 1842 Academie des Sciences.

[41] "Linear-Resistance Meters for Liquid Flow," R. C. Souers and R. C. Binder; *Trans. ASME,* vol. 74, July 1952, p. 837.

[42] "Study of Linear Risistance Flow Meters," F. W. Fleming and R. C. Binder; *Trans. ASME,* vol. 73, July 1951, p. 621.

[43] "Marks' Mechanical Engineers' Handbook," McGraw-Hill Book Co., New York, 6th ed., 1958, p. 3—73.

[44] "Formulation of Equations for Orifice Coefficients," H. S. Bean; ASME paper no. 70—WA/FM-2.

Chapter I-6
Area Meters

I-6-1 In area meters means are employed to cause a variation of the cross-sectional area of the measuring aperture as a function of the flow rate, while the differential pressure producing the flow through the section is maintained substantially constant or at a fixed relationship to the flow rate. In most of the meters of this class, the operation of the area-changing element is automatic; the area variation is produced by the positioning of an area-changing element in such a manner that forces on the element due to fluid flow are exactly balanced by gravity or a mechanical restraining means. Since the position of the area-changing element will be an indication or measure of the rate of flow through the meter, these meters give a direct indication of the *rate* of flow at the instant of observation. This is the outstanding feature of this type of meter.

The most common forms of the area meter are the tapered tube and float, the cylinder and piston, and the orifice and plug. The general theory of operation is the same for all these types, and the working equations have the same form for each. Since more theoretical work has been done with the tapered tube and float type than with the others and since this is probably the most widely used form of area meter, a more thorough theoretical treatment will be given it than the other two forms. It should be remembered that all forms operate essentially in the same manner.

I-6-2 Tapered Tube and Float. In this form, flow is directed vertically upward through a tapered tube, of which the inside diameter increases with elevation. Inside the tube, the floating element is free to move vertically in response to the rate of flow of the fluid until a position is reached where the weight of the float, less its buoyancy in the fluid, is balanced by the drag force of the fluid on the float. Such a device is illustrated schematically in Fig. I-6-1. The variable flow area is the area of the annulus between the tube and the largest diameter of the floating element.

The shape and weight of the float, the range of the annular area variation, and the variation of the tube area with elevation determine the characteristics of the meter for a particular set of fluid properties. It has been found, for example, that a

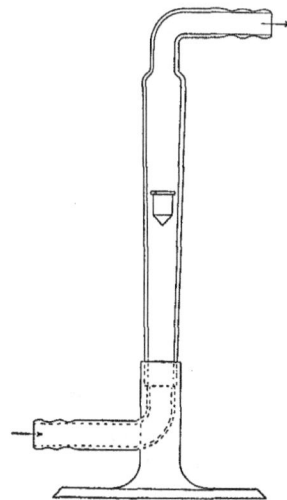

FIG. I-6-1 AREA METER, TAPERED TUBE AND FLOAT

81

thin disc float is nearly independent of fluid viscosity effects over a wide range, while a spherical or cylindrical shape may show significant viscosity sensitivity. Generally, tubes of a constant taper angle are employed, and annular areas are kept less than the area of the floating element so that flow rate is approximately linear with elevation.

I-6-3 The most generally used material for tapered tubes is glass, and accurate methods for forming are now available. Other tube materials such as metal and plastics are used; however, an opaque material requires providing means whereby the float position can be observed or indicated. This is normally accomplished through a magnetic coupling and follower or by having the floating element alter a property such as inductance or capacitance of an electrical current. Also, such methods are useful in connection with recording or transmitting the position of the floating element.

In most early designs the floating elements were made with notches or other features to cause rotation and thus increase the self-centering tendency. The modern design trend is to use guided or mechanically centered floating elements. The two common methods are guiding by means of ribs or flutes formed into the tube bore or by means of a center rod or wire.

I-6-4 Tapered tube and float meters are available in sizes ranging from less than 1/16 in. diam to over 12 in. Fluid velocities at the tube inlet normally range from 1/2 to 10 fps at maximum design flow rate. Normal scale ranges provide a minimum indication of 10 per cent of maximum. For increased readability, scale ranges are sometimes shortened so that the minimum graduation is as much as 20 per cent of maximum flow.

The overall pressure loss will depend upon the friction loss of the particular meter design, plus the practically constant pressure drop across the floating element. The pressure loss depends upon the annular area, and some of the pressure drop across the floating element may be recovered. This pressure drop across the floating element usually will not exceed that calculated from

$$\Delta p_f \le (F/A_f)$$

where

Δp_f = Pressure drop across floating element

F = Weight of floating element less buoyancy, a force

A_f = Area of floating element at largest section

I-6-5 Cylinder and Piston. Figures I-6-2 and I-6-3 show the essential features of the metering elements of this type. The one shown in Fig. I-6-2 has a cylinder with a large number of equally sized orifices, or reamed holes spaced in a uniform helical

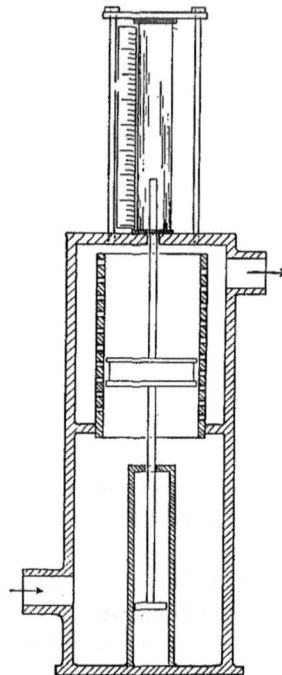

FIG. I-6-2 MULTIPLE-HOLED CYLINDER AND PISTON

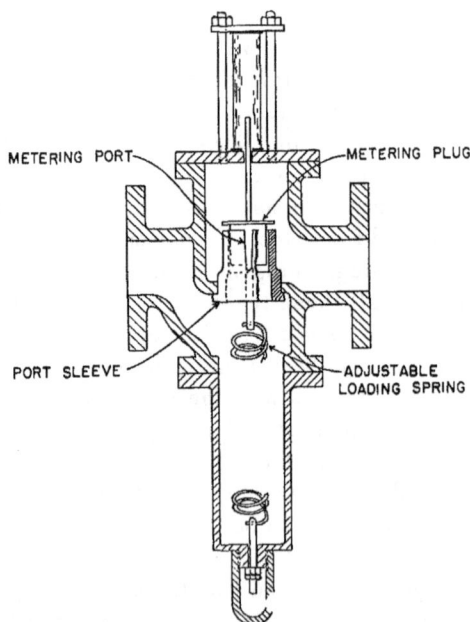

FIG. I-6-3 SLOTTED CYLINDER AND PISTON

pattern. The moving element is a loose-fitting piston. The movement of the piston may be regulated by a weight and dashpot, as shown, or by a spring as in Fig. I-6-3. The number of holes exposed is in direct proportion to the travel of the piston and therefore to the rate of flow. A rod attached to the top of the piston moves inside a sight glass mounted on top of the cylinder. The rate of flow is indicated by the position of the top of the piston rod with respect to a suitable scale mounted beside the sight glass.

A variation of this design replaces the multiple orifices with one or more pairs of longitudinal slots or metering ports in the cylinder, as shown in Fig. I-6-3. The upward movement of the metering plug, as the flow increases, is resisted by a loading spring for high velocities, or rates of flow, or by a calibrated weight for low rates. The movement of an armature attached to the metering-plug guide rod changes the inductance in the bridge circuit of a recorder, the pointer or pen of which shows the rate of flow.

I-6-6 Orifice and Plug. An orifice and plug type of area meter is shown schematically in Fig. I-6-4. A tapered plug, the floating element, rides vertically within the bore of an orifice. At the "no flow" position, the tapered plug nearly completely closes the orifice. Under flowing conditions, calling for an increase in area, the plug rises, increasing the annular

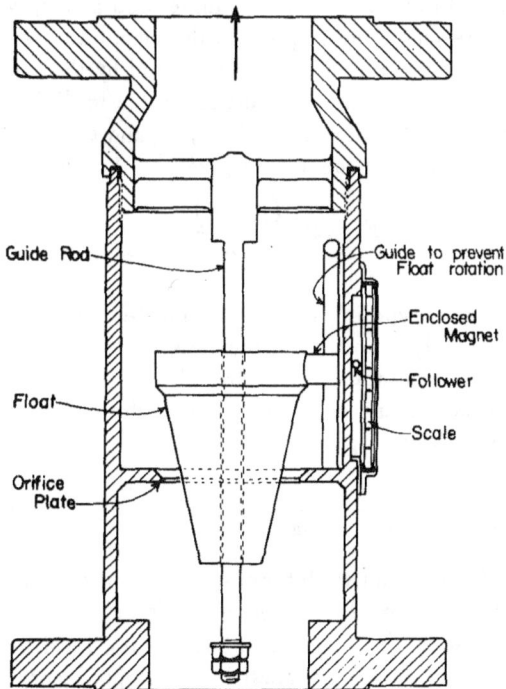

FIG. I-6-4 ORIFICE AND PLUG TYPE OF
AREA METER

area between it and the orifice. In a design such as shown, in which the fluid must pass up through the annulus between the float and the meter body, the shape of the float should be such that the flow through this annulus does not produce significant lift (e.g., as on an airfoil) on the floating element or plug.

I-6-7 Letter Symbols and Theoretical Equations for Area Meters. The letter symbols used in this chapter are:

A =	Area of tube corresponding to diameter, D	Notes 1 & 2
A_i =	Area of tube at inlet end	Notes 1 & 2
a =	Variable metering area (in Fig. I-6-5 the annular area between tube and float at D)	Notes 1 & 2
D =	Tube diameter at elevation of reading surface of float	Note 1
D_f =	Float diameter at reading surface	Note 1
D_i =	Tube diameter at inlet end	Note 1
F =	Force (buoyed weight of float)	lb_f
G =	Specific gravity of the fluid	ratio
g =	Acceleration due to gravity, local	ft/sec^2
g_c =	Conversion factor for units in the force-mass relation	32.174
K =	Flow coefficient	ratio
L =	A characteristic length; for the tapered tube and float, $D - D_f$	Note 1
M =	Mach number	ratio
M_f =	Mass of float	lb_m
m =	Mass rate of flow of fluid being metered	lb_m/sec
p =	Pressure	Note 1
q =	Volume rate of flow of fluid being metered	cfs
R =	Reynolds number	ratio
r =	Pressure ratio. p_2/p_1; p_1 pressure at A_i, p_2 pressure at outlet side of a	ratio
V =	Fluid velocity	fps

α (alpha) = Ratio of the major to minor diameter of the variable annular metering areas, D/D_f ratio

Λ (lambda) = Elevation above some datum   ft

μ (mu) = Absolute viscosity of fluid lb_m/ft-sec

ρ (rho) = Density of fluid lb_m/ft^3

γ (gamma) = Ratio of specific heats of a gas   ratio

Subscript 1 refers to the inlet area, A_i.

Subscript 2 refers to the outlet side of area a.

Note 1: For dimensional continuity diameters should be expressed in ft, areas in ft^2, and pressures in lb_f/ft^2; then in the final working equations the necessary conversion factors must be introduced to provide for using the customary units of inches, in.^{2} and psi.

Note 2: In most of the development of equations it suffices to use $\alpha = D/D_f$, or $A/a = \alpha^2$; however, in places where an area ratio is used the relationship $a/A = [(\alpha^2 - 1)/\alpha^2]$ is implied, since $a = (\pi/4)(D^2 - D_f^2)$.

I-6-8 For the development of equations it is helpful to use an elemental tapered tube and float meter, represented schematically by Fig. I-6-5, for depicting some of the basic quantities. The development of theoretical equations for the flow of liquids through area meters is similar to that for the differential pressure meters. However, with these meters the flow is always vertical; hence differences in elevation above some datum should be taken into account.

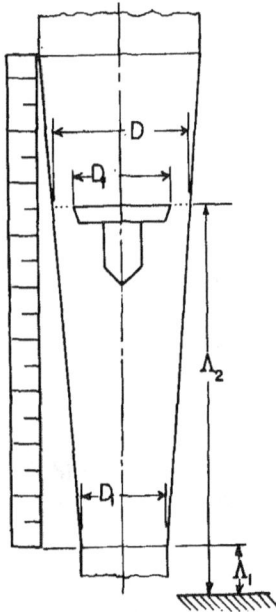

FIG. I-6-5 METERING ELEMENT OF A TAPERED TUBE AND FLOAT AREA METER

Thus, using equation (I-5-6), the general energy relation for area meters is

$$V_2^2 - V_1^2 = 2g\left[\left(\frac{p_1 - p_2}{\rho \, g/g_c}\right) - \left(\Lambda_2 - \Lambda_1\right)\right] \quad \text{(I-6-1)}$$

Referring to Fig. I-6-5, A_i is the area at the inlet (lower) end of the tapered tube and a, the minimum variable annular area between the float and tube; then assuming there are no leaks and the (liquid) density constant, the theoretical volume rate of flow, q_t, is

$$q_t = A_i V_1 = a V_2 \quad \text{(I-6-2)}$$

or

$$\begin{aligned} V_1 &= q_t/A_i \\ V_2 &= q_t/a \end{aligned} \quad \text{(I-6-3)}$$

Using these relations in equation (I-6-1)

$$V_2^2 = 2g\frac{\left[\left(\dfrac{p_1 - p_2}{\rho \, g/g_c}\right) - \left(\Lambda_2 - \Lambda_1\right)\right]}{1 - a^2/A_i^2} \quad \text{(I-6-4)}$$

and

$$q_t^2 = 2g\frac{\left[\left(\dfrac{p_1 - p_2}{\rho \, g/g_c}\right) - \left(\Lambda_2 - \Lambda_1\right)\right]}{1/a^2 - 1/A_i^2} \quad \text{(I-6-5)}$$

I-6-9 In any float-type variable area meter, where gravity is employed to provide the complete restoring force, F, the value of F is determined by the mass of the float, M_f, and the buoyant force due to differences between the density of the float, ρ_f, and that of the fluid, ρ. The relationship is given by the equation

$$F = gM_f\frac{(\rho_f - \rho)}{g_c \rho_f} \quad \text{(I-6-6)}$$

In any other type of variable area meter where gravity provides only a portion of the total restoring force while the remainder, F_s, is provided by some mechanical means such as a spring, the complete restoring force is given by

$$F = gM_f\frac{(\rho_f - \rho)}{g_c \, \rho_f} + F_s \quad \text{(I-6-7)}$$

I-6-10 In order to determine the theoretical rate of flow, the relationship between the force, F, and the pressure drop, $(p_1 - p_2)$, accelerating the fluid from section A to section a, must be developed. For tapered tube and float-type meters, this is done by establishing the force and momentum balance for all material contained within the tube between sections

A and *a*. Neglecting the tube taper angle and all viscous forces this becomes

$$p_1 A + \frac{q_t \rho}{g_c} V_1 = p_2 A + \frac{q_t \rho}{g_c} V_2 + \frac{g}{g_c} M_f$$

$$+ \left[\left(\Lambda_2 - \Lambda_1 \right) A - \frac{M_f}{\rho_f} \right] \rho \frac{g}{g_c} \qquad (I\text{-}6\text{-}8)$$

$$\left(p_1 - p_2 \right) A = \frac{g}{g_c} M_f \left(1 - \frac{\rho}{\rho_f} \right) + \frac{q_t \rho}{g_c} \left(V_2 - V_1 \right)$$

$$+ \left(\Lambda_2 - \Lambda_1 \right) A \rho \frac{g}{g_c}$$

$$= F + \frac{q_t \rho}{g_c} \left(V_2 - V_1 \right)$$

$$+ \left(\Lambda_2 - \Lambda_1 \right) A \rho \frac{g}{g_c} \qquad (I\text{-}6\text{-}9)$$

or

$$\frac{\left(p_1 - p_2 \right)}{\rho (g/g_c)} - \left(\Lambda_2 - \Lambda_1 \right) = \frac{F}{A \rho (g/g_c)} + \frac{q_t}{gA} \left(V_2 - V_1 \right)$$

$$= \frac{F}{A \rho (g/g_c)}$$

$$+ \frac{q_t^2}{gA} \left(\frac{1}{a} - \frac{1}{A} \right) \quad (I\text{-}6\text{-}10)$$

Substitution in (I-6-5) gives

$$q_t^2 = \frac{2g \left[\dfrac{F}{A \rho (g/g_c)} + \dfrac{q_t^2}{gA} \left(\dfrac{1}{a} - \dfrac{1}{A} \right) \right]}{\left(\dfrac{1}{a^2} - \dfrac{1}{A^2} \right)} \qquad (I\text{-}6\text{-}11)$$

so that

$$q_t^2 \left[\left(\frac{1}{a^2} - \frac{1}{A^2} \right) A - 2 \left(\frac{1}{a} - \frac{1}{A} \right) \right] = \frac{2 g_c F}{\rho} \qquad (I\text{-}6\text{-}12)$$

or

$$q_t^2 = \frac{1}{\left(\dfrac{1}{a} - \dfrac{1}{A} \right) \left[\left(\dfrac{1}{a} + \dfrac{1}{A} \right) A - 2 \right]} \times \frac{2 g_c F}{\rho}$$

$$\qquad (I\text{-}6\text{-}13)$$

$$= \left[\frac{A}{\left(\dfrac{A}{a} - 1 \right)^2} \right] \left(\frac{2 g_c A F}{\rho} \right)$$

I-6-11 "Theoretical Hydraulic Equations" for area meters in general are obtained from equation (I-6-13) and from the theoretical mass rate of flow relationship

$$m_t = q_t \rho \qquad (I\text{-}6\text{-}14)$$

Using $a/A = \left(\alpha^2 - 1 \right) / \alpha^2$

$$q_t = \left(\alpha^2 - 1 \right) \sqrt{\frac{2 g_c A F}{\rho}} \qquad (I\text{-}6\text{-}15)$$

$$m_t = \left(\alpha^2 - 1 \right) \sqrt{2 g_c A F \rho} \qquad (I\text{-}6\text{-}16)$$

In area meters of the tapered tube and float design, *a* is the annular area between the tube of area *A* and the float of area A_f. Then

$$A - a = A_f \qquad (I\text{-}6\text{-}17)$$

If *D* is the diameter of area, *A*, and D_f the diameter of area, A_f,

$$\frac{A}{A_f} = \frac{D^2}{D_f^2} = \alpha^2 \qquad (I\text{-}6\text{-}18)$$

and the theoretical hydraulic equations for the tapered tube and float become

$$q_t = \left(\alpha^2 - 1 \right) \alpha \sqrt{\frac{\pi}{2}} \; D_f \sqrt{\frac{g_c F}{\rho}} \qquad (I\text{-}6\text{-}19)$$

$$m_t = \left(\alpha^2 - 1 \right) \alpha \sqrt{\frac{\pi}{2}} \; D_f \sqrt{g_c F \rho} \qquad (I\text{-}6\text{-}20)$$

I-6-12 The Theoretical Equation for a Compressible Fluid is derived in a manner similar to that used for differential pressure meters. As in that case, it is assumed that the fluid is an ideal gas and that the flow from section A through section a is along a reversible isentropic path. Thus this derivation may be started by writing equation (I-5-20) in the form

$$\frac{V_2^2 - V_1^2}{2g_c} = \frac{\gamma}{\gamma-1}\frac{p_1}{\rho_1}\left(1 - r^{(\gamma-1)/\gamma}\right) \quad (I\text{-}6\text{-}21)$$

in which γ = ratio of specific heats, c_p/c_v. Since the mass rate of flow across both sections is the same,

$$m_t = A\,V_1\,\rho_1 = a\,V_2\,\rho_2 \quad (I\text{-}6\text{-}22)$$

As before, $a/A = (\alpha^2 - 1)/\alpha^2$, and $\rho_2/\rho_1 = r^{1/\gamma}$; and combining equations (I-6-21) and (I-6-22) together with these relations gives

$$\frac{m_t^2}{2g_c\,\rho_1\,A^2}\left[\frac{\alpha^4}{(\alpha^2-1)^2\,r^{2/\gamma}} - 1\right] = \frac{\gamma}{\gamma-1}p_1\left(1 - r^{(\gamma-1)/\gamma}\right) \quad (I\text{-}6\text{-}23)$$

Replacing p_1 with $(p_1 - p_2)/(1-r)$,

$$m_t^2 = 2g_c\left(p_1 - p_2\right)A^2\rho_1\left(\frac{\gamma}{\gamma-1}\right)\left[\frac{1 - r^{(\gamma-1)/\gamma}}{1-r}\right]\left[\frac{1}{\frac{\alpha^4}{(\alpha^2-1)^2\,r^{2/\gamma}} - 1}\right] \quad (I\text{-}6\text{-}24)$$

Neglecting any pressure difference due to elevation differences, a force balance for a tapered tube and float meter, by par. I-6-9, is:

$$A\left(p_1 - p_2\right) = F + \frac{m_t}{g_c}(V_2 - V_1) \quad (I\text{-}6\text{-}25)$$

Using values of V_2, V_1 and $(V_2 - V_1)$ derived from equation (I-6-22) gives

$$A\left(p_1 - p_2\right) = F + \frac{m_t^2}{g_c\,A\rho_1}\left[\frac{\alpha^2}{(\alpha^2-1)\,r^{1/\gamma}} - 1\right] \quad (I\text{-}6\text{-}26)$$

Using equation (I-6-26) to eliminate $(p_1 - p_2)$ from equation (I-6-24) gives

$$m_t^2 = 2g_c\,AF\,\rho_1\left(\frac{\gamma}{\gamma-1}\right)\left(\frac{1-r^{(\gamma-1)/\gamma}}{1-r}\right)\left[\frac{1}{\frac{\alpha^4}{(\alpha^2-1)^2\,r^{2/\gamma}} - 1}\right]$$

$$+ 2m_t^2\left(\frac{\gamma}{\gamma-1}\right)\left(\frac{1-r^{(\gamma-1)/\gamma}}{1-r}\right)\left[\frac{1}{\frac{\alpha^2}{(\alpha^2-1)\,r^{1/\gamma}} + 1}\right] \quad (I\text{-}6\text{-}27)$$

from which

$$m_t = \sqrt{2g_c\,AF\,\rho_1}\left\{\frac{1}{\left(\frac{\alpha^2}{(\alpha^2-1)\,r^{1/\gamma}} - 1\right)\left(\frac{\alpha^2}{(\alpha^2-1)\,r^{1/\gamma}} + 1\right)}\frac{1}{\left(\frac{\gamma-1}{\gamma}\right)\left(\frac{1-r}{1-r^{(\gamma-1)/\gamma}}\right) - 2}\right\}^{\frac{1}{2}} \quad (I\text{-}6\text{-}28)$$

Equation (I-6-28) is the theoretical equation for the mass rate of flow of a compressible fluid through section a in terms of the fluid density, ρ_1.

Equation (I-6-28) may be written in the form

$$m_t = (\alpha^2 - 1)\,Y\,\sqrt{2g_c\,AF\,\rho_1} \quad (I\text{-}6\text{-}29)$$

where

$$Y = \frac{1}{(\alpha^2 - 1)}\left\{\frac{1}{\left(\frac{\alpha^2}{(\alpha^2-1)\,r^{1/\gamma}} - 1\right)\left(\frac{\alpha^2}{(\alpha^2-1)\,r^{1/\gamma}} + 1\right)}\frac{1}{\left(\frac{\gamma-1}{\gamma}\right)\left(\frac{1-r}{1-r^{(\gamma-1)/\gamma}}\right) - 2}\right\}^{\frac{1}{2}} \quad (I\text{-}6\text{-}30)$$

Since the pressure ratio, r, is not readily obtainable from the operating conditions in an area meter, it can be determined with equation (I-6-26) and in which $(p_1 - p_2)$ is replaced with $p_1 (1 - r)$

$$A p_1 (1 - r) = F + \frac{m_t^2}{A g_c \rho_1} \left[\frac{\alpha^2}{(\alpha^2 - 1) r^{1/\gamma}} - 1 \right] \quad (I-6-31)$$

Substituting the value of m_t from equation (I-6-29), dividing by $A \gamma p_1$, and rearranging:

$$\frac{F}{A \gamma p_1} = \frac{1 - r}{\gamma} \left[\frac{1}{1 + 2Y^2 (\alpha^2 - 1)^2 \left(\frac{\alpha^2}{(\alpha^2 - 1) r^{1/\gamma}} - 1 \right)} \right] (I-6-32)$$

Thus the compressibility parameter, $(F/A \gamma p_1)$, is a function of r, γ and α, the same quantities that are involved in the expression for Y. Consequently, specific values of these variables define a value of $(F/A \gamma p_1)$ and also define a corresponding value of Y. For any value of γ, this makes it possible to present values of Y as a function of $(F/A \gamma p_1)$ at constant α.

I-6-13 Flow Coefficients and Equations for Actual Rates of Flow. Since the actual rate of flow through an area meter is very seldom exactly equal to the theoretical rate, being less usually, it is necessary to apply a flow coefficient, K, to the theoretical to give the actual rate. Thus, for the general case of an area meter, with an incompressible fluid, the actual rate of flow is, from equation (I-6-16)

$$m = K(\alpha^2 - 1) \sqrt{2 g_c A F \rho} \quad (I-6-33)$$

and for the tapered tube and float the actual rate, by equation (I-6-20), is

$$m = K(\alpha^2 - 1) \alpha \sqrt{\frac{\pi}{2}} D_f \sqrt{g_c F \rho} \quad (I-6-34)$$

For compressible fluids the actual rates of flow are obtained by multiplying the right-hand sides of equations (I-6-33) and (I-6-34) by the expansion factor, Y, equation (I-6-30).

I-6-14 At the present time, in this country, most manufacturers and users of area meters are using a hybrid system of units in the production and use of these meters. Thus, diameters and lengths are in inches, and float weights and forces are in grams, while flow rates are in pounds (m) per hour, gallons per minute, or cu ft per minute. When metering liquids other than water, the density of the liquid is de-

scribed by its specific gravity, G, referred to the density of water at 60 F, the value of which is 62.3664 lb_m per cu ft (in vacuo). It is common practice to use $g = 32.174$ ft/sec^2 without regard to local variations. Also, as used in this edition, $g_c = 32.174$. The conversion factor of grams per pound is 453.59. Using these relations and values, the working equations are:

1. for liquids

$$m \text{ (lb}_m/\text{hr)} = 892.3(\alpha^2-1) K \sqrt{A F G} \quad (I-6-35)$$

$$q \text{ (gal/min)} = 1.784(\alpha^2-1) K \sqrt{\frac{A F}{G}} \quad (I-6-36)$$

2. for compressible fluids

$$m \text{ (lb}_m/\text{hr)} = 113.0 (\alpha^2-1) K \, Y \sqrt{A F \rho_1} \quad (I-6-37)$$

$$q_o \text{ (cu ft/min at } p_o, T_o)$$
$$= \frac{1.883}{\rho_o} (\alpha^2-1) \, K \, Y \quad A F \rho_1 \quad (I-6-38)$$

In these equations, A is in sq in, and F in grams; gallon is the U.S. gal of 231 cu in, ρ_o (lb_m/cu ft) is at the conditions p_o, T_o, and ρ_1 at the conditions p_1, T_1.

I-6-15 For the special case of the tapered tube and float type, it may be convenient to make a substitution for F based on equation (I-6-6). Let G_f be the specific gravity of the float referred to water at 60 F, and as before, G is that of the fluid. Then, for liquids

$$F = \frac{g}{g_c} M_f \frac{G_f - G}{G_f} \quad (I-6-39)$$

and for gases, assuming G/G_f to be very small,

$$F = M_f \frac{g}{g_c} \quad (I-6-40)$$

Note: As used in this place, G is the specific gravity of the flowing fluid, either liquid or *gas*, referred to *water*. The possible alternative in the case of a gas would be to use air as the reference fluid for *both* G and G_f. The result would be the same so far as the use of equation (I-6-40) would be involved.

Also, some of the factors in equation (I-6-34) and the flow coefficient, K, may be combined to give the modified coefficient

$$K' = K (\alpha^2 - 1) \alpha \sqrt{\frac{\pi}{2}} \quad (I-6-41)$$

Thus, the working equations are:

1. for liquids

$$m \ (\text{lb}_m/\text{hr}) = 631.0 \ K' \ D_f \sqrt{\frac{gM_f}{g_c} \frac{(G_f - G)}{G_f}} \ G \quad (\text{I-6-42})$$

$$q \ (\text{gpm}) = 1.261 \ K' \ D_f \sqrt{\frac{gM_f}{g_c} \frac{(G_f - G)}{G_f G}} \quad (\text{I-6-43})$$

2. for gases

$$m \ (\text{lb}_m/\text{hr}) = 79.90 \ K' \ Y \ D_f \sqrt{\frac{g \ M_f \ P_1}{g_c}} \quad (\text{I-6-44})$$

$$q_o \ (\text{cfm}) = \frac{1.332}{\rho_o} K' \ Y \ D_f \sqrt{\frac{g \ M_f \ \rho_1}{g_c}} \quad (\text{I-6-45})$$

in which D_f is in inches and M_f is in grams. [1]

I-6-16   As with differential pressure meters, the Reynolds number and Mach number are useful in comparing and correlating experimentally determined coefficients of geometrically similar area meters [2]. From equation (I-6-33) the velocity of the fluid through area, a, is

$$\frac{m}{a\rho} = \frac{K(\alpha^2 - 1)\sqrt{2g_c \ A \ F \ \rho}}{a \ \rho} \quad (\text{I-6-46})$$

If L is some characteristic dimension of area, a, then

$$R = \frac{L \ K(\alpha^2 - 1)\sqrt{2g_c \ A \ F \ \rho}}{a\rho} \frac{\rho}{\mu} \quad (\text{I-6-47})$$

$$= \frac{L}{a} K \ (\alpha^2 - 1)\left(\frac{\sqrt{2g_c \ A \ F \ \rho}}{\mu}\right)$$

Using L as a characteristic dimension of the area, a, the ratio L/a will be a constant for each particular geometry, and its use in the expression for R provides a proper order of magnitude so the Reynolds numbers for different geometries may be compared.

For the tapered tube and float-type, the characteristic length is taken as the difference between the diameters of the tube and float, that is, $L = D - D_f = D_f (\alpha - 1)$. From equation (I-6-34) the velocity across area a is

$$\frac{m}{\rho a} = \frac{K(\alpha^2 - 1) \alpha \sqrt{\frac{\pi}{4}} D_f \sqrt{2g_c \ F \ \rho}}{\rho \ a} \quad (\text{I-6-48})$$

$$R = D_f(\alpha-1) \frac{K(\alpha^2-1)\alpha\sqrt{\frac{\pi}{4}} D_f \sqrt{2g_c \ F \ \rho}}{\rho \frac{\pi}{4} D_f^2 (\alpha^2-1)} \frac{\rho}{\mu} \quad (\text{I-6-49a})$$

$$= K(\alpha-1)\alpha\sqrt{\frac{4}{\pi}}\left(\frac{\sqrt{2g_c \ F \ \rho}}{\mu}\right) \quad (\text{I-6-49b})$$

$$= K(\alpha^2 - \alpha)\sqrt{\frac{8g_c}{\pi}}\left(\frac{\sqrt{F \ \rho}}{\mu}\right) \quad (\text{I-6-49c})$$

In equation (I-6-49) the term, $(\sqrt{F\rho}/\mu)$, is called the "viscosity influence number," since for a particular value of F the value is determined by the density and viscosity of the fluid.

I-6-17   The Mach number for an area meter is

$$M = \frac{V_2}{\sqrt{\left(\frac{P_2}{\rho_2}\right)g_c \gamma}} \quad (\text{I-6-50})$$

Since $V_2 = q/a = (m/a\rho_2)$,

$$M = \frac{m}{a\sqrt{P_2 \ \rho_2 \ g_c \ \gamma}} \quad (\text{I-6-51})$$

Using the value for m from equation (I-6-33) and including the expansion factor, Y,

$$M = \frac{K \ Y}{a}(\alpha^2 - 1)\sqrt{\frac{2 \ A \ F \ \rho_1}{P_2 \ \rho_2 \ \gamma}}$$

$$= K \ Y \ \alpha^2 \sqrt{\frac{2 \ F \ \rho_1}{A \ P_2 \ \rho_2}}$$

$$= K \ Y \ \alpha^2 \sqrt{\frac{2 \ F}{r^{1/\gamma} \ A \ \gamma \ P_1}} \quad (\text{I-6-52})$$

I-6-18   From an inspection of equations (I-6-47) or (I-6-49) and (I-6-52), it can be seen that whenever the Reynolds number alone can be used for the correlation of test data, the "viscosity influence number," $(\sqrt{F \ \rho}/\mu)$, may be used in its place. Whenever the Mach number alone will correlate data, the compressibility parameter, $F/(A \ \gamma \ p_1)$, may be used instead.

I-6-19   Operating Adjustment Factors. [3] Usually area meters are calibrated for use at a particular set of operating and fluid conditions. There may be times when this is either not possible or the set of conditions changes, and it then becomes useful to apply an adjustment factor. The basic form of such a factor is developed by writing equation (I-6-33) with the inclusion of Y for each condition and ratioing the results:

$$\frac{m_2}{m_1} = \frac{K_2 Y_2}{K_1 Y_1} \left[\frac{\alpha_2^2 (\alpha_2^2 - 1)}{\alpha_1^2 (\alpha_1^2 - 1)} \right] \sqrt{\frac{A_2 \; F_2 \; \rho_2}{A_1 \; F_1 \; \rho_1}} \qquad \text{(I-6-53)}$$

Generally it is not advisable to apply an adjustment factor *unless* the meter tested and the meter used are geometrically similar, the value of K is constant at the values of the viscosity influence numbers involved, and any changes in Y are small. Under these limiting conditions the adjustment factor reduces to

$$\frac{m_2}{m_1} = \sqrt{\frac{F_2 \; \rho_2}{F_1 \; \rho_1}} \qquad \text{(I-6-54)}$$

For volume at operating conditions, the adjustment factor is

$$\frac{q_2}{q_1} = \sqrt{\frac{F_2 \; \rho_1}{F_1 \; \rho_2}} \qquad \text{(I-6-55)}$$

For reference condition volumes,

$$\frac{q_{o2}}{q_{o1}} = \frac{\rho_{o1}}{\rho_{o2}} \sqrt{\frac{F_2 \; \rho_2}{F_1 \; \rho_1}} \qquad \text{(I-6-56)}$$

For the tapered tube and float-type equation (I-6-39) may be used; then with no change in the mass of the float or if the floats in two meters have equal masses,

$$\frac{F_2}{F_1} = \sqrt{\frac{G_f - G_2}{G_f - G_1}} \qquad \text{(I-6-57)}$$

Thus, for liquids the adjustment factors become

$$\frac{m_2}{m_1} = \sqrt{\frac{(G_f - G_2) \; G_2}{(G_f - G_1) \; G_1}} \qquad \text{(I-6-58)}$$

and

$$\frac{q_2}{q_1} = \sqrt{\frac{(G_f - G_2) \; G_1}{(G_f - G_1) \; G_2}} \qquad \text{(I-6-59)}$$

and for gases where the effect of buoyancy on the float is negligible,

$$\frac{m_2}{m_1} = \sqrt{\frac{\rho_2}{\rho_1}} \qquad \text{(I-6-60)}$$

and

$$\frac{q_{o2}}{q_{o1}} = \frac{\rho_{o1}}{\rho_{o2}} \sqrt{\frac{\rho_2}{\rho_1}} \qquad \text{(I-6-61)}$$

References

[1] "Coefficients of Float-Type Variable-Area Flowmeters," V. P. Head; *Trans. ASME*, vol. 76, Aug. 1954, p. 581.

[2] "Correcting for Density and Viscosity of Incompressible Fluids in Float-Type Flowmeters," M. R. Shafer, E. F. Fiock, H. L. Bovey and R. B. VanLone; *Journal of Research*, National Bureau of Standards, vol. 47, Oct. 1951, RP ee47, p. 227.

[3] "Process Instruments and Controls," McGraw-Hill Book Co., New York, 1957, p. 457.

Chapter I-7
Fluid Velocity Measuring Instruments and Meters

I-7-1 There are two basic groups of fluid velocity measuring instruments. One group has a rotating primary element which is kept in motion by the direct movement (velocity) of the fluid stream. With the second group the fluid velocity is deduced from the conversion of fluid velocity energy into pressure as with the Pitot tube or from the dissipation of heat to the fluid as with the hot-wire anemometer. One or more of the velocity measuring devices in each of these groups can be used to measure fluid velocity under the following three conditions:

1. Fluid velocity in an open area.
2. The velocity of fluid flowing in an open channel.
3. The velocity of fluid flow in a closed channel. In the first case a common objective is the measurement of wind velocity or the current in a large stream. In the second and third cases the velocity measurement is usually of secondary interest, since the primary objective is the volume rate or total volume flow.

I-7-2 Terminology. Rotating-cup-, -vane-, or -propeller meters are usually termed *anemometers* when they are designed to measure wind speed; they are usually termed *current meters* when they are designed to measure velocity in an open channel, such as a sluice or stream; and they are usually given the name *flowmeters* when they are designed to measure fluid velocity in a pipe or duct. Turbine-type meters belong to this latter class of flowmeters.

I-7-3 Letter Symbols. The letter symbols used in this chapter are:

C = A coefficient, determined by calibration	ratio
d = Diameter of an instrument element	in. (or ft)
m = Mass rate of flow	lb_m/sec
N = Ideal or theoretical rotational speed	rpm
n = Actual rotational speed	rpm
p = Pressure (absolute or gage as noted)	psi
q = Volume rate of flow	cfs
u = Local instantaneous velocity or local instantaneous deviation of mean velocity	ft/sec
V = Mean area speed of fluid	ft/sec
ρ (rho) = Density of fluid	lb_m/ft³
μ (mu) = Viscosity of fluid	lb_m/ft-sec

Other letter symbols are defined as used in discussing the different instruments, as some are not defined the same in all places. Also, the special subscripts are defined as used.

I-7-4 Anemometers: Cup Type. Cup-type anemometers have hemispherical or conical cups mounted on the spokes of a wheel with a vertical shaft. Three- and four-cup designs are common (Fig. I-7-1). The lips of the cups may be beaded for greater strength and reduced sensitivity to flow angle in a vertical plane, i.e., in a plane through the axis of rotation. Rotor diameters range from 3 to 18 in., and the corresponding cup diameters range from 1 to 5 in.

FIG. I-7-1 THREE-ARM CUP ANEMOMETER

I-7-5 The peripheral speed of the center of the cups is only about one-third of the linear speed of the fluid stream. The ratio, (linear fluid speed)/(linear peripheral speed of cup center), is termed the *factor* of the anemometer. This factor may range from 2.8 to 3.1, depending on design. It *must* be determined by a wind-tunnel calibration. The practical lower limit of the range is 2 to 3 miles per hour. The upper limit of the range is usually 60 to 100 mph. Inaccuracy in steady, horizontal flow is on the order of 1/2 to 1 mph, after application of systematic corrections stated by the manufacturer. This systematic correction may be about 2 per cent at 60 mph and less at lower speeds.

In gusty winds the indicated air speed is considerably higher than the true mean air speed. Turbulence, $\sqrt{(\bar{u})^2}/V$, of 3 per cent may produce an indication that is 3 to 5 per cent too high. Here u is the instantaneous deviation of the actual air speed from the mean area air speed, V.

Because of symmetry, there is no effect of flow angle variation in a horizontal plane. The effect of vertical flow angles will depend on the cup, wheel and support design. To indicate the order of magnitude of possible error, an indication that was 6 per cent too high has been reported for an anemometer whose vertical axis was tilted 30 deg. Vertical flow angles exceeding 45 deg must not be permitted, since an abrupt, drastic change in meter performance occurs at this angle.

A 30 per cent reduction in static pressure produces a reduction of about 6 per cent in indicated speed.

I-7-6 It must be recognized that the anemometer indicates only the air speed at its own location and this local air speed may be strongly influenced by the proximity of other objects such as a thermometer well, a branch connection or even its own support when used within ducts and by buildings or other structures when used in free-field measurements. Hence, to give a representative indication of free-field airspeed, the instrument must be mounted on as tall and slender a support as consistent with mechanical sturdiness.

I-7-7 Vane and Propeller Types: Figures I-7-2 and I-7-3. The adjectives "vane type" and "propeller type" are used indiscriminately when referring to these designs of either anemometers or current meters. Ordinarily these meters have a multibladed rotor whose axis of rotation is approximately parallel to the direction of the stream. The blade faces are inclined to this direction, so that axial fluid velocity

FIG. I-7-2 A SHROUDED VANE ANEMOMETER

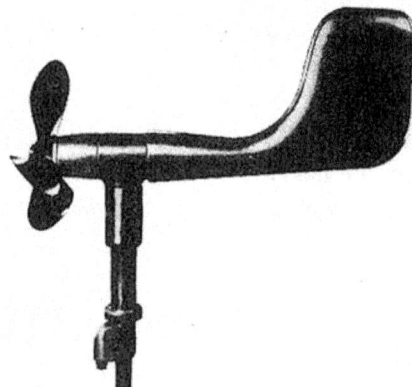

FIG. I-7-3 MAST-MOUNTED PROPELLER ANEMOMETER

is converted into rotor rotation. In the most efficient propeller design, the blades are twisted and tapered in cross section so that each radial element of blade operates at the most efficient angle of attack for that fluid speed at which maximum accuracy is desired, and the blades are of airfoil cross section. Most practical designs, however, forego some of these features in order to realize simplicity and economy of construction:

1. The taper may be omitted, or restricted to a taper in width only.

2. The blades may be of constant-thickness sheet metal or plastic.

3. The blades may be of rectangular, trapezoidal, quasicircular, or oval planform.

4. The blades may be short lengths of a helical screw, with axial extension a small fraction of one turn.

I-7-8 Vane-type anemometers range in size from less than 3 in. to over 15 in. in outside diameter and cover airspeeds from about 30 to 10,000 fpm (about 0.34 to 114 mph). However, any one meter is usable with accuracy only between a 10:1 to 20:1 airspeed range. In contrast to cup-type anemometers, the vane type have a peripheral velocity, at some radial location on the blade, that is substantially equal to the air velocity multiplied by the tangent of the angle of the blade to the stream. Anemometers that are shrouded are substantially insensitive to flow angles up to 20 deg or more. Calibrations are generally established in a wind tunnel. When the calibration is certified by a governmental laboratory, uniform, axial, nonturbulent flow is assumed, normally. Since the calibration thus obtained will depend upon the friction in the instrument, recalibration may be necessary from time to time. The effect of small changes of air density, such as due to changes of elevation, may be evaluated as follows.

Let V = Mean airspeed
V_b = Peripheral speed of the center of pressure of one vane
ρ = Density of the air

Subscript 1 refers to the calibration conditions.
Subscript 2 refers to some later condition of use.
Then [1, 2],

$$\frac{V_{b2}}{V_{b1}} = \frac{V_2}{V_1} = \sqrt{\frac{\rho_1}{\rho_2}} \qquad (I\text{-}7\text{-}1)$$

It is important to note that equation (I-7-1) is applicable only to the use of anemometers in low-pressure ducts, tunnels and free-field measurements of wind speeds and cannot be applied to pipeline flowmeters, e.g., turbine meters, where viscosity effects and large changes in density are involved. Also, in free-field velocity measurements, the same considerations apply as given for cup-type anemometers in Par. I-7-6.

When an anemometer is used to measure airflow rate in a duct, the meter calibration may be appreciably in error if the smallest transverse dimension of the duct is less than 10 times the outside diameter of the meter.

I-7-9 Current Meters: Cup Type. Much that was given regarding cup-type anemometers applies to current meters, also. For open-channel measurements, a type commonly used is the Price meter, with five or six conical cups (Fig. I-7-4). The *factor* of these meters is about 2.8. For these meters and the cup-type anemometers also, it is possible to develop an analytical analysis of the operation — such a relation is extremely complex and of little practical value. However, such an analysis as well as results of tests show that these meters tend to over-run. That is, read high. Thus, it is the general practice with these meters to calibrate each instrument. These calibrations may be had at government or university laboratories. Usually these calibrations

FIG. I-7-4 A FIVE-ARM CUP-TYPE CURRENT METER

are made by towing the meter, at several uniform speeds, through a long channel of still water. Alternately, the meter may be mounted on a long (about 20 ft) arm that can be rotated at various constant velocities over a body of still water.

The range of fluid speeds covered by a conventional meter with a 5-in. outside diam bucket wheel is about 0.1 to 15 fps. A smaller meter with a 2-in. diam bucket wheel has a lower limit of about 0.05 fps.

I-7-10 Horizontal flow angles up to 45 deg produce negligible change in the *factor* of the Price meter; the meter continues to indicate the magnitude of fluid velocity. If a meter is rigidly attached to the bottom of a support rod and is subjected to flow with a vertical component directed downward, the indicated water speed should be divided by the cosine of the flow angle to obtain the true water speed. For flow with a vertical component directed upward, the indicated speed should be divided by 1.5 times the cosine of the flow angle to obtain the true water speed.

For a rms (root-mean-square) fluctuation, $\sqrt{\bar{u}^2}$, of fluid velocity along the direction of mean velocity, V, the meter indication will be too high. The fractional error in indication is on the order of $0.2(\bar{u}^2/V^2)$ when the oscillation frequency is between 5 and 50/min, if $\bar{u}^2$ is appreciably smaller than V and if V is averaged over at least 2 min.

The Price current meter is usable in air, also. Its factor is a single-valued function of $V/\sqrt{\rho}$, where ρ is the density of the fluid over at least the range of densities between water and air. The factor is substantially constant when $V/\sqrt{\rho}$ exceeds (10 ft/sec) $\sqrt{\mathrm{lb}_m/\mathrm{ft}^3}$ [3].

I-7-11 Propeller Type. For open-channel measurement, a commonly used propeller-type current meter (Fig. I-7-5) has two or three blades, of helical-screw shape, on a horizontal-axis rotor. Diameters range from 1 to 5 in. Various fluid velocity ranges

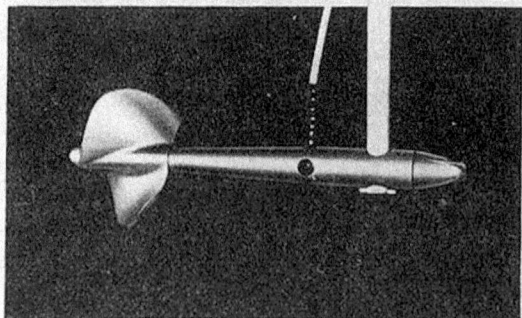

from 0.1 to 15 fps can be covered by using propellers with different pitches. Calibrations are performed in the same manner as described for the cup-type current meters.

For oscillations of velocity in the direction of the mean fluid velocity, in the range of 5 to 50 oscillations/min, negligible error in indication results when the rms oscillation velocity is much smaller than the mean stream velocity.

In isotropic turbulence, the propeller-type current meter is affected much less than the cup-type current meter.

Ordinarily, the assumption of a cosine law yields results accurate to about 1 per cent for flow-misalignment angles up to 15 deg; however, special designs are obtainable in which the angle limit is raised to 45 deg. When effect of flow misalignment on any design of current meter is important, the effect should be determined specifically for that design of meter by calibration in the towing tank.

Concepts Applicable to Velocity Meters with Rotating Primary Elements, Particularly Anemometers and Current Meters

I-7-12 Range. In any blade-type mechanical meter designed to provide approximate proportionality between rotational speed and fluid velocity, there exists a balance between the driving torque exerted by the hydrodynamic impact of the fluid upon the meter blades and the opposing torques due to bearing friction and to hydrodynamic drag upon the rotor. The hydrodynamic torques vary approximately as the square of fluid velocity. The bearing torque, above a certain rotational speed, may be almost constant or may increase slowly with speed; however, as speed approaches zero, the bearing torque generally approaches a rather high value, the "sticking" or "static-friction" value. Figure I-7-6 shows a typical relationship between these quantities. The

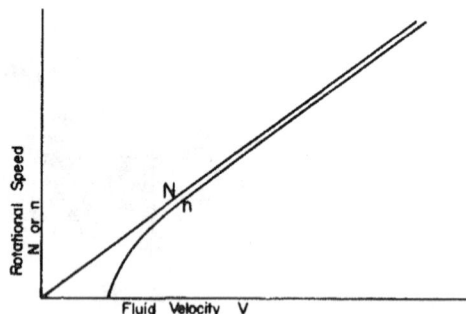

FIG. I-7-5 PROPELLER-TYPE CURRENT METER

FIG. I-7-6 RELATION BETWEEN ROTATIONAL SPEED AND FLUID VELOCITY

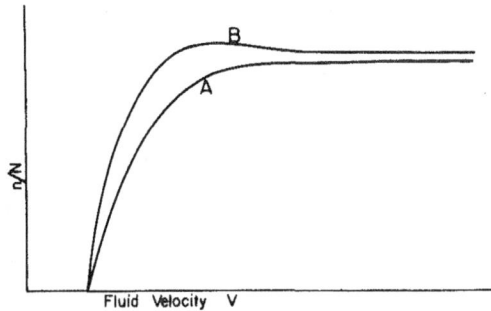

FIG. I-7-7 TYPICAL FLOWMETER PERFORMANCE
CURVES

angle of attack of the fluid upon the blades adjusts itself to maintain equality of driving and opposing torques. As the fluid velocity increases, the angle of attack ordinarily approaches an asymptotic value close to zero; and the slip between actual rotational speed, n, and ideal rotational speed, N, decreases as fluid velocity increases. Thus, typically, the ratio n/N has the appearance shown by curve A of Fig. I-7-7. However, secondary effects associated with Reyuolds-number variation may alter this curve slightly so that it has the appearance of curve B.

The lower limit, $n = 0$, of the range is established when driving torque is equal to sticking torque of the bearing. Since, in most bearings, the static friction is greater than the running friction, the fluid velocity, V_1 – at which the meter first begins to turn as V is increased – is greater than the fluid velocity, V_2 – at which the meter stops turning as V is decreased. Thus, the region near the lower limit of the range is not only the region of greatest inaccuracy but also the region of poorest reproducibility. The *practically useful lower limit* of any meter is the value n_{min} at which the systematic calibration corrections Δn ($\equiv N - n$) themselves have an acceptably small uncertainty, $\delta(\Delta n)$. Ordinarily, $\delta(\Delta n)$ is the Gaussian probable error of measurement. (In establishing Δn by calibration, it is not always necessary to know N, from a theoretical analysis; it is often sufficient to choose arbitrarily an N_{max} as the value of n_{max} corresponding to a fluid velocity V_{max} near the upper limit of the range and then to define $N = N_{max} (V/V_{max})$.)

The upper limit of the range of a meter is generally set by one of the following considerations:

1. The mechanically safe upper limit of rotation of the bearings and the rotor structure.

2. The occurrence of cavitation at the blades in the case of liquids, or of sonic flow in the case of gases.

3. A too high pressure drop across the meter.

I-7-13 Alignment and Nonalignment.

In the following discussion the angles referred to are:

θ = Angle between meter axis and mean path of fluid

ϕ = Angle between meter axis and channel center line

When the axis of the meter is at an angle, θ, with the mean direction of the fluid flowing at speed V (Fig. I-7-8), the meter's indication is not always $V \cos \theta$. The deviation from this value can be determined by calibration and hence can be applied as a correction which is a systematic function of θ. However, if θ is not known when the meter is used, the resulting random error can be minimized only by choosing a meter design that is either self-aligning or else that has other design provisions for minimizing effects of nonaligned flows.

In open-channel gaging, a current meter may be used in one of three ways to determine the equivalent velocity V_o in the direction of the geometric center line of the channel:

1. $\theta = \phi$. The meter may be rigidly attached to a vertical rod which is held so that the meter axis is parallel to the channel center line. The angle ϕ between local stream direction and channel center line must be estimated or independently determined, and the correction for flow misalignment must be made from empirical knowledge, by calibration, of the meter's flow-angle characteristic, $(V_{indicated}/V_{true})$ vs θ.

2. $\theta = 0$. The meter may be rigidly attached to a vertical support which is rotated until a maximum (or minimum) indication is obtained from the meter. The method is convenient only when substantially continuous indication of rotation rate is available. The velocity V_o is then $V_{ind} \cos \phi$, where ϕ is the angle between meter axis and channel center line and also the angle between the local stream direction and channel center line.

3. $\theta = 0$. The meter may be fitted with an aligning tail vane and either pivoted about the end of a vertical rod or else hung from a chain, so that the meter aligns itself with the local stream direction. Then $V_o = V_{ind} \cos \phi$.

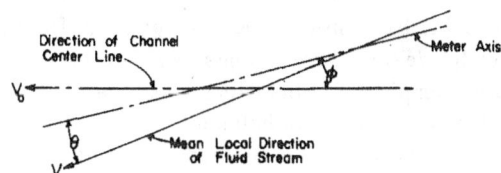

FIG. I-7-8 ANGLES INVOLVED IN NONALIGNMENT

I-7-14 Turbulence. Oscillations of velocity may be superposed on the mean velocity of the fluid. These oscillations may be resolved into a component u_x along the line of the mean-fluid velocity, V, and components, u_y and u_z, perpendicular to that line. Eddies may also be resolved into such components. The common name *turbulence* is conveniently assigned to all such oscillations. An appropriate measure of this quantity is the rms amplitude of the oscillation divided by the mean velocity.

A distinction is to be made between turbulence, as discussed here, and pulsations, as defined in Chapter I-3. Turbulence develops as a result of the fluid movement, aggravated in most cases by objects or unevennesses in the path or channel of the fluid stream [4]. On the other hand, pulsations generally arise from the action of some body or force (e.g., a pump) upon the fluid. Moreover, pulsations can be present in a fluid stream even when the flow rate is zero.

When turbulent eddies exist in a stream, with turbulence $|\Delta V/V| \equiv \left(\overline{u_x^2} + \overline{u_y^2} + \overline{u_z^2}\right)^{\frac{1}{2}}/V$, the amplitude distribution is Gaussian at points remote from the stream boundaries but necessarily is skewed near the boundaries. Fortunately, in those situations where high accuracy is needed, the value of $|\Delta V/V|$ is usually less than 15 per cent. An upper limit of possible error in the measurement may be taken as $(1/2)|\Delta V/V|^2$, so that the limit of error would then be about 1 per cent.

1. When velocity fluctuations are so slow that the meter can follow them, the average value of the meter's indication correctly represents the mean fluid velocity.

2. When velocity fluctuations are so rapid the meter indication remains substantially stationary, the meter's indication is too high because of the square-law relation between hydrodynamic torque and velocity. The magnitude of the error depends on the relative magnitudes of hydrodynamic torque and bearing torque. If these torques are known and the fluid is a liquid, it may be possible to compute the error analytically for propeller and vane-type meters, provided the fluctuations are small and along the line of the meter axis. (The error is a function of the ratio of rms fluctuation amplitude, $\sqrt{\overline{u_x^2}}$, to the mean velocity, V.) For cup-type meters, for large fluctuation or for cases where torques have not been measured, empirical calibration is necessary to determine the error for turbulence.

3. When the fluid is a gas and the velocity fluctuations due to either turbulence or pulsations are too rapid for the meter element to follow faithfully, the meter indication will be incorrect by an indeterminable amount. In such a case, some means of making the flow steady and with a normal velocity distribution must be provided upstream of the meter.

I-7-15 Fluid Density and Viscosity. The meter calibration may be different for different fluids, especially near the lower limit of the range, because hydrodynamic torques vary directly with fluid density as a first-order effect and with kinematic viscosity or Reynolds number as a second-order effect.

I-7-16 Temperature. Change in temperature alters the effect of bearing friction upon rotational speed and also alters the density, viscosity and Reynolds number of the fluid. The overall effect of temperature, which is generally slight in a velocity meter, is most noticeable near the lower limit of the range and must be determined empirically if systematic correction is desired.

I-7-17 Spatial Distribution of Velocities. When used for local-velocity measurement (as distinguished from the case where the entire flow passes through the meter), the meter's indication is generally intended to permit inference of the mean linear speed of the stream or of its bulk velocity. An area-wide survey with one or several meters is made and the indications averaged or integrated; or else an assumption is made concerning the relative velocity distribution across the cross section of the stream. The surveying techniques that permit use of the fewest number of meters or that minimize the time or effort of the survey are treated in a later section.

I-7-18 Effect of Boundaries. The boundaries of the fluid may be solid, as at the sides and bottom of a duct or sluice, or may be a free surface, as at the top of a sluice or river. When the distance of a local-velocity meter from any such boundary becomes nearly the same as a transverse dimension of the meter itself, interference occurs, so that the meter's indication may no longer represent that fluid velocity which would exist if the meter were absent. The effect on any spatial average or integral becomes greater as the meter dimension becomes a greater fraction of the transverse dimensions of the stream. Unless empirical calibrations of these proximity effects are available, it is important to use a meter whose dimensions are very small compared to the channel dimensions.

Similar interference due to proximity may exist when two or more meters are used in a rake or array to provide an areal survey. Interference is lessened when adjacent meters have opposite directions of rotation, so that buildup of large eddies is avoided.

Nevertheless, the transverse dimension of each meter must be many times smaller than the distance between adjacent meters, unless careful laboratory determination of the magnitude and sign of the interference effects has been made.

I-7-19 Dynamic Response. In estimating the ability of a velocity meter to follow a fluctuating flow, it is sufficient to consider it as following a first-order differential equation, characterized by a time constant, τ. This time constant is dependent on rotor mass and dimensions and is inversely proportional to fluid density and to fluid velocity.

I-7-20 Secondary Instrumentation. The rate of primary-element rotation may be deduced from:

1. Counting and timing of a succession of contact closures.

2. Counting and timing of electromagnetically- or photoelectrically-generated pulses.

3. Timing of the interval between appropriate readings of a mechanical revolution counter geared to the fluid-driven rotor.

4. A continuous ac or dc voltage, proportional to rate of rotation, generated by a tachometer element.

5. Deflection of a magnetic-drag type of element using eddy currents induced in an aluminum disc. The counting and timing of pulses, or direct indication of pulse rate with a frequency meter, or measurement of the proportional voltages can be performed by a variety of conventional, readily available instruments, whose listing and description are outside the scope of this text. However, where time-varying velocities are to be followed, the period or the time constant of any voltmeter or frequency meter used must be sufficiently small so that fidelity of dynamic measurement is not seriously impaired.

I-7-21 Pipeline Flowmeters. This general name is applied to those velocity-responsive meters that are connected into a pipeline so that the entire stream of fluid in the pipe passes through the meter. Various designs of pipeline flowmeters are used extensively for metering both liquids and gases. The more specific names, *turbine meter* and *propeller meter*, are applied to namy of these meters by both makers and users.

Because the area of the pipe is known, it is common practice to treat these flowmeters as volumetric-velocity rather than linear-velocity meters. Also, some of them, particularly for liquids, resemble very closely volumetric displacement, or quantity meters (Chapter I-4). These differ from some quantity meters mainly in having much greater clearances between the casing and rotor so that the slip may be several per cent rather than 0.2 per cent

or less, as in some quantity meters. The advantage of the greater clearances is the much lower susceptibility to jamming and damage by foreign matter; indeed, stopping the rotor does not stop the flow as can be done with some quantity meters. Also, the pressure drop through the pipeline flowmeters is much lower than with quantity meters. A disadvantage of the larger clearances is the larger systematic corrections for flow rate, especially at the lower rates, and for Reynolds number or fluid viscosity effects.

I-7-22 Design. Most turbine-type flowmeters consist of a casing or housing with end flanges or other fittings for connecting into a pipeline and a rotor suitably mounted. The secondary element may be a revolution counter (i.e., a register) or other means of sensing and totalizing the revolutions of the rotor, as with a pulse generator.

One method of enclosing a propeller-type flowmeter for liquids is shown in Fig. I-7-9. Here the direction of fluid flow relative to the rotor is axial. Two other turbine meters for liquids are shown in Figs. I-7-10 and I-7-11. In the first the fluid flows radially outward through the rotor blades, while in the second the meter is so mounted that the flow is axially downward. Figure I-7-12 represents, in a general way, what is probably, at the present time, the most common arrangement of housing and rotor of turbine meters for liquids.

It is to be noted that in each of the above-mentioned meters the rotor is preceded by a section of vanes. In the case of the axial flowmeters, the purpose of the vanes is to insure that the fluid stream reaching the rotor is free of rotational currents (i.e., swirls). In the radial-flow-type meters,

FIG. I-7-9 PROPELLER METER MOUNTED IN A SPECIAL SECTION OF PIPE. THIS METHOD OF INSTALLATION SUITABLE ONLY FOR PIPES 4-IN. AND LARGER. FLOW IS FROM RIGHT TO LEFT.

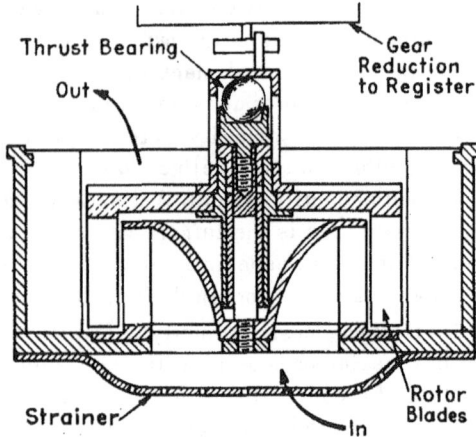

FIG. I-7-10 DISTINCTIVE FEATURES OF A RADIAL-
FLOW TURBINE METER FOR LIQUIDS

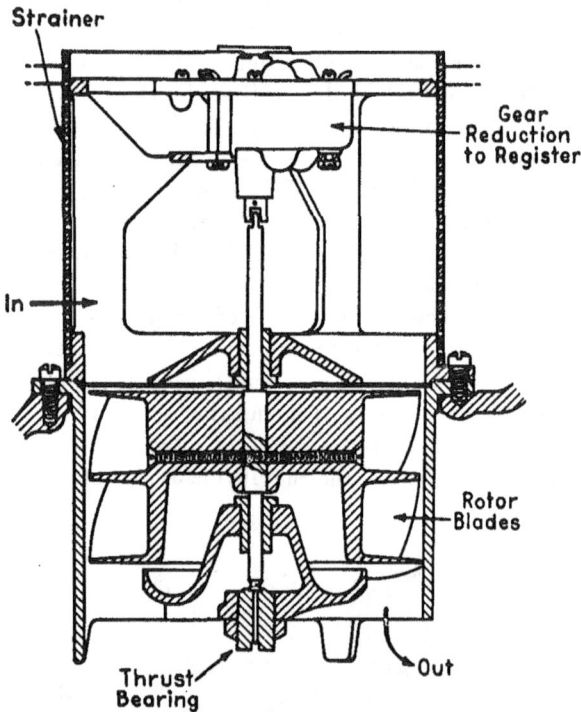

FIG. I-7-11 DISTINCTIVE FEATURES OF AN AXIAL-
FLOW TURBINE METER

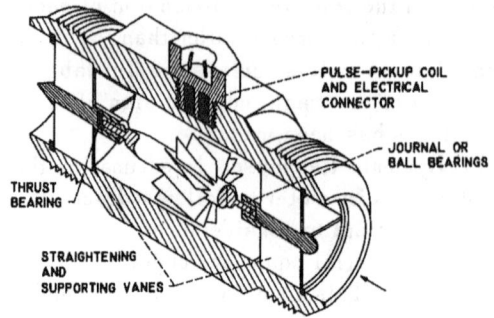

FIG. I-7-12 AN AXIAL-FLOW TURBINE METER FOR
LIQUIDS WITH AN ELECTROMAGNETIC-
PULSE PICKUP COIL

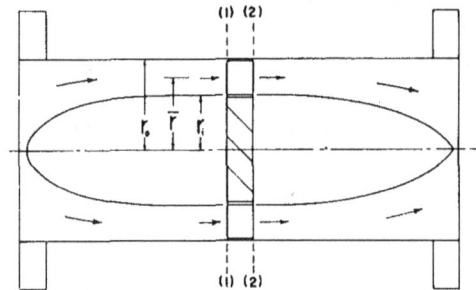

FIG. I-7-13 SCHEMATIC DRAWING OF AN AXIAL-FLOW
TURBINE FLOWMETER

the vanes direct the stream segments radially outward or inward.

I-7-23 The design of turbine meters for metering
gases is somewhat different from that of a meter for
liquids, due to the differences in the densities of
liquids and gases. The driving torque depends on the
density and the square of the fluid velocity. If the
fluid channel through the meter, shown schematically
in Fig. I-7-13, is made annular, so that the flow area
is less than that of the pipe in which the meter is
connected, the fluid velocity and therefore the
driving torque will be increased. In a turbine meter
for gas this annular area may be considerably less
than the pipe area so as to magnify the driving
torque, as in Fig. I-7-14. Also, it is necessary to
have the resisting torque very low; yet it is important
that this resisting torque should be as constant as
possible throughout the operating range of the meter.

I-7-24 In another design of turbine meter for both
liquids and gases, the fluid flow is tangential to the
rotor, and the rotor axis is normal to the axis of the
pipe (Fig. I-7-15). The rotor axis is displaced from
the channel center line, possibly by as much as the
radius of the pipe; and a housing, or "vortex
chamber," shrouds the part of the rotor outside the
line of the channel wall. The flow of the mainstream
against the part of the rotor within the channel produces a strong driving torque. An opposing torque is
produced from the centrifugal and angular momentum
imparted by the rotor to the fluid that is carried
around within the chamber, supplemented by frictional drag of this fluid against the chamber. Both of
these torques increase in approximately a constant

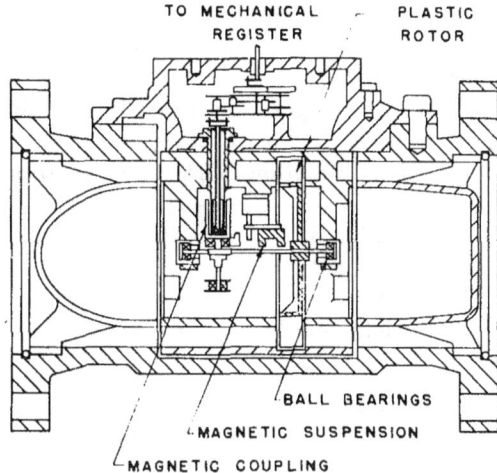

FIG. I-7-14 A GAS TURBINE METER

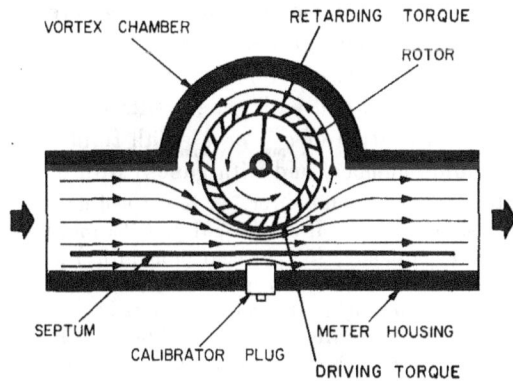

FIG. I-7-15 A TANGENTIAL-FLOW TURBINE METER
FOR LIQUIDS AND GASES

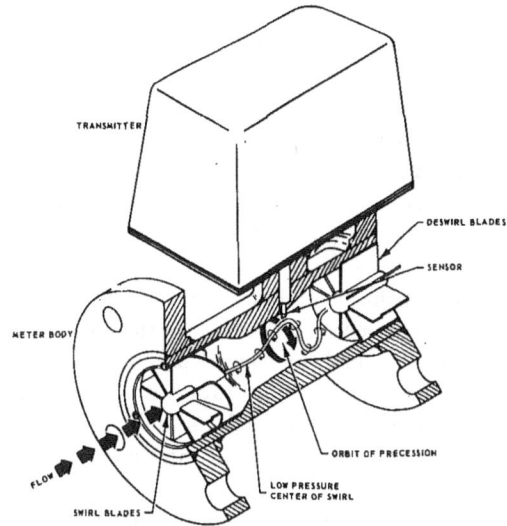

FIG. I-7-16 A PRECESSING VORTEX FLOWMETER

ratio with increases of either fluid velocity or density. Changes of fluid viscosity have little or no effect on the meter operation. The normal meter capacity range is about 10 to 1 for either liquid or gas. With water at maximum flow, the pressure loss is about 6 psi. With gas at a density of 1 lb_m/ft^3, the pressure loss at maximum rated flow is about 12.5 psi.

I-7-25 When a gas in concentric swirling flow within a circular pipe enters a sudden enlargement, the center of rotation precesses about the center line of the pipe. The angular velocity, or rpm, of these precessions has a direct relation to the velocity of swirl upstream of the enlargement and therefore to the volume rate of flow. A meter (Fig. I-7-16), designed to make use of such behavior of a gas stream, has a set of curved vanes at the inlet that impart a swirling motion to the gas stream which is

concentrated by a short constricted section a little downstream of the vanes. Following the constricted section there is an enlarged section in which the center of the swirling flow precesses. A suitable sensor located at the inlet end of the enlarged section responds to the small pressure effects of the precessions, so that a voltage pulse is given for each precession. The frequency of these pulses is proportional to the gas velocity at the meter inlet and thus to the volume rate of flow. For pipe Reynolds numbers above about 5000 and Mach numbers below about 0.12, the frequency output ranges from about 10 to 1000 cycles/sec [4, 5].

I-7-26 Pressure Loss. For any given meter of the velocity type, the pressure loss is proportional to the density of the fluid and the fluid velocity through the meter. With most pipeline flowmeters for liquids, the pressure loss is usually between 5 to 10 psi at maximum capacity. With a compressible fluid, the density and therefore the pressure loss increases as the operating pressure is increased, for the same velocity through the rotor section. Turbine meters, for gases operating at a line pressure of 1000 psig, will have an overall pressure loss of about 0.5 to 1.0 per cent of the line pressure. When operating on gas at near atmospheric pressure, the pressure loss may be only a few inches of water. With meters such as shown in Fig. I-7-16, the pressure loss is equivalent to about $10V^2/2g_c$, where V is the mean area velocity at the meter inlet.

I-7-27 Range. The flow range of a pipeline flow-meter depends on the size of the meter, the density and the viscosity of the fluid. The smallest sizes of meters may have only a 3 to 1 range, but for most flowmeters, such as discussed above, the normal range is about 10 to 1 within an accuracy band of about ± 1/2 per cent. With flowmeters for liquids, the higher viscosity generally makes for a wider flow range. Viscosity has little effect on the range of turbine meters for gases. For any one gas, however, the increase of density with increasing pressure will increase the range in proportion to the square root of the pressure. Maximum flow rate through any one meter is limited by the safe speed of rotor rotation, and the overall pressure loss acceptable. Very little overspeeding is permissible in the larger meters, but as much as 100 per cent overspeeding may be safe for short, infrequent periods in the smallest meters.

Turbine meters such as represented by Figs. I-7-9 through I-7-14, range from about 1/4 in. equivalent pipe size to about 20 in. The corresponding full-scale flow rates with liquids range from about 0.7 to about 30,000 gpm, while for gases the full-scale flow rates range from about 150 scfm or less at a line pressure of 0.25 psig to about 230,000 scfm at a line pressure of 1440 psig.

I-7-28 Calibration. For design purposes it may be possible, in some cases, to estimate the approximate relationship between rotational speed of the rotor and the average velocity of the fluid stream, from the blade angle and dimensions of the meter [6]. However, for definite values of this relationship, a calibration of a velocity-type pipeline flowmeter is necessary. Desirably, a meter should be calibrated under the same conditions, with respect to fluid and installation arrangement, as it will be subjected to in actual use. Where this is not possible, the conditions of use should be simulated as closely as possible [7].

For calibrating flowmeters for liquids, the reference measurements may be made with a calibrated volumetric tank and timer, a weigh-tank and timer, plus a density determination or a calibrated displacement type flow prover [8, 9]. For gas turbine meters the reference measurements may be made with a bell-type prover, an orifice meter of known characteristics, a sonic flow nozzle, a displacement flow prover, or one of the supplemental methods described in Chapter I-9. It should be remembered that the factors that can change the density of a gas can also affect the volumetric flow rate calibration [10-11-12].

If it is desired to associate a single calibration factor, C, with a given meter, the characteristics that must be determined are:

1. C_{max}, the calibration factor at the nominal full-scale volume flow rate, q_{max}, or at the equivalent rotor speed, n_{max}.

2. q_{min} or n_{min}, the practical lower limit of flow rate or rotor speed at which C does not deviate from C_{max} by more than the allowable percentage error in q.

3. The variation of C_{max} with fluid viscosity (or Reynolds number) and for gases, with density.

4. The variation of q_{min} or n_{min} with fluid viscosity (or Reynolds number) and for gases, with density.

If it is feasible to use a calibration curve, C vs n, for a given meter, the characteristics that must be determined are:

1. C as a function of n.

2. q_{min} or n_{min}, the practical lower limit of flow rate or rotor speed at which the calibration factor, C, does not show a random *probable error* greater than the allowable random probable error.

3. The variation of $C(n)$ with fluid viscosity (or Reynolds number) and with gases, with density.

4. The variation of q_{min} or n_{min} with fluid viscosity (or Reynolds number) and with gases, with density.

Note 1: The probable error, e_p, is defined as follows: regardless of the type of statistical distribution, there exists an even chance (50 per cent probability) that any one error will not exceed e_p. In a Gaussian distribution, $e_p = 0.67 e_\sigma$, where e_σ is the *root-mean-square error*.

Note 2: It is usually difficult to obtain a reliable estimate of the possible systematic errors that could be present in transferring a calibration, especially if the fluid, pressure, temperature or flow disturbance in use deviates greatly from the calibration conditions.

For turbine meters that will be used over a range of fluid temperatures of several hundred degrees, a temperature correction is generally necessary to account for the expansion or contraction of the meter housing, if measurement accuracies on the order of 1 per cent are to be achieved.

I-7-29 Turbulence and Pulsations. Velocity fluctuations, whether due to turbulence or pulsations, will cause a pipeline flowmeter of the turbine type to register too high. When the fluid is a liquid and the velocity fluctuations or pressure pulses are small and of low frequence (e.g., less than 10 cps), it may be possible to compute the error analytically. If the fluid is a gas, the error is indeterminable, regardless of the magnitude or frequency, and the velocity fluctuations or pressure pulses must be eliminated by some means upstream of the meter [13].

I-7-30 Installation. Designs, such as shown in Figs. I-7-10 and I-7-11, generally include strainers which also act to reduce effects of swirl and of asymmetric velocity distribution that may exist upstream of the flowmeter and to increase pressure drop. When sufficiently fine strainers are incorporated in the meter, long straight lengths of pipe or straightening vanes may not be needed to ensure applicability of the meter calibration.

Meters represented by Figs. I-7-9, I-7-12 and I-7-14, which are of the straight-through type, ordinarily do not include strainers. For this reason and also because the designs do not involve several changes of flow direction, they have a lower pressure drop for a given volumetric flow rate. However, under these conditions effects of upstream swirl or asymmetric velocity distribution must be minimized by adequate provisions for straightening the flow and flattening the velocity profile.

If the pressure drop across the meter at full-scale flow rate is rather high, e.g., 5 psi or more, flow-straightening provisions are necessary only if the entering flow has very strong swirl or velocity asymmetry, such as that produced by pumps or partially opened valves very close to the flowmeter inlet. If the flowmeter is designed for very low pressure drop, e.g., 1 psi or less, flow-straightening provisions are generally necessary if errors are to be limited to approximately 1 per cent. The exact installation requirements should be obtained from the flowmeter manufacturer for the specific situation. However, the following typical requirements indicate the order of magnitude of the necessary precautions:

1. For a meter that has built-in straightening vanes, at least 5-pipe diameters upstream and 2-pipe diameters downstream of straight pipe.

2. For a meter without built-in straghtening vanes, at least 5-pipe diameters of inlet piping with straightening vanes near the inlet end.

3. For an installation with upstream strainer or upstream flow expander, at least 4-pipe diameters of straight pipe between the meter entrance and the strainer or expander.

In the case of flowmeters metering liquids, the following are suggested to ensure that the flowmeter is always filled with liquid even at very low flow rates:

1. The flow should be directed upward in a vertical run.

2. A horizontal run should be followed by a long-radius elbow and a vertical riser, 4- or more pipe diameters after the flowmeter.

3. A "horizontal" run of pipe should actually slope upward for a sufficient distance downstream of the flowmeter so that the flowmeter will always

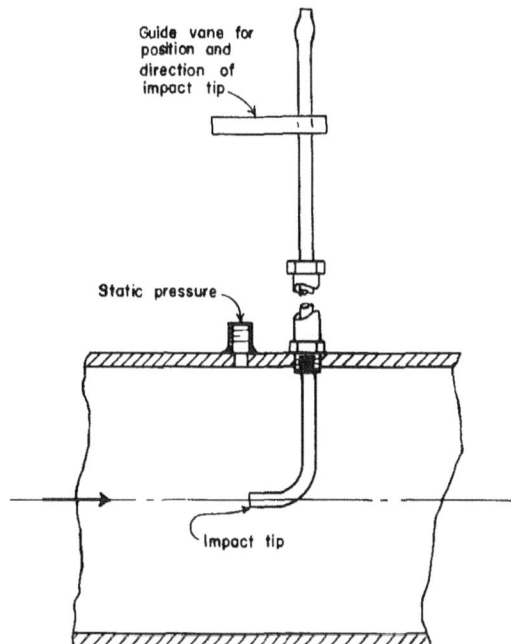

FIG. I-7-17 A TYPE OF BASIC PITOT TUBE OR IMPACT TUBE

be filled with liquid.

Control valves that produce appreciable pressure drop should, if possible, be located downstream of the meter.

Instruments for Determining Point-Velocity and Mean-Velocity

I-7-31 **Pitot and Pitot-Static Tubes.** The Pitot tube or impact tube (Fig. I-7-17) has a short section of one end bent at 90 deg to the main body or stem of the tube. In use the tube is held so that the open end of the short section faces directly into the moving fluid. Thus a pressure gage connected to the outer end of the stem will register the total or stagnation pressure of the stream element intercepted by the open end of the bent tip. When such a tube is used in a closed conduit, it is necessary to measure the static pressure with one or more side-wall pressure taps, as shown in Fig. I-7-17.

Note: The terms, Pitot tube and "impact" or "total pressure" tube, are used interchangeably in this report.

Pitot-static tubes (Fig. I-7-18) have two coaxial tubes so proportioned and assembled that the smaller inner tube senses the total or stagnation pressure. The static pressure is sensed through one or more rings of small holes in the wall of the outer tube located well back from the tip. An instrument coefficient of 1.00 has been found to apply to Pitot-static tubes similar to the two shown in Fig. I-7-18.

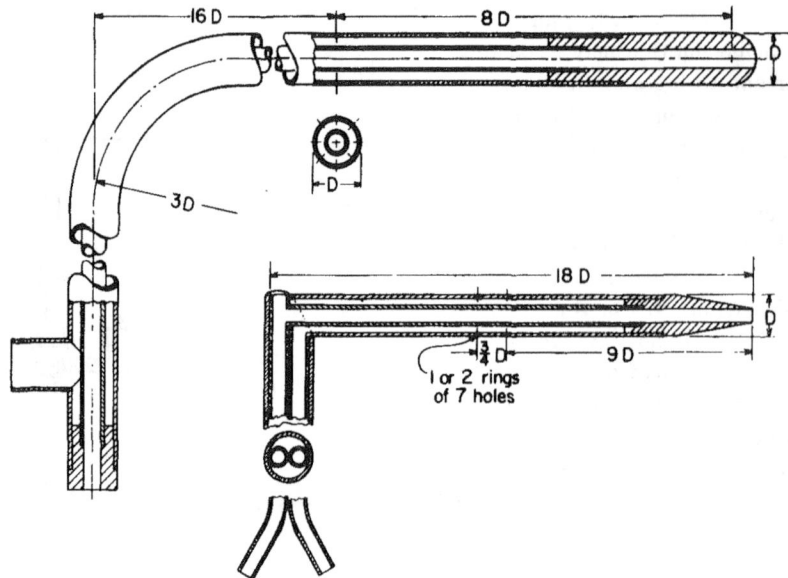

FIG. I-7-18 TWO COMMENDABLE DESIGNS OF PITOT-
STATIC TUBES. VALUES OF D BETWEEN
3/16 IN. AND 5/16 IN. INCLUSIVE, ARE
SUITABLE

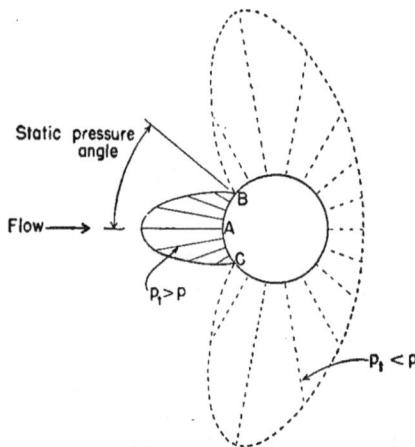

FIG. I-7-19 DISTRIBUTION OF PRESSURE AROUND A
CYLINDER THAT IS NORMAL TO THE
DIRECTION OF FLOW OF A STREAM. THE
LENGTHS OF THE RADIAL LINES
REPRESENT QUALITATIVELY THE
MAGNITUDE OF THE DIFFERENCE BE-
TWEEN THE STATIC PRESSURE AND OB-
SERVED PRESSURES AROUND THE
CYLINDER

However, a tube, such as the upper one, with a
hemispherical tip, is sensitive to any lack of smoothness at the junction of the tip with the cylindrical
section. An ellipsoidal tip has been reported to be
less sensitive [14].

I-7-32 Several other designs of Pitot-static
tubes are useful for special purposes. One of these,
which is sometimes called a directional Pitot tube,
is useful for measuring fluid velocity in both magnitude and direction. This instrument can be explained
by considering the two-dimensional flow around a
circular cylinder, the axis of which is perpendicular
to the flow. Let p_t be the pressure at any point on
the cylinder, and let p be the static pressure in the
approaching stream. Experiments have shown that
the distribution of the pressure difference, p_t-p, is
represented approximately by the radial ordinates in
Fig. I-7-19. At the stagnation point, A, the pressure
difference is a maximum and decreases toward each
side, becoming zero at points B and C. If there were
an opening in the cylinder at either point B or C, the
pressure transmitted to a gage would be the true
static pressure. Experiments have shown that the
static-pressure angle increases slightly as the
velocity of the stream increases, the value for
average conditions being about 39¼ deg [15].

One possible construction of the direction-finding
Pitot is shown by Fig. I-7-20(a) and consists of a
cylindrical tube with two holes and compartments.
The axis of the tube is perpendicular to the stream.
Each opening, i.e., each compartment, can be connected to one side of a differential gage, as shown
schematically in Fig. I-7-20(c). If the Pitot is
placed in a stream and rotated until the pressure

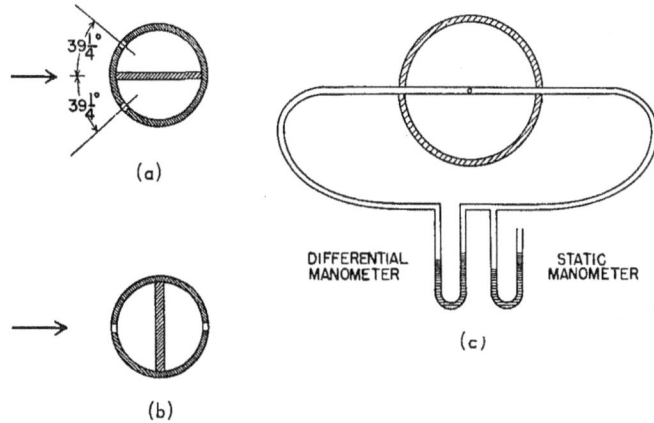

FIG. I-7-20 (a) SECTION OF A DIRECTIONAL PITOT
(b) SECTION OF A REVERSED PITOT
CYLINDRICAL TYPE
(c) METHOD OF USING A STRAIGHT-
THROUGH PITOT OF EITHER DIREC-
TIONAL OR REVERSED TYPE

difference is zero, the pressure at each hole will be
the static pressure, and the bisector of the angle
between the holes gives the line of flow. Thus, a
small-diameter directional tube may be used to
study the flow pattern in a fluid stream where helical
or other nonaxial patterns exist.

Again, if the tube is rotated until the axis of one
hole is parallel to the line of flow, the pressure at
this hole will be full-impact pressure, i.e., the static
pressure plus the velocity equivalent pressure. At
the other hole, since its axis is 78½ deg from the
first, the pressure will be less than the static pressure by an amount that may be from one to over two
times the magnitude of the velocity equivalent pressure. This is shown qualitatively by the dotted part
of Fig. I-7-19. In this case the differential pressure
may be as much as three times that which would be
obtained with the simple impact tube and sidewall
static hole in Fig. I-7-17 [16].

I-7-33 Another modification of this form of tube is
to locate the small pressure holes 180 deg apart,
thus making a form of reversed tube, as illustrated
by Fig. I-7-20(b). In this case, if one hole faces
directly into the stream it will measure the full impact pressure. The pressure measured by the other
hole will be less than the static pressure by an
amount between 25 and 50 per cent of the velocity
equivalent pressure and about 20 per cent of the
maximum possible amount mentioned in the preceding paragraph [17].

A design of combined-reversed Pitot tube is illustrated in I-7-21. In this tube the static is a

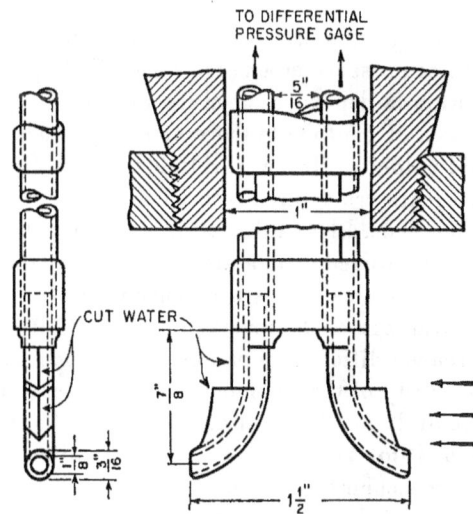

FIG. I-7-21 PITOT TUBE, COMBINED TYPE WITH
REVERSED STATIC TUBE

reversed duplicate of the impact tube. The instrument
coefficient of a tube of this design will be lower
(about 0.86) than for the regular Pitot-static tube or
the impact-side wall tap combination [18, 19].

One other design of an impact tube combined with
a reversed static, or suction pressure tube, is shown
by Fig. I-7-22. Here some averaged value of the
velocity equivalent pressure is picked up by a
special inner tube from four openings in the outer
tube that extends across the pipe.

The main disadvantage of these modified designs
of Pitots, as compared to the conventional designs

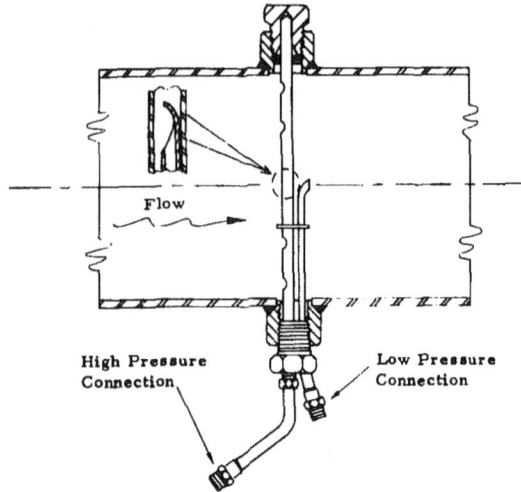

FIG. I-7-22 A REVERSED TYPE OF IMPACT TUBE IN-
CORPORATING A METHOD FOR OBTAIN-
ING AN AVERAGED IMPACT PRESSURE

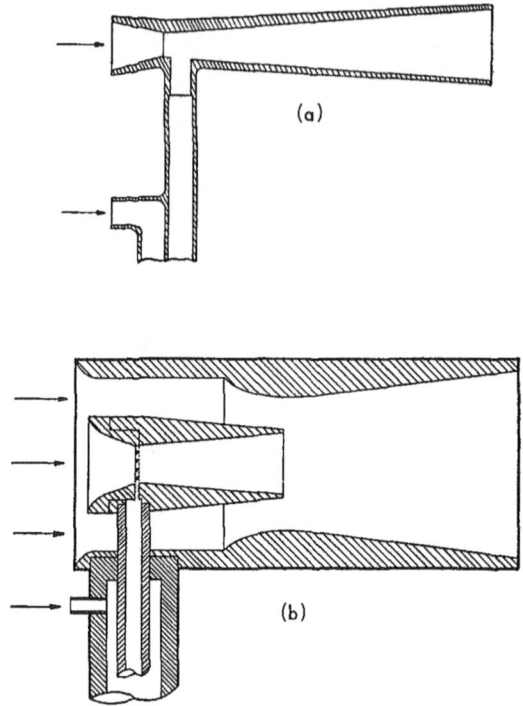

FIG. I-7-23 TYPICAL PITOT-VENTURI ELEMENTS
(a) SINGLE VENTURI TYPE
(b) DOUBLE VENTURI TYPE

for measuring rates of flow, is the variable distribution of pressure around the tubes. This pressure
distribution is dependent upon the ratio of the projected area of the tube to the area of the channel and
the fluid velocity. Hence, if reliable rate-of-flow
measurements are desired with Pitots of the reversed
type, they should be calibrated in the channel where
they are to be used or in a similar one, over the
range of velocities that will be encountered.

An advantage of the reversed type of Pitot tube is
the increased differential pressure corresponding to a
given velocity of the fluid under measurement. This
is particularly helpful in streams of low velocity.

I-7-34 Pitot-Venturi. As the name implies, this
instrument combines a simple Pitot or impact tube
with a Venturi. The objective of the combination is to
obtain a greater differential pressure than would be
obtained with a conventional Pitot-static tube or a
Venturi with a simple sidewall static. As shown by
Fig. I-7-23, the high pressure is obtained from the
Pitot or impact tube, while the low pressure is
taken from the Venturi throat.

At one time Pitot-Venturi heads were used as the
sensing element for aircraft airspeed indicators.
More recently Pitot-Venturi heads have been used as
flow-sensing elements in channels where the velocity may be low or the pressure loss due to other
types of differential producers would be excessive.
For both types of use it has been found necessary to
calibrate each such flow element individually since
slight variations in dimensions may have appreciable
effects on the calibration characteristics [20].

I-7-35 The equation to be used for computing the
fluid velocity from the pressure measurements made
with Pitot and Pitot-static tubes is similar to that
used for Venturi tubes and other differential pressure
producing elements [21].

If

p_1 = Static pressure of the stream in a plane
normal to the axis of the impact tube
and immediately upstream of the tip
thereof

p_s = Total or stagnation pressure sensed by
the impact tip

V_1 = Velocity of an elemental segment of the
fluid stream coaxial with the impact tip
and immediately upstream of the tip

V_s = Velocity of the same elemental segment
just within mouth of impact tip

= 0 (by assumption)

ρ_1 = Density of the fluid in the plane of, and
under, the pressure, p_1

ρ_s = Density of the fluid in the entrance of
the impact tip and under pressure, p_s

then with liquids $\rho_s = \rho_1$, and

$$V_1 = \sqrt{\frac{2(p_s - p_1)}{\rho_1}} \qquad \text{(I-7-2)}$$

If p_1 and p_s are measured in psi, ρ_1 in lb_m/ft^3, then

$$V_1 = 96.26 \sqrt{\frac{p_s - p_1}{\rho_1}} \quad \text{ft/sec} \qquad \text{(I-7-3)}$$

If the fluid is compressible, i.e., a gas, the elemental segment that is stopped at the mouth of the impact tip undergoes an isentropic compression as the pressure changes from p_1 to p_s. For this condition

$$V_1 = \sqrt{\frac{2k}{\gamma - 1} \frac{p_1}{\rho_1} \left[\left(\frac{p_s}{p_1} \right)^{\frac{\gamma-1}{\gamma}} - 1 \right]} \qquad \text{(I-7-4)}$$

Using the units as given above

$$V_1 = 96.26 \sqrt{\frac{\gamma}{\gamma - 1} \frac{p_1}{\rho_1} \left[\left(\frac{p_s}{p_1} \right)^{\frac{\gamma-1}{\gamma}} - 1 \right]} \quad \text{ft/sec} \qquad \text{(I-7-5)}$$

Here, as elsewhere in this report, γ is the ratio of the specific heats of the gas.

Equation (I-7-5) can be written

$$V_1 = 96.26 \sqrt{\frac{p_s - p_1}{\rho_1}}$$

$$\left\{ \frac{\gamma}{\gamma - 1} \frac{p_1}{p_s - p_1} \left[\left(\frac{p_s}{p_1} \right)^{\frac{\gamma-1}{\gamma}} - 1 \right] \right\}^{\frac{1}{2}} \qquad \text{(I-7-6)}$$

which is the same as equation (I-7-3) modified by the last factor, (in braces $\{\quad\}^{\frac{1}{2}}$), which accounts for the compression of the gas at the impact tip. At low velocities of the stream (e.g., below about 50 ft/sec for air as sea level) the magnitude of this factor is so small that usually it is neglected.

For use with metric units, with p_1 and p_s in kg_f/cm^2, ρ_1 in kg_m/m^3, and V_1 in m/sec, then for liquids

$$V_1 = 140048 \sqrt{\frac{p_s - p_1}{\rho_1}} \qquad \text{(I-7-7)}$$

and for gases

$$V_1 = 140048 \sqrt{\frac{\gamma}{\gamma - 1} \frac{p_1}{\rho_1} \left[\left(\frac{p_s}{p_1} \right)^{\frac{\gamma-1}{\gamma}} - 1 \right]} \qquad \text{(I-7-8)}$$

I-7-36 Velocity measurements — made with Pitot and Pitot-static tubes similar in construction to those of Figs. I-7-17 and I-7-18 — are correct to 0.3 per cent if the angle between the mean stream path and axis of the impact tip is not greater than 12 deg.

In a stream with isotropic turbulence, the value of V_1^2 deduced from the pressure measurements is too high by a fraction on the order of $V_t^2/(3V_1^2)$, where V_t is the rms value of the turbulence velocity components. In pipe flow, the turbulence may not be isotropic, but this correction may still be of the right order of magnitude.

I-7-37 The Hot-Wire Anemometer. This instrument is used extensively in detecting and measuring the velocities of gas streams, especially where there may be rapid changes in velocity (i.e., pulsations) and direction of the local stream (i.e., turbulence). This is because the single wire element, stretched between two needle-like prongs, is equally sensitive to all changes of velocity in directions perpendicular to the wire's length. In the absence of electrical lag compensation, reasonably faithful reproduction will be obtained of velocity fluctuations, the period of which is about six times the time constant, τ, of the wire. If higher frequency response is desired, electrical compensation, as described below, must be used.

I-7-38 Probe Design. In a typical hot-wire probe, such as shown in Fig. I-7-24, the wire length between prongs should be at least 100 times the wire diameter. The recommended wire diameter is between 0.001 and 0.003 in. It should have a high temperature coefficient of resistivity, such as provided by tungsten, nickel, platinum, or some of the nickel alloys. Alternately, a very small bead thermistor may be used, when suspended on stretched wires smaller in diameter than the bead, in which case the

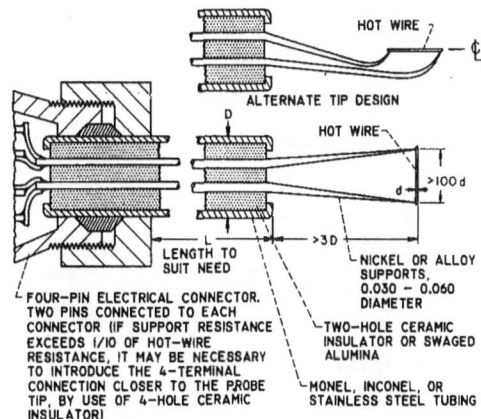

FIG. I-7-24 TYPICAL HOT-WIRE ANEMOMETER PROBE DESIGN

distance between supports may be as small as 15 times the wire diameter. However, a thermistor can be used only in a constant-temperature circuit, as described below.

For metallic wires within the range of diameters given above and for gas velocities above 5 ft/sec but below sonic velocity, the wire time constant, τ, is

$$\tau = \frac{c_w(\pi d_w^2/576)\rho_w L}{\eta(\pi d_w/12)L}$$

$$= \frac{c_w d_w o_w}{48\eta} \tag{I-7-9}$$

where

c_w = Specific heat of the wire $\quad$ Btu/lb$_m$ F
d_w = Diameter of the wire $\quad$ in.
L = Length of wire $\quad$ ft
ρ_w = Density of the wire $\quad$ lb$_m$/ft^3
η = Coefficient of convective heat transfer of gas stream, a value of which may be derived from

$$\eta = \frac{12k}{d_w}\left[\left(\frac{c_p\mu}{k}\right)^{0.3}\right]$$

$$\left[0.35 + 0.47\left(\frac{d_w\rho V}{12\mu}\right)^{0.52}\right] \tag{I-7-10}$$

where

c_p = Specific heat of gas at constant pressure
k = Coefficient of heat conduction (of the gas)
V = Velocity of the gas stream $\quad$ ft/sec
ρ = Density of the gas stream $\quad$ lb$_m$/ft^3
μ = Viscosity of the gas stream $\quad$ lb$_m$/ft sec

Equation (I-7-10) is applicable within the range [22]

$$0.1 \leq \left(\frac{d_w\rho V}{12\mu}\right) \leq 1000$$

Note: For dimensional consistency, the wire diameter should be expressed in feet, but usually wire diameters are given in inches (or gage number), hence the use of $d_w/12$.

Equation (I-7-9) may be converted to

$$\tau = N\sqrt{d_w^3/p_1 V} \tag{I-7-11}$$

or

$$\tau = N\sqrt{\frac{d_w^3(T/530)}{p_1 V}} \tag{I-7-12}$$

in which p_1 is in psia and T in deg R.

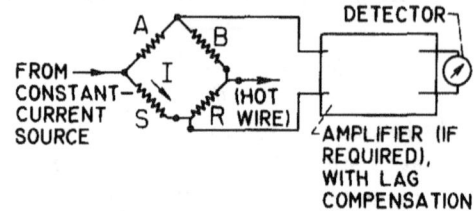

FIG. I-7-25 CONSTANT-CURRENT HOT-WIRE ANE-MOMETER CIRCUIT

Depending on the kind of metal wire within the range of diameters given above and the properties of the gas in which the hot-wire anemometer is being used, the value of N may range from about 6550 to about 7660.

I-7-39 Circuits. The hot-wire anemometer is commonly used in one of two basic types of electric circuits: the *constant-current* and the *constant-temperature* circuits. Variations of these basic circuits have been constructed with relatively rugged wires and only slight adaptations of standard commercial components, which have provided response to several hundred cycles per second.

I-7-40 Constant-Current Circuit. The basic concept of this circuit is shown in Fig. I-7-25. The hot wire, R, is fed with a constant current, I, while it is in a bridge circuit with a fixed series resistor, S. The ratio arms, A and B, may be 1 to 100 times the values of S and R, respectively. (If S >> R, a constant voltage applied to the bridge will approximate the constant-current condition.) For convenience, the bridge may be balanced at the mean gas speed over the wire, so that the alternating component of detector indication represents gas-speed fluctuations.

The upper frequency limit of response established by the wire time constant may be extended several hundredfold by appropriate electrical compensation within the detector. This is termed "lag compensation" and is accomplished by designing the detector to have a transfer function that is the inverse of the transfer function of the wire. Such compensation is accomplished adequately only when the wire time constant, τ, is known (i.e., when the mean gas speed is known) and when the gas-speed-fluctuation velocity is only a small fraction of the mean gas speed.

I-7-41 Constant-Temperature Circuit. The basic concept of this circuit, shown in Fig. I-7-26, is to

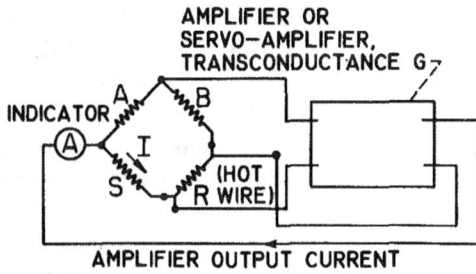

FIG. I-7-26 CONSTANT-TEMPERATURE HOT-WIRE
ANEMOMETER CIRCUIT

maintain bridge balance, and hence constant wire
temperature, by appropriate continuous adjustment
of the heating current, I. The fluctuation in I is a
measure of gas-speed fluctuation. For small fluctuations about a mean gas speed, the required fractional
change in I is one-fourth the fractional change in air
speed. The wire time constant need not be known,
nor need it be constant; but the amplifier must have
sufficient frequency response and gain, while
stability is maintained of the complete feedback
loop, including both amplifier and bridge.

In this circuit, the series resistor, S, is on the
order of wire resistance, R, while ratio arms, A and
B, are equal to each other but from 1 to 100 times the
value of S, depending on the power available and the
sensitivity desired. Because of the finite amount of
feedback, there is actually a very small unbalance
signal at the amplifier-input terminals; this signal
varies with the gas speed.

The constant-temperature circuit is particularly
effective when the turbulence is large, say over
20 per cent. The upper frequency limit of response
of the wire itself, $(\approx 1/(6\tau))$, is multiplied by a factor
on the order of RG, where R is wire resistance and G
is amplifier transconductance, provided RG is very
much greater than unity. However, the amplifier
proper must be flat to a frequency that is two or
three times higher than $RG/(6\tau)$.

I-7-42 Circuit Refinements. Although Figs. I-7-24
and I-7-25 show Wheatstone bridge circuits to
clarify the exposition, the actual circuit must be of
the 3-wire, Kelvin, or multiple-bridge type, in order
to minimize effects of changing lead and contact
resistances associated with the hot-wire element,
which is itself generally of very low resistance.

I-7-43 Calibration. In most cases, theoretical
computation of anemometer calibration is adequate,
based upon the electrical-circuit equations and the
known laws of heat transfer to wires transverse to a
gas stream. Where high accuracy is desired, calibration in an air jet of known velocity is performed to

obtain the mean-velocity calibration. The calibrating
air jet must be exceptionally free of all turbulence.
Such calibration permits computation of wire time
constant and of wire response to turbulence.

I-7-44 Averaging. Bulk-volumetric-velocity
measurement in an open channel, or in a duct too
large to allow convenient use of a pipeline flowmeter, can be performed by surveying the cross
section of the flow passage. The survey may be
accomplished by one of two basic methods or by a
combination of them:

1. Continuous traverse with a single meter,
simultaneous, continuous recording of fluid velocity
versus traverse distance, and subsequent integration
of the record, with appropriate weighting.

2. Recording of the simultaneous readings of a
number of meters placed at fixed locations across the
flow passage and subsequent summation of the
readings, with appropriate weighting.

Whatever averaging technique is used, each
meter's indication must be corrected for its calibration before any averaging is performed.

I-7-45 Continuous-Traverse Method. This method
has the advantage that only one or a few accurately
calibrated meters are needed, that velocity is
determined at every point of the traverse path, and
that minimum disturbance of the stream is produced
by the presence of the meter or meters.

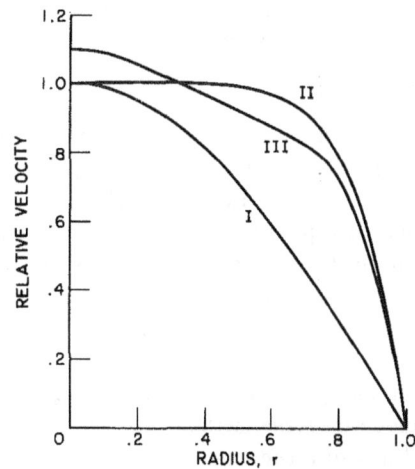

(a) VELOCITY DISTRIBUTIONS

	DISTRIBUTION		
	I	II	III
CENTROID OF EQUAL AREAS	-0.26	+1.16	+0.77
NEWTON-COTES	-0.00_5	-0.20	-0.08
CHEBYSHEF	-0.00_4	+0.02	+0.01
GAUSS	-0.00_2	-0.00_1	-0.00_2

(b) ERROR, %, IN 4-POINT APPROXIMATION

FIG. I-7-27 ERROR IN 4-POINT AVERAGING OF SOME
ARBITRARY AXIALLY-SYMMETRIC
VELOCITY DISTRIBUTIONS

The method has the disadvantages that the survey takes considerable time, that fluid velocities throughout the channel cross section must remain substantially constant during the survey, and that rather complex equipment may be needed to effect the traverse and to record the meter readings continuously.

In order to check constancy of all local velocities during the survey, a second meter should be used as a reference in some suitable fixed location in the stream. This second meter should preferably be of the same type and should be of the same sensitivity as that used for the traverse; however, it need not be of equally high accuracy. The reading of the reference meter should be recorded continuously while the traverse is being made. If the reference meter shows no change as large as the acceptable inaccuracy of the survey, the criterion of time-invariant velocities has been met. If the reference meter shows slight changes during the course of the survey, the recorded readings of the traversing meter should be corrected in the same proportion as the simultaneous velocity change shown by the reference meter. The postulate made here is that small changes in bulk velocity may occur without a comparable accompanying change in local-velocity distribution; thus, the correction procedure is valid only when the changes in reference-meter indications are on the order of just a few per cent and when there is no reason to expect locally severe changes of velocity (amplitude or direction) along the path of the survey.

The speed of traverse must be low enough so that negligible error is introduced by the lag in response to the velocity meter. A rule of thumb is that the time to traverse a passageway should be at least a thousand times the equivalent time constant of the velocity meter. For some current meters, this time constant is no greater than the time for one revolution of the rotor.

I-7-46 In an open channel or a rectangular duct, the continuous-traverse method requires traverses along the lines of a rectangular Cartesian grid transverse to the mean flow direction. Sometimes, an array of meters is located on a rod to obtain flow integration — by the second method of Par. I-7-41 — along one coordinate direction; and the entire array is moved, in the orthogonal coordinate direction, in a continuous traverse. The disposition of meters along the rod is treated under "Location of Measuring Stations."

In a circular duct surveyed with a Pitot tube, a polar-coordinate grid is used. Traverse should be made along several equally spaced radii, all lying in the same plane. A minimum of four radii, 90 deg apart, is recommended. The reference Pitot tube may conveniently be placed on the center line of the duct, sufficiently far from the plane of the traverse so that there is no hydrodynamic interference with the traversing probe; or else the reference Pitot tube may be placed about $0.7R$ from the duct center (where R is duct radius), midway in azimuth between adjacent radii of traverse and nearer to the plane of traverse.

If $\overline{V}$ is the average of readings, all at radius r, on each of the lines of traverse, then bulk velocity is given by $2\pi_o \int^R r(\overline{V} + \Delta V) \cdot dr$ where ΔV is the correction for the meter at reading $\overline{V}$.

I-7-47 **Fixed-Array Method.** This method has the advantages that the entire measurement can be made quickly, because simultaneous measurements are made at all measuring points of the flow-passage cross section, and that elaborate mechanical traversing apparatus is not necessary.

The method has the disadvantages that many accurately calibrated meters are required in order to obtain good averaging and that the simultaneous presence of these meters may appreciably disturb the velocity distribution of the stream being studied.

To retain the advantage of simultaneity, the meters should be of the remote-reading (telemetering) type and the signals transmitted to a multichannel recorder. This instrument will generally be a recording voltmeter if meter output is a continuous voltage proportional to fluid speed. A few high-speed multipoint instruments, each of which records a group of data serially, may be sufficient to record all of the data quickly enough to meet the ideal of "simultaneity" for all practical purposes. For meters whose output is a pulse for every fraction or multiple of a revolution, an "events recorder" or "operation recorder" that provides simultaneous marking of pulse occurrence in many channels, as a function of time, is usable. One-hundred channel recorders of this type are commercially available.

I-7-48 In an open channel, the measuring stations should be located on a rectangular Cartesian grid. In general, the accurate determination of the velocity distribution along any one vertical line requires measurements at at least five depths. However, preliminary trials may show that fewer measurements are sufficient. Many open-channel distributions are parabolic. In such cases the mean velocity along a vertical is close to that at 0.6 times the depth, and a closer mean is obtained by averaging the velocities at 0.2 and 0.8 times the depth.

Table I-7-1 Station Locations and Weights for Averaging

(Averaging for linear interval $0 \leq x \leq 1$)

(Averaging in a circular duct, in interval $0 \leq r \leq 1$)

Number of Stations	Method											
	Centroid of Equal Areas (a)			Newton-Cotes			Chebyshef (a)			Gauss		
n	x	r	w	x	r	w	x	r	w	x	r	w
2	0.2500 .7500	0.5000 .8660	1/2	(b) 0 1	(b) 0 1	(b) 1/2	(e) 0.2113 .7887	(e) 0.4597 .8881	(e) 1/2	(e) 0.2113 .7887	(e) 0.4597 .8881	(e) 1/2
3	0.1667 .5000 .8333	0.4082 .7071 .9129	1/3	(c) 0 0.5 1	(c) 0 0.7071 1	(c) 0.1667 .6667 .1667	0.1464 .5000 .8536	0.3827 .7071 .9239	1/3	0.1127 .5000 .8873	0.3357 .7071 .9420	0.2778 .4444 .2778
4	0.1250 .3750 .6250 .8750	0.3536 .6124 .7906 .9354	1/4	(d) 0 0.3333 0.6667 1	(d) 0 0.5774 .8165 1	(d) 0.1250 .3750 .3750 .1250	0.1027 .4072 .5928 .8973	0.3203 .6382 .7699 .9473	1/4	0.0694 .3300 .6700 .9306	0.2635 .5745 .8185 .9647	0.1739 .3261 .3261 .1739
5	0.1000 .3000 .5000 .7000 .9000	0.3162 .5477 .7071 .8367 .9487	1/5	0 0.25 .50 .75 1	0 .5000 .7071 .8660 1	0.0778 .3556 .1333 .3556 .0778	0.0838 .3127 .5000 .6873 .9162	0.2891 .5592 .7071 .8290 .9572	1/5	0.0469 .2308 .5000 .7692 .9531	0.2166 .4804 .7071 .8771 .9763	0.1185 .2393 .2844 .2393 .1185
6	0.0833 .2500 .4167 .5833 .7500 .9167	0.2887 .5000 .6455 .7638 .8660 .9574	1/6	0 0.2 .4 .6 .8 1	0 0.4472 .6325 .7746 .8944 1	0.0660 .2604 .1736 .1736 .2604 .0660	0.0669 .2887 .3667 .6333 .7113 .9331	0.2586 .5373 .6057 .7958 .8434 .9660	1/6			
7	0.0714 .2143 .3571 .5000 .6429 .7857 .9286	0.2673 .4629 .5976 .7071 .8018 .8864 .9636	1/7	0 0.1667 .3333 .5000 .6667 .8333 1	0 0.4082 .5774 .7071 .8165 .9129 1	0.0488 .2571 .0321 .3238 .0321 .2571 .0488	0.0581 .2352 .3381 .5000 .6619 .7648 .9419	0.2410 .4849 .5814 .7071 .8136 .8745 .9705	1/7			
8	0.0625 .1875 .3125 .4375 .5625 .6875 .8125 .9375	0.2500 .4330 .5590 .6614 .7500 .8292 .9014 .9682	1/8	0 0.1429 .2857 .4286 .5714 .7143 .8571 1	0 0.3780 .5345 .6547 .7559 .8452 .9258 1	(f) 0.0435 .2070 .0766 .1730 .1730 .0766 .2070 .0435	(f) 0.0513 .2036 .2964 .4487 .5513 .7036 .7964 .9487	(f) 0.2266 .4513 .5444 .6698 .7425 .8388 .8924 .9740	(f) 1/8			
9	0.0556 .1667 .2778 .3889 .5000 .6111 .7222 .8333 .9444	0.2357 .4082 .5270 .6236 .7071 .7817 .8498 .9129 .9718	1/9	0 0.1250 .2500 .3750 .5000 .6250 .7500 .8750 1	0 0.1250 .5000 .6124 .7071 .7906 .8660 .9354 1	0.0349 .2077 −.0327 .3702 −.1601 .3702 −.0327 .2077 .0349	0.0442 .1995 .2356 .4160 .5000 .5840 .7644 .8005 .9558	0.2103 .4466 .4854 .6450 .7071 .7642 .8743 .8947 .9776	1/9			
10	0.05 .15 .25 .35 .45 .55 .65 .75 .85 .95	0.2236 .3873 .5000 .5916 .6708 .7416 .8062 .8660 .9220 .9747	1/10	0 0.1111 .2222 .3333 .4444 .5556 .6667 .7778 .8889 1	0 0.3333 .4714 .5774 .6667 .7454 .8165 .8819 .9428 1	0.0319 .1757 .0121 .2159 .0645 .0645 .2159 .0121 .1757 .0319	(g) 0.0419 .1564 .2500 .3436 .4581 .5419 .6564 .7500 .8436 .9581	(g) 0.2046 .3954 .5000 .5862 .6768 .7361 .8102 .8660 .9185 .9788	(g) 1/10			
n	x	r	w	x	r	w	x	r	w	x	r	w
	Centroid of Equal Areas			Newton-Cotes			Chebyshef					

Notes

(a) All measurements of equal weight

(b) Trapezoidal rule

(c) Parabolic rule (Simpson's rule)

(d) Three-eighths rule

(e) 0.2 − 0.8 rule

(f) Two 4-station intervals

(g) Two 5-station intervals

109

The average velocity along each vertical line should first be computed; then these values should be used to obtain the average for all the vertical lines. The number of vertical lines required and their locations are treated under "Location of Measuring Stations."

In a circular duct, the stations at which meters are located should lie on several equally spaced coplanar radii. A minimum of four radii, 90 deg apart, is recommended. The number of stations required on each radius and their locations are discussed below.

I-7-49 Location of Measuring Stations. To achieve high accuracy in bulk-volumetric-velocity determination, with greatest economy in the number of measuring stations, the location of the measuring stations and the weight assigned to each measurement must be carefully selected. The two basic channel types involved in the selection of meter locations and assignment of weights are:

1. *Linear:* The averaging of readings along a straight line of unit length. The measuring station is located at distance x from one end point of the interval $0 \leq x \leq 1$. For example, in open-channel gaging, the mean value of velocity at any abscissa is first determined by such an averaging along an ordinate; then these mean values are themselves used in such an averaging.

2. *Circular:* the averaging of readings along the radius of a circular duct. Each reading, at any given radius, may itself be the average of m readings at angles $360/m$ deg apart. The measuring station is located at distance r from the center line of a duct of unit inside radius.

For each channel type several combinations of meter location and weight may be chosen. The more common of these are:

1. *Centroid of Equal Areas:* Equally weighted measurements at equally spaced areal increments. Utilizing repeated application of the trapezoidal rule, this mode of averaging is the simplest and least accurate of those listed here.

2. *Newton-Cotes:* Appropriately weighted measurements at equally spaced areal increments.

3. *Chebyshef:* equally weighted measurements at appropriately spaced locations. If all measurements are of equal probable error, this method yields the smallest probable error of the mean of equally weighted observations.

4. *Gauss:* appropriately weighted measurements at appropriately spaced locations. If all measurements are of equal probable error, this method yields the smallest probable error of the mean and is generally the most accurate of the four methods.

Table I-7-1 lists the values of n, x, r, and w for these four distributions. Here, n is the number of measuring stations, x and r are respective locations for linear and circular averaging, and w is the weight assigned to the measurement at that location.

For values of n other than those listed in the table, divide the interval into sufficiently smaller subdivisions, so that each subdivision can be treated by direct application of the table.

The relative effectiveness of the various distributions is suggested by the examples shown in Fig. I-7-26, in which are tabulated the errors in making four-point approximations to the arbitrarily drawn, axially symmetric, linear-velocity profiles shown. The error is expressed as per cent of the true bulk velocity for the respective curve.

References

[1] "Theory of the Vane Anemometer," E. Ower; *Philosophical Magazine*, 7th series, vol. 2, Nov. 1926, p. 881.

[2] "Aircraft Speed Instruments," H. K. Beij; National Advisory Committee for Aeronautics, Annual Report 1932, Report 420, p. 416.

[3] "Performance Characteristics of a Water Current Meter in Water and in Air," G. B. Schubauer and M. A. Mason; *Journal of Research*, National Bureau of Standards, vol. 18, 1937, RP 981, p. 351.

[4] "Uber den Mechanismus des Flussigkeit und Luftwiderstandes (About the Mechanism of the Resistance of Liquids and Air)," Th. von Karman and H. Rubach; *Physialische Zeitschrift*, 49, 1912. Also, Nachr. d. Wiss. Ges. Gottingen, (Reports of the Scientific Society of Gottingen), Th. von Karman; *Math. Phys. Klasse*, 509, 1911 and 547, 1912

[5] "A Digital Flowmeter without Moving Parts," A. E. Rodely, D. F. White and R. C. Chanaud; ASME Paper 65-WA/FM-6, Abstr. in *Mechanical Engineering*, Mar. 1966, p. 90.

[6] "Driving Torques in a Theoretical Model of a Turbine Meter," M. Rubin, R. W. Miller and W. G. Fox; *Trans. ASME, Journal of Basic Engineering*, vol. 87, No. 2, June 1965, p. 413.

[7] API Standard 1101, "Measurement of Petroleum Liquid Hydrocarbons by Positive Displacement Meter," and 1962 Supplement.

[8] "Instant Meter Proving," G. W. Swinney; ASME 65-WA/FM-4, Abstr. in *Mechanical Engineering*, Mar. 1966, p. 90.

[9] "Mechanical Prover Eases Doubt in Gas Measurements," A. W. Jasek; *The Oil and Gas Journal*, June 20, 1966.

[10] "The Measurement of Liquid Hydrocarbons by Turbine Meter Systems," (proposed) API Standard 2534, Nov. 1967.

[11] "Density Effect and Reynolds Number Effect on Gas Turbine Flowmeters," W. F. Z. Lee and H. J. Evans; *Trans. ASME, Journal of Basic Engineering*, vol. 87, No. 4, Dec. 1965, p. 1043.

[12] "Experimental Analysis of Gas Turbine Performance," H. J. Evans and C. R. Sparks; *Gas*, Oct. 1967, p. 78.

[13] "Pulsation Errors in Turbine Flowmeters," A. Haalman; *Control Engineering*, vol. 12-5, May 1965, p. 89.

[14] "Review of the Pitot Tube," R. G. Folsom; *Trans. ASME*, vol. 78, Oct. 1956, p. 1447.

[15] "Wind Pressure on Circular Cylinders and Chimneys," H. L. Dryden and G. C. Hill; *Journal of Research*, National Bureau of Standards, vol. 5, Sept 1930, p. 653.

[16] "A Three-Dimensional Spherical Pitot Probe," J. C. Lee and J. E. Ash; *Trans. ASME*, vol. 78, Apr. 1956, p. 603.

[17] "Influence of Compressibility on Cylindrical Pitot-Tube Measurements," L. W. Thrasher and R. C. Binder; *Trans. ASME*, vol. 72, July 1950, p. 647.

[18] "Investigation of Errors of Pitot Tubes," C. W. Hubbard; *Trans. ASME*, Aug. 1939, p. 477.

[19] "The Pitot Tube in Current Practice," E. S. Cole; *Civil Engineering*, vol. 5, Apr. 1935, p. 220

[20] "Pitot- Venturi Flow Element," H. W. Stoll; *Trans. ASME*, vol. 73, Oct. 1951, p. 963.

[21] "The Theory of the Pitot and Venturi Tubes," E. Buckingham; First Annual Report of the NACA, 1915, Report No. 2, Part 2.

[22] "Heat Transmission," McAdams; McGraw-Hill Book Co., New York, Second Edition, 1942, p. 222.

Chapter I-8
Head-Area Meters

I-8-1 General Features. In head-area meters both the area of the stream and the head of fluid producing the flow are variable but not independently, the area being a function of the head. These meters are also distinguished from other classes of rate meter in that they are used in open conduits or in conduits which do not flow full, i.e., in which there is a free surface. Consequently they are applicable to the metering of liquids only. Examples of this type of meter are the weir, which is in effect a dam over which the liquid is made to flow, and the Venturi flume, which is a formed section in a channel of very slight slope. These meters are used almost exclusively for measuring water flows, although small weirs have been used for metering other liquids.

Note: With these meters the pressure producing the flow is measured by the *height* of the free upstream surface of the stream above the crest of a weir or flume. The use of the name *head* for this height of fluid is almost universal, and hence it is used here. It is to be noted that the measurement of this "head" represents a single pressure in contrast to the measurement of two pressures or a pressure difference as required with most of the meters discussed in preceding chapters.

The inclusion of weirs in a report on meters may appear in contrast with such forms of enclosed meters, as described in Chapter I-5. However, since the weir and Venturi flume are recognized and widely used in the measurement (i.e., metering) of liquids, they come within the classification of meters as that term is used in this report.

I-8-2 The following letter symbols are used in the discussion of weirs and flumes:

b	= Adjustment to be added to crest length, L_w	ft
h	= Height of a column of flowing fluid producing the flow over a weir or in a flume	ft
h_1	= Depth of liquid in the approach channel to a weir	ft
h_v	= Pressure equivalent of the fluid velocity expressed as a column of the fluid	ft
h'	= Adjusted head of a triangular notch weir	ft
L_c	= Width of approach channel upstream of a weir	ft
L_w	= Horizontal length of a weir crest	ft
P	= Elevation of a weir crest or apex of a triangular-notch weir above bottom of approach channel	ft
q	= Volume rate of flow	cfs
θ (theta)	= Angle included between sides of a triangular-notch weir	deg
Δ (delta)	= Adjustment to be added to observed head of a triangular-notch weir	ft

C and g_c are the same as used throughout this report (sixth edition of *Fluid Meters*).

Weirs

I-8-3 Weir Classification. Sharp-crested weirs can be classified according to the form of notch or opening, as follows: rectangular notch, the original form; the *V* or *triangular* notch, which is used to provide higher head readings at low rates of flow; and special notches, such as the trapezoidal and Cippoletti weirs, and the hyperbolic and parabolic notches, which are intended to have a constant discharge coefficient or to have the head directly proportional to the rate of flow.

I-8-4 Thin-Plate Weirs. For rectangular notch weirs, the basic form, the weir plate should be vertical and the crest horizontal. The upstream edge of the crest should be sharp and square, free from burrs and wire-edge; and its thickness, perpendicular to the plate face, should be about 0.05 in. (1.3 mm). Usually the thickness of the plate will be greater than this, and the downstream edge of the crest should then be beveled at an angle of not more than 45 deg with the downstream face of the plate. The sides of the notch should be at 90 deg with the crest, and their height such as to be 1 to 2 in. (25 to 50 mm) greater than the maximum expected height of fluid surface, or head. The upstream edges of the sides should be square and their thickness the same as at the crest.

I-8-5 Special Terms.

Crest: Referring to Figs. I-8-1 and I-8-4, the bottom edge of the weir notch is called the sill or crest.

Crest Length, L_w: The horizontal distance along the crest between the sides. Obviously, the triangular notch has zero crest length.

Head, h: The head on the weir is the vertical distance from the level of the crest to the surface of the liquid just upstream from the point where it starts to curve over the weir crest. With the triangular notch the head is measured from the level of the apex. (See Fig. I-8-7.)

Nappe, Free Flow, Submerged Flow: The sheet of liquid flowing over the weir crest is termed the nappe. When the nappe falls free of the downstream face of the weir plate, the flow is said to be free or aerated. If the surface of the liquid on the downstream side of the weir plate is not far enough below the level of the crest to permit full aeration, the flow over the weir is said to be submerged. By referring to Fig. I-8-1 it is seen that the depth of the stream, z, over the crest is less than the head, h. Also, if the stream falls free of the downstream face of the weir plate, that is, if the nappe is aerated, the under surface of the nappe will rise slightly above the level of the crest, as represented by the height, z_1, in Fig. I-8-1. Free flow, independent of the elevation of the downstream surface, will

not take place over rectangular weirs at very low heads. The recommended minimum value of h to insure free flow is 0.1 ft (0.03 m).

Contracted Weir: When the width of the approach channel, L_c (Fig. I-8-2), is greater than the crest length, L_w, the nappe will contract, so that below the crest it will have a minimum width less than L_w. This narrowing of the nappe is termed the weir contraction, and the weir has been called a contracted weir, or a weir with end contractions. In order that the liquid drag against the channel sides will not affect the side contractions of the nappe, the difference, $(L_o - L_w)$, should be at least four times the maximum value of h that is expected.

Suppressed Rectangular Weir: With a rectangular weir, if the width of the approach channel is equal to the crest length, that is, $L_o = L_w$, (Fig. I-8-3), there will be no side contractions of the nappe. This has been called a suppressed rectangular weir.

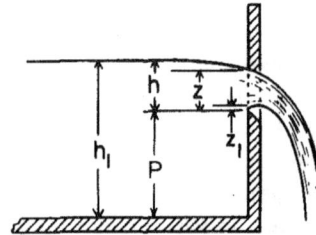

FIG. I-8-1 VERTICAL CONTRACTION OF A
STREAM OVER A WEIR

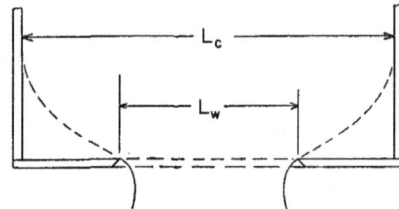

FIG. I-8-2 WEIR WITH SIDE CONTRACTIONS
(RECTANGULAR WEIR) PLAN
VIEW

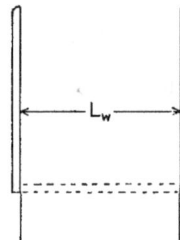

FIG. I-8-3 WEIR WITH SIDE CONTRACTIONS
SUPPRESSED (FULL-WIDTH
WEIR) PLAN VIEW

Note: The terms "contracted weir" and "suppressed weir" as defined above, have been used for many years in connection with rectangular weirs. In a tentative draft of the report by the committee on the Measurement of Liquid Flow in Open Channels, Technical Committee 113 of the International Standard Organization (ISO/TC-113), the proposed terms are, respectively: "rectangular-notch weir" and "full-width weir."

I-8-6 Measurement of Head. The most common and one of the most accurate methods of measuring the head over a weir is with a hook gage. To use it, the hook is submerged and then raised slowly until the point just begins to break the surface. The scale to which the hook is attached can be mounted with a vernier index. The zero reading is that obtained when the hook point is level with the bottom of the notch.

Floats are sometimes used to measure the head, in which case a stem can be attached to the float and held so as to move vertically beside a scale or the float can be connected to a pointer by suitable linkage. In some cases, this pointer may be the pen of a recording mechanism, thereby obtaining a continuous record of the head and of the rate of flow.

Regardless of the method used to measure the head, the measurement should be made at a distance sufficiently back (upstream) from the weir to insure that the surface is unaffected by the flow. To secure this condition it is recommended that the distance of the head-measuring section should be 3 to 4 times the expected maximum value of the head, h. Also, the channel above the weir should be relatively wide and deep. In some instances, baffles or screens are placed across the channel to break up currents produced by the incoming stream. Another method used to secure a still liquid surface is the still well, which is a box or pipe by the side of the weir chamber, at the desired distance upstream of the notch and connected to the weir chamber by a

small pipe or opening near the bottom. The liquid will rise in this well to the same height as in the weir chamber and will be practically undisturbed by currents in the chamber.

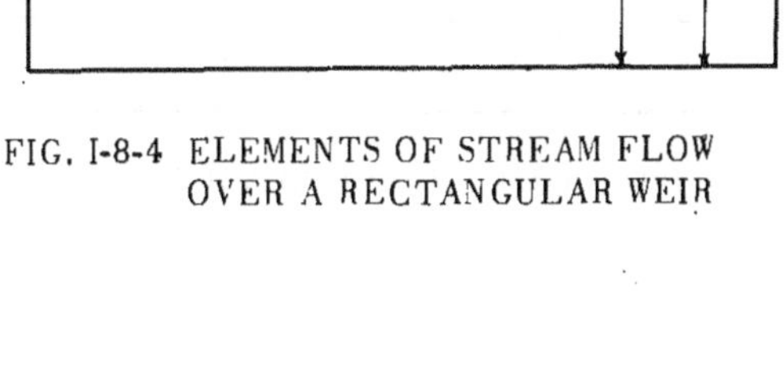

I-8-7 Equations of Flow: Rectangular Notch. Referring to Fig. I-8-4, the theoretical velocity of a stream filament or layer issuing from a rectangular notch weir at a distance, y, below the surface is

$$V = \sqrt{2g_c y} \qquad \text{(I-8-1)}$$

If the thickness of this layer is dy and its width, L_w, the elemental volume rate of flow is

$$
\begin{aligned}
dq &= L_w V \, dy \\
&= L_w \sqrt{2g_c y} \, dy
\end{aligned}
\qquad \text{(I-8-2)}
$$

Integrating equation (I-8-2) between the limits $y = 0$ and $y = h$ gives the total theoretical rate of flow over the weir, which is

$$
\begin{aligned}
q_t &= L_w \sqrt{2g_c} \int_0^h y^{1/2} \, dy \\
&= \frac{2}{3} L_w \sqrt{2g_c h^3}
\end{aligned}
\qquad \text{(I-8-3)}
$$

Due to several causes, the actual rate of flow, q, is less than the theoretical, so that a coefficient of discharge is used, thus giving

$$q(\text{cfs}) = \frac{2}{3} C L_w \sqrt{2g_c h^3} \qquad \text{(I-8-4)}$$

where, as before

$$C = q/q_t \qquad \text{(I-8-5)}$$

Note 1: Throughout this chapter both h and L_w are in feet.

Note 2: In equation (I-8-4) if h and L_w are in meters and $g_c = 9.80665$ m/sec^2, then q will be in m^3/sec.

In the development of equation (I-8-4) no attention is given to the possible effects of the fluid properties, viscosity and surface tension. Also, it has long been known that the relative proportions of the weir notch and weir channel have an effect on the value of C. Procedures for taking account of these effects in measurements of water have been developed from tests made at the Georgia Institute of Technology [1]. By these procedures equation (I-8-4) is written

$$q = \frac{2}{3} C' L_w \sqrt{2g_c (h')^3} \qquad \text{(I-8-6)}$$

FIG. I-8-4 ELEMENTS OF STREAM FLOW OVER A RECTANGULAR WEIR

where

$$C' = f(L_w/L_c,\ h/P)$$

$$L_w' = L_w + b$$

$$h' = h + 0.003 \text{ ft (or } h + 0.9 \text{ mm)}$$

The adjustment term, b, is a function of L_w/L_c, as given by Fig. I-8-5 and when $L_w/L_c = 1$, i.e., $L_w = L_c$, $b = -0.003$ ft (or -0.9 mm). The values of the coefficient, C', are given in Fig. I-8-6.

While theoretically the values of L_w and h of a rectangular weir have no limits, experimental work has seldom extended beyond crest widths of about 4 ft and heads between 2.50 and 2.75 ft. For weirs with end contractions, L_w should be greater than 0.5 ft (152 mm) so as to avoid the effects of mutual interference from the end contractions. With full-width weirs, i.e., no side contractions, the crest

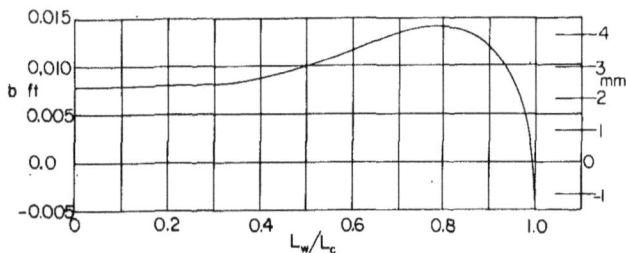

FIG. I-8-5 ADJUSTMENTS TO BE APPLIED (ADDED) TO THE CREST LENGTH, L_w

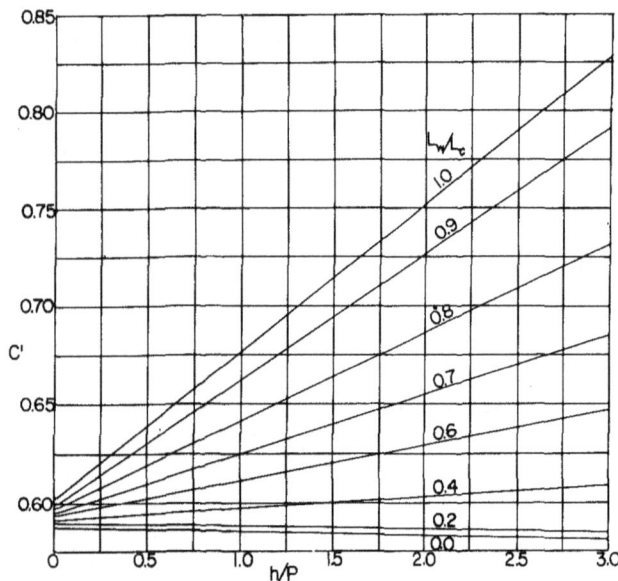

FIG. I-8-6 COEFFICIENTS, C', TO BE USED IN EQUATION (I-8-6) FOR RECTANGULAR WEIRS, AS A FUNCTION OF L_w/L_c AND h/P

width, L_w, is not limited to the 0.5 ft minimum. However, if the stream to be measured is so small as to require small values of L_w and h, greater accuracy will be obtained by using a triangular-notch weir.

I-8-8 Triangular-Notch Weir. The triangular-notch weir is used with the apex down, and L_h is the width between the sides at a distance, h, above the apex (Fig. I-8-7). The line bisecting the notch angle should be vertical, and the notch should be centered in the channel. The upstream edge of the notch and the thickness of the crest, i.e., the notch faces, should conform to the requirements for the rectangular notch, as given in Par. I-8-4. The nappe should be fully aerated so that the discharge is unaffected by variations in the tailwater elevation. Since at very low rates of flow the nappe tends to cling to the weir plate, it is recommended that heads below 0.2 ft (61 mm) should be avoided.

Following the same procedure used in considering the rectangular weir and referring to Fig. I-8-7, the theoretical rate of flow through the elemental segment, dy, is

$$dq_t = \sqrt{2g_c\,y}\ L_{h-y}dy \qquad (I\text{-}8\text{-}7)$$

or, since $L_{h-y}/L_h = (h-y)/h$, $L_{h-y} = (L_h/h)(h-y)$, and

$$dq_t = \sqrt{2g_c\,y}\ \frac{L_h}{h}(h-y)\,dy \qquad (I\text{-}8\text{-}8)$$

Integrating equation (I-8-8) between the limits $y = 0$ and $y = h$ gives

$$q_t = \frac{4}{15}L_h\sqrt{2g_c\,h^3} \qquad (I\text{-}8\text{-}9)$$

If $\theta =$ the angle between the sides,

$$L_h = 2h\tan\frac{\theta}{2} \qquad (I\text{-}8\text{-}10)$$

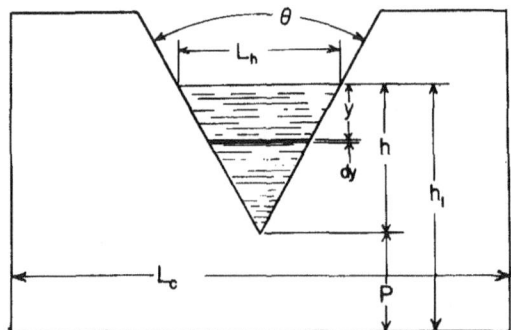

FIG. I-8-7 TRIANGULAR-NOTCH WEIR

and including the coefficient of discharge the actual rate of flow is

$$q(\text{cfs}) = \frac{8}{15} C \, \tan \frac{\theta}{2} \sqrt{2g_c \, h^5} \qquad \text{(I-8-11)}$$

Note: In equation (I-8-11), if h is in meters and $g_c = 9.80665$, q will be in m³/sec.

If $\theta = 60$ deg, $L_h = 1.1547 \, h$, and equation (I-8-11) becomes

$$q(\text{cfs}) = 2.470 \, C \, h^{5/2} \qquad \text{(I-8-12)}$$

If $\theta = 90$ deg, $L_h = 2 \, h$, and

$$q = 4.278 \, C \, h^{5/2} \qquad \text{(I-8-13)}$$

The value of the coefficient, C, to be used in the equations for a V-notch weir is dependent mainly on the notch angle, θ, and only slightly on the head, h, or rather the ratio, h/P (Fig. I-8-7). The kinematic viscosity and surface tension of the liquid may affect the value of C slightly, but data to define such effects are inadequate [2, 3]. From a study of a number of test programs on V-notch weirs, it was found that the influence of h/P on C become negligible when the observed head, h, was replaced with

$$h' = h + \Delta \qquad \text{(I-8-14)}$$

where the correction term, Δ, was related to the notch angle, θ. Thus equation (I-8-11) became

$$q(\text{cfs}) = \frac{8}{15} C' \, \tan \frac{\theta}{2} \sqrt{2g_c h'^5} \qquad \text{(I-8-15)}$$

Values of C' and Δ are given by the curves in Fig. I-8-8.

In the test programs reviewed, only a few values of h exceeded 1.25 ft; the values of P ranged from about 0.3 to 4.5 ft; and the ratio, h/P, up to about 0.35. While there were a few tests reported for notch angles below 20 deg and above 100 deg, the results were scattered and too few to provide a

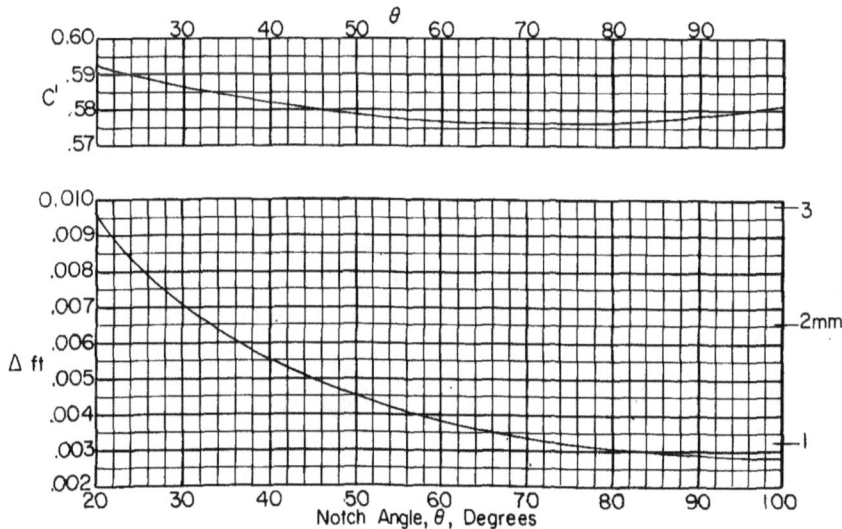

FIG. I-8-8 COEFFICIENTS, C', AND HEAD ADJUSTMENT VALUES, Δ, FOR TRIANGULAR NOTCH, THIN-PLATE WEIRS

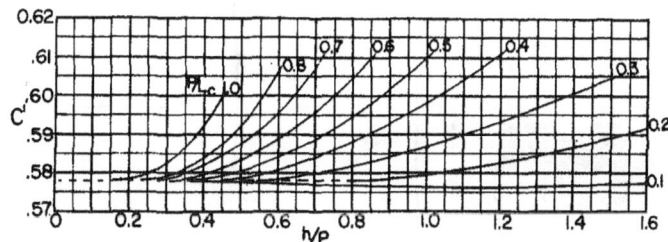

FIG. I-8-9 COEFFICIENTS FOR A 90° NOTCH WEIR WHEN A CONSTANT HEAD ADJUSTMENT OF +0.0028 FT (0.85mm) IS USED. (THE ORIGINAL CURVES WERE PREPARED BY SHEN FROM THE TEST DATA OF NUMACHI, KUROKAWA AND HUTIZAWA[5])

reliable evaluation of C or C'. Therefore, the curves for C' and Δ in Fig. I-8-8 are limited to notch angles between 20 and 100 deg.

For the special case of the 90-deg notch weir, Fig. I-8-9 gives values of C' for a wide range of values of h/P and P/L_c. For use with these values of C', a constant correction to h of $\Delta = + 0.0028$ ft (+0.85 mm) is to be used for the full range of conditions.

I-8-9 Trapezoidal Notch, Cippoletti Weir. The trapezoidal notch is a combination of the rectangular and triangular notches; the ends of the notch slope outward so that the notch is wider at the top than at the sill. In the form known as the Cippoletti weir, the slope of the ends has a value such that the additional discharge through the added triangular portions of the notch will exactly compensate for the effects of end contractions. Furthermore, it is assumed that with a rectangular weir the effect of an end contraction is equivalent to reducing the crest length by 0.1 h, so that with two fully developed end contractions the effective length of the crest becomes $L_w - 0.2\, h$. Thus, the requirement for the Cippoletti weir is met when

$$\frac{0.4}{3} h \sqrt{2 g_c h^3} = \frac{8}{15} h \ \tan \frac{\theta}{2} \sqrt{2 g_c h^3} \qquad \text{(I-8-16)}$$

from which

$$\tan \frac{\theta}{2} = 0.25 \qquad \text{(I-8-17)}$$

that is, the slope of one end of the notch will be 0.25, measured as the tangent of the angle formed by an end and a line normal to the sill, as shown in Fig. I-8-10. Thus, the equation for the rate of discharge through a Cippoletti weir is

$$q(\text{cfs}) = \frac{2}{3} C\, L_w \sqrt{2 g_c\, h^3} \qquad \text{(I-8-18)}$$

in which L_w is the length of the crest. One report on the Cippoletti weir suggests using $C = 0.63$ for those conditions where $h/L_w \leqq 1/5$ and $L_w \gtreqless 0.5$ ft [6].

I-8-10 Special Notches. Several other special notches, such as circular, proportional, exponential, parabolic, hyperbolic, and cycloidal, are available. All of these special forms have a common objective, namely, to simplify the relationship between the rate of flow and the head. Figure I-8-11 illustrates one of these special forms in which the rate of discharge is directly proportional to the head. Since these special forms are used to such a limited extent, no space will be devoted here to the equations for their rates of discharge.

I-8-11 Installation and Operation of Sharp-Crested Weirs. The weir should be preceded by a straight channel of uniform cross section, long enough to secure uniform velocity distribution. In some of the test programs the lengths of the well-formed approach channels were 25 to 30 ft. It would seem appropriate for the length to have some relation to the channel width, L_c, such as 4 to 6 times L_c. Shorter lengths of approach channel may be used by placing baffles across the channel near the inlet end. In some cases it may be necessary to place a screen across the approach channel to catch litter carried by the stream.

On the downstream side of the notch plate, the edges of the channel end wall, to which the plate is mounted, should be cut away so that the stream will fall freely without adhering to the end wall. Also, the width of the downstream channel at the weir should be sufficiently greater than the crest length, L_w, so that the sides of the nappe fall free. This is particularly important with full-width or supressed weirs.

Where a single value for the coefficient is used, as suggested above for the Cippoletti weir, it may be desirable to correct the observed value of the head, h, for the effect of the approach velocity, when that is appreciable. To do this, let

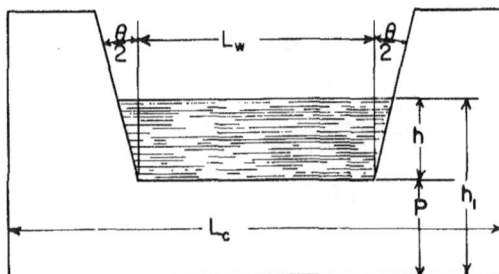

FIG. I-8-10 TRAPEZOIDAL NOTCH WEIR

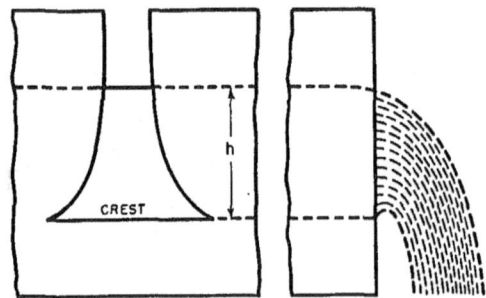

FIG. I-8-11 HYPERBOLIC NOTCH (IN THE RETTGER WEIR, THE SHAPE OF THIS NOTCH IS MODIFIED SLIGHTLY TO GIVE A CONSTANT COEFFICIENT)

h = The observed head

h_t = The total or true head

h_v = The head corresponding to the mean velocity of approach = $V_1^2/2g_c$

Then

$$h_t = h + h_v$$
$$= h + V_1^2/2g_c \qquad \text{(I-8-19)}$$

The approximate rate of flow is first obtained by using the observed head, h; and dividing this approximate flow by the cross-sectional area of the approaching stream, $L_c h_1$, gives a first value of the velocity of approach, V_1. With the use of this value of V_1 in equation (I-8-19), a value of h_t is obtained. Usually one such determination is sufficient, and the value of h_t is used in computing the rate of flow.

I-8-12 Tolerances. The accuracy of flow measurements made with thin-plate sharp-crested weirs will depend upon the sharpness of the weir edge, the exactness with which the weir dimensions, L_w or θ, are determined, and the method — and therefore the accuracy of measuring or recording the head, h. Then if the flow measurement involves the use of coefficients read from curves such as those of Figs. I-8-6 or I-8-8, there will be the question of how closely the weir system corresponds to those from which the originating data came. Estimates of the possible errors from each of the several individual items can be made, and the possible overall error, or uncertainty, obtained as the square root of the sum of the squares of the individual errors.

Most of the reports on tests of weirs place estimates of the error in flow measurements to be between 0.5 per cent and 3 to 4 per cent. It is probable that these estimates correspond most nearly to a single standard deviation, if such were or had been determined. If this is so, then on the basis used in other parts of this report, the *tolerance* to be applied to flow measurements with thin-plate weirs will be between ±1.0 per cent to ±6 to 8 per cent.

I-8-13 Broad-Crested Weirs. Broad-crested weirs have considerable thickness of crest as measured along (parallel to) the channel. This thickness of the crest should be such that the nappe will not spring free at the upstream edge. This will require a thickness equal to at least twice the (maximum) head if the upstream edge of the crest is square. By rounding the upstream edge this thickness may be decreased somewhat, possibly as much as one-fourth.

The basic assumption usually made for computing the flow over a broad-crested weir is that the depth,

h' (Fig. I-8-12), of the stream over the crest will be equal to the critical depth.

Note: The critical depth for a stream in an open channel is that depth at which its velocity energy is equal to one-half its depth. With such a depth the total energy of the stream is a minimum. In a channel of unit width and neglecting turbulence losses, this condition is attained if

$$h_o = q^{2/3}/g^{1/3} \qquad \text{(I-8-20)}$$

For this condition $h' = 2/3\ h$, so that the head producing the flow is $1/3\ h$ and, applying this in equation (I-8-4),

$$q(\text{cfs}) = 3.09\ C L_w h^{3/2} \qquad \text{(I-8-21)}$$

The value of C for use in equation (I-8-21) will depend upon the type of upstream edge — whether square or rounded — upon the thickness of the crest, and upon the depth of h'. It may range from about 0.85 for a square-cornered crest to nearly 1.00 for a crest with a rounded upstream corner and a slope downstream of about 1 in 30 [3, 7].

I-8-14 Round-Crested Weirs: Figure I-8-13. The curvature of the crest surface of a round-crested weir may be radial or that of some other geometric curve such as a parabola. The upstream face of these weirs may be vertical or sloping, while the downstream face is always sloping. Both the degree of the crest rounding and the amount of slope to the faces affect the rate of discharge. Furthermore, weirs of this type are seldom built for the purpose

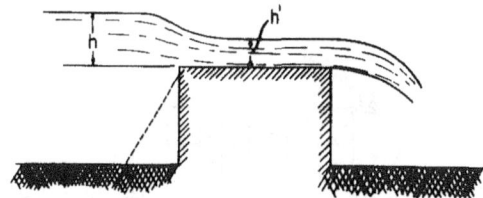

FIG. I-8-12 SECTION OF A BROAD-CRESTED WEIR (THE DOTTED LINE INDICATES A POSSIBLE MODIFICATION OF THE SECTION OUTLINE)

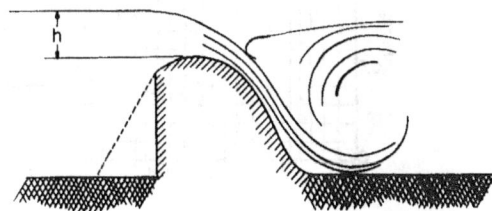

FIG. I-8-13 SECTION OF A ROUND-CRESTED WEIR

of measurement only; indeed, it is probable that measurement of flow is a secondary consideration in most places where they are used. An example would be the use of a dam spillway for obtaining the approximate rate of overflow. In most such cases the relation between head and discharge will be derived from a test of a model.

An important feature of both broad-crested and round-crested weirs is that they can be operated under a considerable degree of submergence without appreciably affecting the rate of discharge. Thus, they may be used in many cases where the overall head available is insufficient to permit the use of a sharp-crested weir [7, 8].

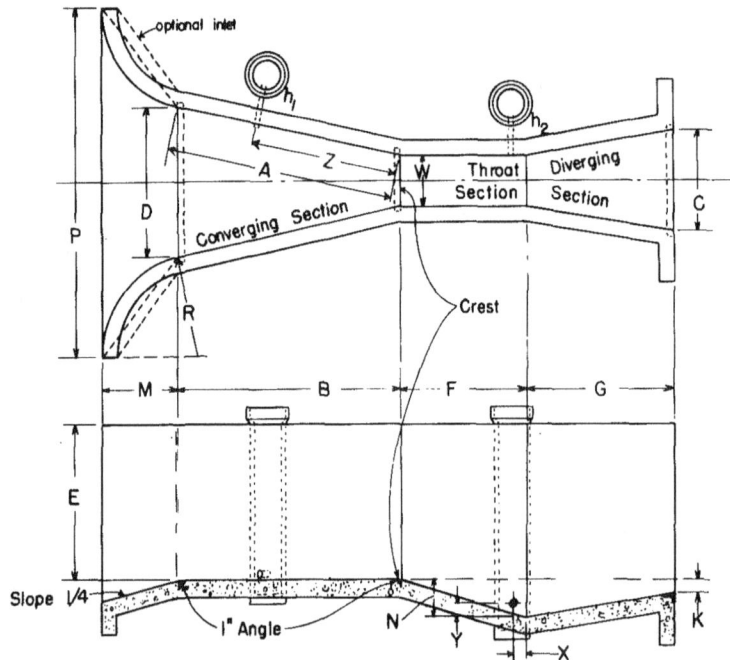

FIG. I-8-14 RATE OF FLOW VERSUS HEAD CURVES FOR PARSHALL FLUMES

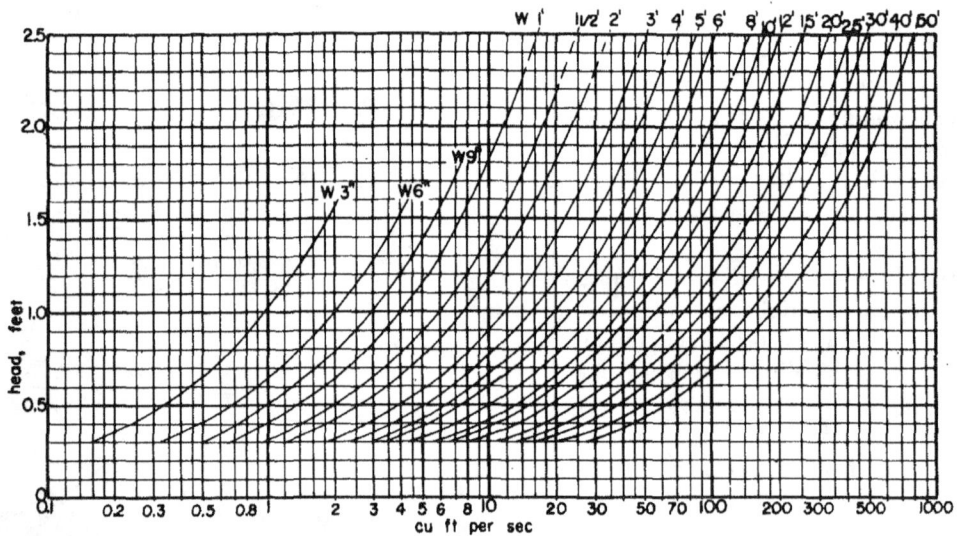

FIG. I-8-15 PLAN AND SECTIONAL ELEVATION OF CONCRETE PARSHALL MEASURING FLUME

Table I-8-1 Dimensions and Capacities of the Parshall Measuring Flume for Various Throat Widths, W

(Letters Refer to Dimensions in Fig. I-8-14)

W		A		Z		B		C		D		E	F	G	K	N	R		M		P		X	Y	Free-flow capacity Mini-mum	Free-flow capacity Maxi-mum
Ft	In.	Ft	In.	Ft	In.	Ft	In.	Ft	In.	Ft	In.	Ft	Ft	Ft	In.	In.	Ft	In.	Ft	In.	Ft	In.	In.	In.	cfs	cfs
0	3	1	6⅜	1	¼	1	6	0	7	0	10 3/16	2	½	1	1	2¼	1	4	1	0	2	6¼	1	1½	0.03	1.9
0	6	2	7/16	1	4 5/16	2	0	1	3½	1	3⅝	2	1	2	3	4½	1	4	1	0	2	11½	2	3	0.05	3.9
0	9	2	10⅝	1	11⅛	2	10	1	3	1	10⅝	2½	1	1½	3	4½	1	4	1	0	3	6½	2	3	0.09	8.9
1	0	4	6	3	0	4	4⅞	2	0	2	9¼	3	2	3	3	9	1	8	1	3	4	10¾	2	3	0.11	16.1
1	6	4	9	3	2	4	7⅞	2	6	3	4⅜	3	2	3	3	9	1	8	1	3	5	6	2	3	0.15	24.6
2	0	5	0	3	4	4	10⅞	3	0	3	11½	3	2	3	3	9	1	8	1	3	6	1	2	3	0.42	33.1
3	0	5	6	3	8	5	4¾	4	0	5	1⅞	3	2	3	3	9	1	8	1	3	7	3½	2	3	0.61	50.4
4	0	6	0	4	0	5	10⅝	5	0	6	4¼	3	2	3	3	9	2	0	1	6	8	10¾	2	3	1.3	67.9
5	0	6	6	4	4	6	4½	6	0	7	6⅝	3	2	3	3	9	2	0	1	6	10	1¼	2	3	1.6	85.6
6	0	7	0	4	8	6	10⅜	7	0	8	9	3	2	3	3	9	2	0	1	6	11	3½	2	3	2.6	103.5
7	0	7	6	5	0	7	4¼	8	0	9	11⅜	3	2	3	3	9	2	0	1	6	12	6	2	3	3.0	121.4
8	0	8	0	5	4	7	10⅛	9	0	11	1¾	3	2	3	3	9	2	0	1	6	13	8¼	2	3	3.5	139.5
10		14	3¼	6	0	14		12	0	15	7¼	4	3	6	6	13½	8	0	6	0	32	0	12	9	6	200
12		16	3¾	6	8	16		14	8	18	4¾	5	3	8	6	13½	9	0	8	0	35	0	12	9	8	350
15		25	6	7	8	25		18	4	25	0	6	4	10	9	18	11	0	9	0	40	0	12	9	8	600
20		25	6	9	4	25		24	0	30	0	7	6	12	12	27	12	0	10	0	48	0	12	9	10	1000
25		25	6	11	0	25		29	4	35	0	7	6	13	12	27	12	0	10	0	55	0	12	9	15	1200
30		26	6¾	12	8	26		34	8	40	4¾	7	6	14	12	27	12	0	10	0	64	0	12	9	15	1500
40		27	7½	16	0	27		45	4	50	9½	7	6	16	12	27	13	0	11	0	80	0	12	9	20	2000
50		27	7½	19	4	27		56	8	60	9½	7	6	20	12	27	13	0	11	0	95	0	12	9	25	3000

For large flumes some dimensions, especially R, M and P, may be varied as necessary to conform to the channel space available.

Flumes

I-8-15 Measuring Flumes. In these flumes the measuring section may be produced by a contraction of the sidewalls, by a raised section or hump of the channel bed (i.e., a very low broad-crested weir) or by both side contractions and hump. A common characteristic of these flumes is the formation of a standing wave close to the outlet from the constricted section, for which reason they sometimes are called standing wave flumes. Also, they will operate with a considerable degree of submergence, which means that the overall head or fall may be very low. Furthermore, since there may be none or very little change to the contour or slope of the channel bottom, the presence of silt or other material in the stream usually will not be detrimental.

As yet, there is no common or generally used design of these flumes; hence, for the purpose of this report, it will suffice to describe briefly three designs used to some extent in this country.

I-8-16 The Improved Parshall Flume [8-10]. A measuring flume that has proved to be well adapted to use in irrigation canals and ditches is the Parshall Flume. This consists of an entrance section with converging walls and level floor, a throat section with parallel walls and a downstream sloping floor, and an outlet section with diverging walls and a rising floor.

The size of the flume is taken as the horizontal distance between the parallel vertical walls of the throat and is identical with the length of the crest. The crest is the line where the level floor of the converging entrance section joins the inclined floor of the throat section. Figure I-8-14, together with Table I-8-1, gives the form and appropriate dimensions for flumes of several throat widths, up to and including 50 ft.

The zeros of the scales in the still wells for measuring the inlet and throat heads, h_1 and h_2, are set at the elevation of the crest. The ratio of h_2/h_1 expressed as a per cent is termed the per cent of submergence. If the per cent submergence is 60 or less with the 9-in. and smaller flumes or less than 70 per cent for the 1-ft and larger flumes, the rate of flow is unaffected by the downstream condition. Under this condition only the inlet head, h_1, need be observed, and the rate of flow can be read directly from a table or from curves such as Fig. I-8-15.

If the degree of submergence is greater than 60 or 70 per cent, both the inlet and throat heads, h_1 and h_2, must be observed. Also, the rate of flow will be less than it would be for the same value of h_1 if h_2 were

lower. For determining the rate of flow under such conditions reference should be made to one of the papers on this flume.

For the larger sizes of Parshall flumes, that is, those having a throat width of 10 to 50 ft, the free flow discharge is given by the empirical equation

$$q(\text{cfs}) = (3.6875 \, L_w + 2.5) \, h_1^{1.6} \qquad \text{(I-8-22)}$$

I-8-17 Palmer-Bowlus Flume [11, 12]. This flume was developed primarily for use in circular conduits such as storm and sanitary sewers, in which the flow only partly fills the conduit, except under abnormal conditions. In general, both sidewall and a bottom contraction or hump are used, as shown by Fig. I-8-16. It is suggested that the length of the throat

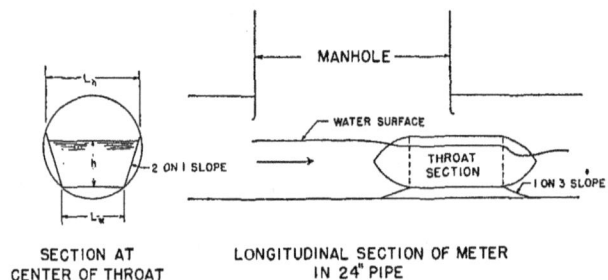

FIG. I-8-16 GENERAL PROPORTIONS OF THE PALMER-BOWLUS FLUME

should be equal to or greater than the diameter of the conduit. The optimum size of throat (i.e., its cross-sectional area) for any given conduit is one in which the energy head, $V^2/2g$, in the throat will be greater than the energy head in the free flowing conduit for all values of q. The general equation for the flow is

$$q(\text{cfs}) = C \sqrt{\frac{A_h^3}{L_h \, g_c}} \qquad \text{(I-8-23)}$$

in which A_h is the cross-sectional area of the stream in the throat and L_h is the width of the free surface of the stream corresponding to any depth, h. Assuming a slope of the sides, as shown in Fig. I-8-16, equation (I-8-23) becomes

$$q(\text{cfs}) = C \, h^{3/2} \left[\frac{(2L_w + h)^3}{8(L_w + h)} \, g_c \right]^{1/2} \qquad \text{(I-8-24)}$$

The value of the coefficient, C, will have to be determined by a calibration with a model or with one of the special methods described in Chapter I-9.

I-8-18 Parabolic Discharge Flume. A third type of measuring flume is the parabolic discharge flume. It is used on the outflow end of a pipe. As shown in

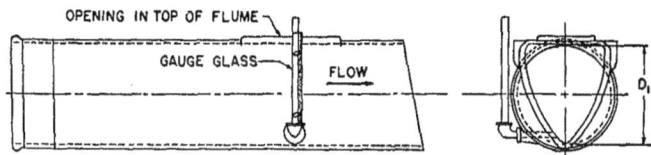

FIG. I-8-17 A FREE DISCHARGING
PARABOLIC FLUME

Fig. I-8-17, it is formed by constricting the sides at the outlet end to form a parabola, the apex of which is in line with the bottom of the pipe. The size of the flume is taken as the diameter of the pipe to which it is attached. Unlike the measuring flumes discussed above, the discharge from this flume must fall free.

The relation between the outflow depth gage reading and rate of flow is established by calibration. Typical quantity-depth curves for several sizes are shown in Fig. I-8-18.

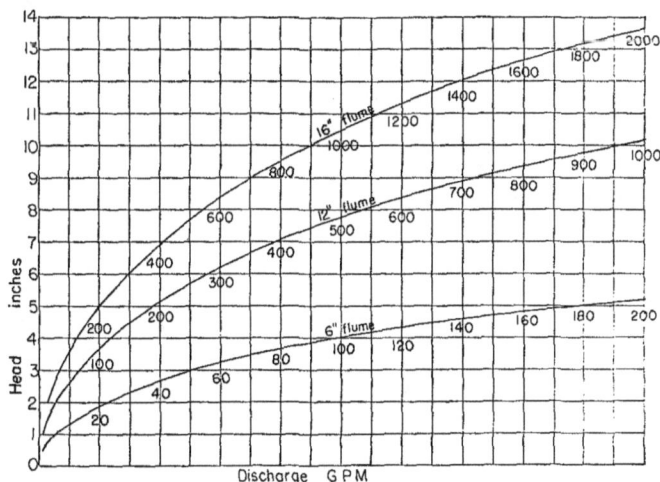

FIG. I-8-18 REPRESENTATIVE HEAD-
DISCHARGE CURVES FOR
PARABOLIC FLUMES

I-8-19 Tolerances for Flumes. No numerical values of the probable accuracy of measurements made with flumes are at hand. One report on Parshall flumes states the measurements therewith are accurate for irrigation purposes. It seems probable the flow measurements with flumes will be somewhat less accurate than with weirs, that is, the tolerance to be used with flumes is greater than with weirs. There is no basis for saying how much greater. Of course, when a calibration can be made of a flume, a value of the tolerance to be applied to measurements with *that* flume can be evaluated, taking into account *all* the sources of error in the observations on both the flume and the reference method of measurement.

References

[1] "Discharge Characteristics of Rectangular Thin-Plate Weirs," Prof. C. E. Kindsvater and R. W. Carter; *Trans. ASCE*, vol. 124, 1959, p. 772.

[2] "The V-Notch Weir for Hot Water," E. S. Smith, Jr.; *Trans. ASME*, vol. 56, 1934, p. 787, and vol. 57, 1935, p. 249.

[3] "Fluid Mechanics," Cox and Germano; D. Van Nostrand Co., New York, 1941, chapter VII.

[4] "A Preliminary Report on the Discharge Characteristics of Triangular Thin-Plate Weirs," J. Shen; unpublished report to Water Resources Div., Geological Survey, U.S. Dept. Interior, 1959; released to Committee on Measurement of Liquid Flow in Open Channels, ISO/TC-113.

[5] "Uber den Uberfallbeiwert eines rechtwinkelig-dreieckigen Messwehrs, (On the Discharge from a 90°-Notch Weir)," F. Numachi, T. Kurokawa, and S. Hutizawa; *Soc. Mech. Engrs. (Japan) Trans.*, vol. 6, Feb. 1940, no. 22.

[6] "Hydraulics," J. N. LeConte; McGraw-Hill Book Co., New York, 1926, p. 46.

[7] "Hydraulic Measurements," H. Addison; Chapman & Hall, Ltd., London, 1949, chapter XI.

[8] "Measurement of Water in Irrigation Channels with Parshall Flumes and Small Weirs;" Circular 843, Soil Conservation Service, U.S. Dept. of Agriculture, May 1950.

[9] "The Parshall Measuring Flume," R. L. Parshall; Colorado Agricultural Experiment Station Bulletin 423, Mar. 1936.

[10] "Parshall Flumes of Large Size," R. L. Parshall; Colorado Agricultural Experiment Station Bulletin 426-A, Mar. 1953. (Reprint of Experiment Station Bulletin 386.)

[11] "Design of Palmer-Bowlus Flumes," J. H. Ludwig; *Sewage and Industrial Waste*, vol. 23, Sept. 1951, p. 1096.

[12] "Adaptation of Venturi Flumes to Flow Measurements in Conduits," H. K. Palmer and F. D. Bowlus; *Trans. ASCE*, vol. 101, 1936, p. 1195.

Chapter I-9

Other Meters
and
Methods of Determining
Rates of Flow

Magnetic Flowmeters

I-9-1 The symbols used are:

B = Flux density   webers/in.^{2}

D = Diameter of pipe   in.

d_e = Diameter of electrodes   cm

e = Strength of induced emf   volt

q = Rate of fluid flow   in.^{3}/sec

V = Velocity of the fluid, area average   in./sec

R_f = Fluid resistance, electrical   ohms

R_s = Resistive impedance in secondary element   ohms

δ (delta) = Distance between electrodes (through fluid)   in.

ξ (xi) = Fluid conductivity   mhos/cm

Note: The units for some of the quantities are not the same as given in the general list of symbols (Table I-2-1). Also, in commercial use the actual unit may be a multiple or submultiple of the basic unit.

I-9-2 The basic principle of the magnetic flowmeter is similar to that of an electric generator. As indicated schematically in Fig. I-9-1, a pipe is so placed with respect to the magnetic field that the path of the fluid flowing in the pipe is normal to the magnetic field. In accordance with the Faraday law, the motion of the fluid through the magnetic field induces an electromotive force across the fluid in a path that is mutually normal to the magnetic field and the average direction of fluid motion. By placing insulated electrodes in the pipe in a diametrical plane normal to the magnetic field and connecting these to a specially designed voltmeter, the strength of the induced emf may be measured.

The magnet may be either a permanent magnet or an electromagnet, and thereby we have the two classifications: dc and ac magnetic flowmeters. The primary element of a magnetic flowmeter consists of the permanent magnet, or the coils and core of an electromagnet, the pipe or flow tube through which

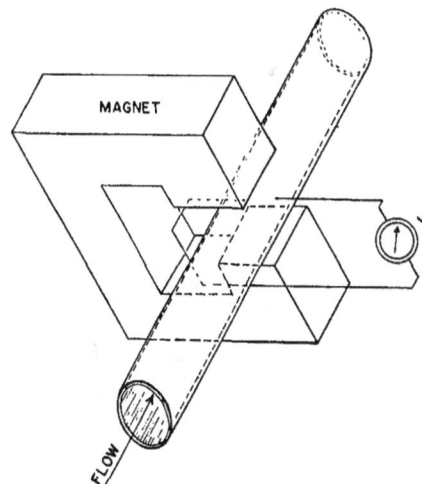

FIG. I-9-1 ELEMENTS OF AN ELECTROMAGNETIC FLOWMETER

the fluid flows, and the electrodes. The secondary element is the instrument for receiving and indicating in some form the magnitude of the induced emf. The leads from the electrodes are a part of the secondary element.

A typical cross section of an ac or electromagnetic flowmeter is shown in Fig. I-9-2. The coils and core structure are arranged to produce a magnetic field. The flow tube is a nonmagnetic material such as plastic, aluminum, brass, or a 300-series stainless steel. With metallic tubes an insulating lining is used to prevent the metal tube from short-circuiting the conducting path of the induced emf through the fluid from one electrode to the other. The lining may be glass or plastic.

I-9-3 The size range of magnetic flowmeters is very extensive. On the small end of the range a meter has been used to measure the flow of blood in blood vessels [1]. On the other extreme, a 9-ft meter has been installed at a sewage plant. The line pressure at which these meters may be operated will depend upon the pressure rating of the flow tube pipe and flanges. The maximum temperature at which a meter may be operated safely will depend, primarily, upon the limiting temperature for the flow tube lining and, to a lesser degree, upon the insulation of the magnetic coils. The safe operating temperature for most insulating materials ranges from about 170 to about 360 F. The minimum temperature of operation will be that at which the fluid to be metered will flow. The primary element of a magnetic flowmeter can be furnished with a housing that will meet the hazardous area requirements of Class I, Group D, Division I.

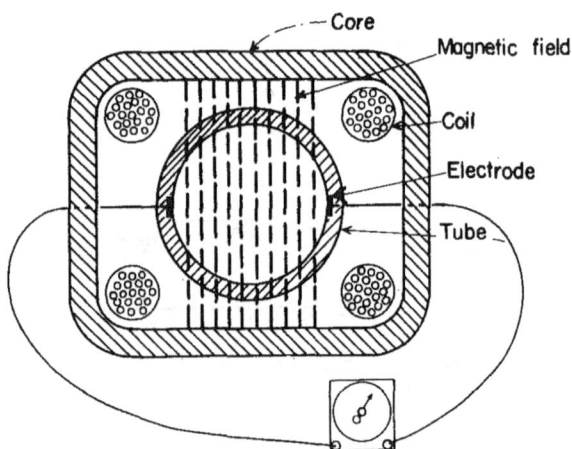

FIG. I-9-2 SECTION THROUGH A MAGNETIC
FLOWMETER AT THE ELECTRODES
(SCHEMATIC)

I-9-4 Elementary Theory of Operation [2, 3].
In accordance with the Faraday law of induction, the strength of the induced voltage is given by

$$e = B \, \delta \, V \qquad \text{(I-9-1)}$$

The volumetric rate of flow is

$$q = \frac{\pi D^2}{4} \, V \qquad \text{(I-9-2)}$$

and combined with equation (I-9-1)

$$q = \frac{\pi}{4} \, \frac{D^2}{\delta} \, \frac{e}{B} \qquad \text{(I-9-3)}$$

(With most actual magnetic flowmeters δ is slightly less than D.)

The flux density produced in a dc magnetic flowmeter, with a given permanent magnet, is constant; hence the generated emf is directly proportional to the velocity or volume rate of flow of the fluid. Thus, the rate of flow can be determined by an instrument that measures this voltage. On the other hand, the flux density of an electromagnet energized by alternating current will be related to the characteristics of this current as supplied. Since these characteristics are not constant, the flux density, B, is not constant. By equation (I-9-3) the rate of flow is proportional to the ratio, e/B, and therefore with ac magnetic flowmeters it is necessary to use an instrument that will measure this ratio as the secondary element.

The dc magnetic flowmeter is less expensive (to produce). It is particularly suited to the measurement of fluid flows subject to very high frequency disturbances (pulsations) and to the measurement of fluids of very high conductivity such as molten metals [4]. However, with fluids of lower conductivity, polarization of the fluid at the electrodes usually occurs. This is due to the dc current flowing between the electrodes causing an insulating layer to be formed at the electrode surfaces, thereby decreasing the signal voltage. Also, any voltages in the electrode circuit generated by galvanic or thermal action must be kept low enough not to decrease the accuracy of the flow measurement below a desired value, or provisions made to cancel such voltages in the secondary element.

I-9-5 The ac magnetic flowmeters are by far the more common type. As indicated above, the secondary element must measure the ratio of e/B. That part of the signal representative of the value of B is usually obtained by one of the following methods:

1. A reference signal is derived which is proportional to the supply voltage to the magnetic coils. It is assumed that the flux density in the conduit is proportional to supply voltage.

2. A reference signal is derived which is proportional to the current to the magnet coils. It is assumed that the flux density in the conduit is proportional to the current to coils.

3. A reference signal is derived from a coil wound around the iron of the magnet assembly. It is assumed that the flux density in the conduit is proportional to the flux density in the iron of the magnet assembly.

4. A reference signal is derived from a coil placed in the air gap of the magnet assembly. It is assumed that the flux density in the conduit is proportional to the flux density in the air gap of the magnet assembly.

The phase relations between the flow signal, e, and the reference signal, B, must be taken into account. If the reference signal used is the supply voltage, (1) above, the flow signal will be approximately 90 deg out of phase with it. If the current to the field coils is the reference signal, (2) above, the flow signal will be approximately in phase with it. However, changes in the conductivity of the fluid may affect the phase of e.

Signals other than flow signals may be picked up by the leads from the electrodes. A signal may be generated by the varying flux intersecting a coil composed of the electrode leads, the electrodes, and the fluid which connects the two electrodes, i.e., a transformer effect. Such a signal will be approximately 90 deg out of phase with the flow signal. The secondary element must separate such extraneous signals from the true flow signal.

I-9-6 The importance of fluid conductivity may be qualitatively assessed as follows. Consider the primary element as a simple generator developing a generated voltage, e, (equation (I-9-1)) which is connected in series with an internal fluid resistance, R_f. The output voltage, e_o, from this generator is received by the secondary element which has an input resistive impedance, R_s. The resistance R_f is that of the fluid between the electrodes and is given approximately by

$$R_f \cong \frac{1}{\xi d_e} \qquad \text{(I-9-4)}$$

Then the ratio of the output voltage to the generated voltage is

$$\frac{e_o}{d} = 1 - \left(\frac{1}{1 + R_s \, \xi \, d_e} \right) \qquad \text{(I-9-5)}$$

To illustrate, if the impedance, R_s, is a megohm, the fluid ordinary service water, with a conductivity of 0.0001 mhos/cm, and the electrode diameter, d_e, is 1 cm, then the voltage ratio, e_o/e, will be approximately 99 per cent. Now if the conductivity of the fluid were *increased* by a factor of 10, the voltage ratio would change to 99.9 per cent. That is, an increase in fluid conductivity of 1000 per cent changed the voltage ratio less than 1 per cent. On the other hand, if the fluid conductivity were *decreased* by a factor of 10, the ratio, e_o/e, will drop to 90 per cent, which is a 10 per cent change.

From this assessment of fluid conductivity, it appears that for a specific combination of primary and secondary elements there is a fluid conductivity value *above* which any increase in conductivity will not change the voltage ratio more than 1 per cent but below which decreasing the conductivity may produce a voltage ratio change greater than 1 per cent. For convenience, this boundary-line value of fluid conductivity may be called the "threshold conductivity." Naturally, this threshold conductivity will be different for different combinations of primary and secondary elements.

I-9-7 **Fluids and Fluid-Flow Characteristics as Affect Magnetic Flowmeters.**

Conductivity: A conductivity of 10 micromhos/cm will meet the threshold conductivity for most magnetic flowmeters, and sometimes a lower value may be considered. Such a value will include many industrial liquids. With special equipment fluids with conductivities ranging down to about 0.1 micromhos/cm may be metered. However, most petroleum products and derivatives are below a practical threshold conductivity, as are also gases, unless the gas is or can be highly ionized.

Density: The magnetic flowmeter is velocity responsive and therefore a volume rate of flowmeter; its indications are independent of variations of density.

Viscosity: The meters are independent of viscosity variations.

Velocity Profile: Changes in velocity profile, from streamline flow to very turbulent flow, have no effect, provided the flow profile is symmetrical about the pipe axis. Swirls or helical flow patterns seem to have little or no effect. A nonsymmetric flow pattern, such as might be produced by up to 50 per cent of the pipe area close to the electrodes being blocked, may produce effects of several per cent, possibly as high as 20 per cent, depending on the orientation of the blockage with respect to the electrodes. Blockages several pipe diameters upstream of the electrodes may have little or no effect.

Direction of Flow: Flows in either direction can be metered, that is, the meters are bidirectional.

Pipeline Deposits: The effects of deposits on the surfaces of the pipe and the electrodes, such as rusts, scale, and dissolved substances that precipitate along the pipe wall, will depend on the nature and amount of buildup. If the deposit has the same conductivity as the flowing fluid, there will be no effect. If the coating on the electrodes is nonconductive, it will have the effect of insulating the generated signal from the electrodes.

I-9-8 Accuracy and Tolerances. When the primary and secondary units of a magnetic flowmeter have been mutually attuned, carefully, it may be possible for the indicated flow, by tests, to stay within about ±0.25 per cent of span from full-rated capacity down to 5 per cent of rated capacity. It is doubtful if this is representative of the average magnetic flowmeter in use. Probably ± 1.0 per cent of span from full-rated capacity to 5 per cent of rated capacity will be more representative of these meters.

Force or Vane Meters

I-9-9 Symbols used are the following:

a = Projected cross-sectional area of a ball or vane, in a plane normal to the average fluid motion   See Note

C_d = A drag coefficient, determined by test

D = Diameter of pipe   See Note

F = Force exerted by the fluid against a ball or vane lb_f

g_c = Proportionality factor in the force-mass-acceleration relation

m' = Mass rate of fluid flow of a stream element having a cross-sectional area of a lb_m/sec

q' = Volume rate of flow of the stream element   cfs

V = Average velocity of the same stream element upstream of the ball or vane   ft/sec

δ (delta) = Angular deflection from the vertical of a vane or a string supporting a ball

ρ (rho) = Average density of the fluid stream lb_m/ft^3

ψ (psi) = Mass of ball or vane lb_m

ψ_w = Mass of an equal volume of water lb_m

Note: In the development of equations it is both convenient and conventional to evaluate linear dimensions in ft and areas in ft^2. Then, at the end, in giving the working equations to be used in the actual computation of flows, numerical conversion factors are introduced to provide for the use of the common units, inches and square inches.

I-9-10 In the general operation of these meters, the force produced by the impact of the moving fluid particles upon a swinging element of the meter deflects this element from its undisturbed position. The degree of deflection is a measure of the force and, hence, of the fluid velocity.

When used in an open stream, a true force meter will cause no loss of head. When applied to a closed channel, some loss of head will be found, the amount depending upon the size of the vane and the proportion of the pipe area it occupies.

I-9-11 Hydrometric Pendulum. This type of force meter may be used in open-channel streams [7]. In its simplest form, it consists of a ball of greater density than water, suspended by a string. As the ball is lowered into the stream, the impact of the water on the ball carries it downstream so that the supporting string is deflected from a vertical line. The velocity of the stream can be computed from the angular deflection of the string in the following manner.

Let ψ be the mass of the ball and ψ_w, the mass of a body of water having a volume equal to that of the ball. Since ψ must be greater than ψ_w, the resultant downward force, when the ball is in the stream, will be ($\psi - \psi_w$). The horizontal force, F, with which a stream element, having a cross-sectional area equal to the projected area of the ball, presses against the ball, is

$$F = C_d m' \frac{(V - 0)}{2g_c}$$

$$= C_d m' \frac{V}{2g_c}$$

(I-9-6)

but

$$m' = a V \rho \qquad (I\text{-}9\text{-}7)$$

hence

$$F = C_d a\ \rho\ \frac{V^2}{2g_c} \qquad (I\text{-}9\text{-}8)$$

The angle δ will be given by

$$\tan \delta = \frac{F}{(\psi - \psi_w)} \qquad (I\text{-}9\text{-}9)$$
$$= \frac{C_d a\ \rho\ V^2}{2g_c\ (\psi - \psi_w)}$$

and therefore

$$V = \sqrt{\frac{2g_c\ (\psi - \psi_w)\ \tan\ \delta}{C_d a\ \rho}} \qquad (I\text{-}9\text{-}10)$$

Replacing a with $(\pi d^2/576)$ and using a numerical value of g_c gives

$$V = 108.6 \sqrt{\frac{(\psi - \psi_w)\ \tan\ \delta}{C_d\ d^2\ \rho}} \qquad (I\text{-}9\text{-}11)$$

This method of obtaining the stream velocity will give only approximate values, since fluctuations of the stream currents will make accurate measurements of the angle of deflection impossible. The principal cause for unsteadiness of the ball and the supporting string is the formation of a train of eddies in the rear of the ball.

I-9-12 Vane Meters. Meters of this type are called by such names as gate, drag, target, force, etc. All are for use in closed channels. Possibly the simplest form of such a meter, represented by Fig. I-9-3, is a plate mounted so that when there is no flow of fluid it will rest against the end of a section of pipe, completely closing it. The force of a fluid stream against the plate causes it to swing away from the end of the pipe. The angle through which the plate swings will be a measure of the quantity rate of flow [8]. Such a meter might be used for a sediment-ladened liquid. Currently, vane-type meters, which may be represented schematically by Fig. I-9-4, are being used extensively. The vane may be a simple disc (as illustrated), a sphere, or a cylinder of a length greater than its diameter. Various methods are used for mounting the support rod or force bar and for measuring and transmitting the magnitude of the force against the vane. In most of

these meters the amount of the movement of the vane, or of the force bar, at the restraining element (indicated only by F in Fig. I-9-4) will be very small, possibly less than 0.05 in.

I-9-13 From Fig. I-9-4 it is seen that the fluid passage is the annular area between the pipe and vane, the area of which is $(\pi/4)\ (D^2 - d^2)$. The force, F, received by the vane is transferred by the force bar to a measuring and transmitting mechanism. The relation between the force, F, and the average fluid velocity upstream of the vane may be derived from equation (I-9-8) in which a is the area of the vane. Thus, an equation for the volume rate of flow is

$$q = C_d\ \left(\frac{D^2 - d^2}{d}\right)\sqrt{\frac{\pi}{4}\ \frac{2g_c\ F}{\rho}} \qquad (I\text{-}9\text{-}12)$$

and for the mass rate of flow

$$m = C_d\ \left(\frac{D^2 - d^2}{d}\right)\sqrt{\frac{\pi}{4}\ 2g_c\ F\ \rho} \qquad (I\text{-}9\text{-}13)$$

If the diameters are expressed in inches instead of feet (see Note, Par. I-9-9), the area term becomes $[(D^2 - d^2)/12d]$ in both equations.

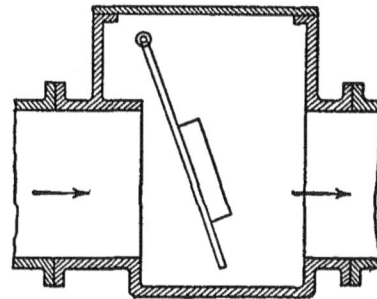

FIG. I-9-3 A SWINGING-PLATE VANE METER

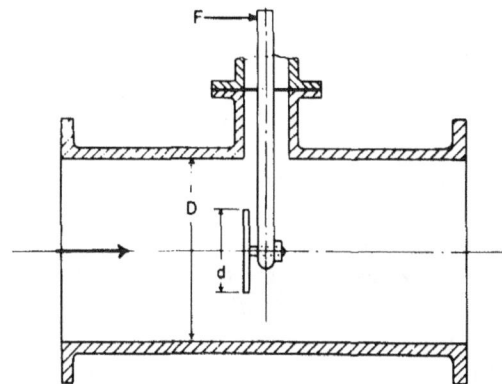

FIG. I-9-4 PRIMARY ELEMENT OF A VANE METER (SCHEMATIC)

The value of C_d will depend upon the shape of the vane, the method of its mounting and that of the force bar, and the force-measuring element. There are no published procedures for computing its value; hence the value to be used with any particular meter must be obtained from the manufacturer or by a calibration.

I-9-14 There is no published literature on tests to determine the effects of installation conditions upon force or vane meters. However, in any specific installation, if the density of the fluid is constant, then from the above equations it is seen that the force, F, and therefore the rate of flow is a function only of the velocity squared. This implies that the proper functioning of these meters can be affected by the velocity distribution profile. Likewise, there are no public test results available by which accuracy and tolerance values may be assigned.

In comparison with differential pressure meters, these meters have the advantage of not requiring pressure taps and lead lines. Thus it may be possible to use these meters for metering liquids carrying suspended materials.

Transverse-Momentum or Mass Flowmeters

I-9-15 In these meters there is superimposed upon the normal velocity of motion of the stream a known velocity in a direction or path that is perpendicular to the first. Thus there is produced a change in the momentum of the fluid. The force required to produce or to overcome this change in momentum will be directly proportional to the mass rate of flow.

A construction feature common to some of the transverse-momentum meters developed thus far is the need for rotating seals. Also, for their operation an external source of power is required to provide a constant rotational speed.

I-9-16 In one of these meters, which has been called the gyroscopic design, the primary element may be a section of pipe bent in a flat circular loop resembling the lower case Greek letter phi, ϕ. This loop is then rotated at constant speed, about the axis of the adjoining pipe, which may be called the x-axis. The rotating mass of fluid in the coil sets up a flywheel effect about an axis normal to the first, which may be called the z-axis. As a result of this flywheel effect, a torque is produced about an axis perpendicular to the other two, that is, about the y-axis. This torque is proportional to the mass rate of flow of the fluid and

the angular velocity at which the loop is rotated. Therefore, by measuring this torque, as with a calibrated spring, a direct indication of the mass rate of flow is obtained [9].

I-9-17 In a second design, which may be termed the radial-flow design, the primary element resembles a centrifugal pump. The fluid, upon entering the unit, encounters a runner like that of a centrifugal pump with radial guide vanes. This runner is connected to the housing by a torque tube. On reaching the periphery of this runner the fluid is guided back to its normal axial flow path through a second runner, in which there are guide vanes also. The second runner is integral with the shell or case. The whole assembly is rotated at constant speed by some suitable source of power. The drag or torque exerted by the case upon the first runner through the torque tube may be measured by strain gages attached to the torque tube. This torque will be proportional to the mass rate of flow of the fluid [10].

I-9-18 In a third type, which may be termed the axial-flow type, the primary element contains two similar, rotatable cylinders. These cylinders are mounted end to end on shafts that are coaxial but independent. The housing, which is an enlarged section of the pipe, is machined on the inside to provide very little clearance around the cylinders. The central area of the cylinders is blanked off. In the annular space thus formed are a number of passages, parallel to the cylinder axes, through which the fluid flows. The first or upstream cylinder is rotated at a constant speed and imparts to the fluid flowing through the passages a constant angular momentum. The second cylinder absorbs this angular momentum and is restrained from rotating by one or more springs. The angular deflection of this cylinder against the restraining force of the springs will be a measure of the mass rate of flow of the fluid [11].

I-9-19 A fourth design does not require external power. As in many axial-flow turbine meters, there is a closed streamlined tube mounted centrally, thus leaving an annular passage for the fluid. Near the inlet end there is a set of radial vanes mounted so the faces make an acute angle with axis of the pipe and thus divert the fluid stream into a helical motion. Close to the outlet edge of these stationary vanes, there is a second set of radial vanes mounted on a hub which rotates freely. Also, the faces of these second vanes are parallel to the pipe axis. Thus the fluid, which has been given a tangential velocity component, causes these second vanes to rotate. The speed of rotation is maintained constant either by an (external) energy-absorbing governor unit or by a governor system that varies the angular mounting of the stationary inlet vanes. Close to the outlet plane of the rotating vanes, there is a third set of vanes in a tubular unit that is mounted in fluid-sealed journals, thus providing for possible rotation of the tubular unit. The faces of the vanes in the tubular unit are parallel to the tube axis, and the axial length is long compared to the radial length. Since the speed of rotation of the second or rotating vanes is constant, the angular component of the velocity of the fluid as it leaves the rotating vanes and enters the third set is independent of the volume rate of flow. Hence, the angular momentum of the fluid entering the tubular unit of vanes is directly proportional to the mass rate of flow. Due to the axial length of these vanes, all of the angular momentum is absorbed from the fluid, and the resulting torque on this journal-mounted unit is a measure of the mass rate of flow. This torque may be measured and indicated by suitable means.

Alternately, the torque of the journal-mounted tubular vanes may be held constant and, through a servomechanism, used to regulate the angular setting of the first, or inlet, set of vanes. In this way the speed of rotation of the second, or middle, set of vanes will be variable and will be directly proportional to the mass rate of flow. The speed of rotation can be determined, as with a pulse pickup system, and indicated in terms of mass rate of flow [12].

Thermal Meters

I-9-20 As applied to the measurement of fluids, thermal meters are those in which the transfer of heat to or from the fluid stream is a basic part of the metering process. By a measurement of the heat supplied to, or received from, the stream, the mass rate of flow may be computed. While theoretically a thermal meter could be used for either liquid or gas flows, practically, they have been applied to the metering of gas streams only. Depending upon the manner in which the heat transfer is utilized they may be grouped as follows:

1. Those in which the effect of a known quantity of heat on a stream of fluid is determined.

2. Those in which the effect of a stream of fluid on a hot body is determined (or compared).

I-9-21 First Group: Requirements and Computation of Flow. For meters in the first group to operate satisfactorily, it would be necessary to have

1. A knowledge, to the required degree of accuracy, of the specific heat at constant pressure of the fluid stream being measured.

2. Efficient means of transferring heat to the fluid stream and of measuring the heat input.

3. Sufficiently accurate means of measuring or controlling the temperature change of the fluid stream due to the transfer of heat.

The relation between the rate of flow of fluid and the heat input is expressed by the following equation:

$$H_t = m c_p ([T_a]_2 - [T_a]_1) \qquad (I\text{-}9\text{-}14)$$

or

$$m = \frac{H_t}{c_p ([T_a]_2 - [T_a]_1)} \qquad (I\text{-}9\text{-}15)$$

where

c_p = Specific heat of the fluid at constant pressure, Btu/lb/F
H_t = Heat energy supplied the heater, Btu/sec
$[T_a]_1$ and $[T_a]_2$ = The entrance and exit temperatures, respectively, F
m = Mass rate of flow of the fluid, lb_m/sec

If the heat energy is measured by the electric energy supplied the heater, so that

$$H_t = 0.000948 \, W_j \qquad (I\text{-}9\text{-}16)$$

in which W_j is in international joules (watt-seconds), then [13],

$$m = \frac{0.000948 \, W_j}{c_p ([T_a]_2 = [T_a]_1)} \qquad (I\text{-}9\text{-}17)$$

I-9-22 A modification of this procedure is provided by a thermal proportioning method which eliminates the need of knowing the specific heat of the stream fluid. Briefly, the method involves dividing the fluid (gas) stream into a small and large stream (e.g., 0.1 and 0.9) and supplying sufficient heat to each stream that the temperature of each is raised a few degrees. This temperature change should not only be the same in degrees but over the same range, also. The amount of heat to each stream must be measured. The small stream is directed through a meter of suitable size and accuracy and thence back into the mainstream line. With this quantity and the ratio of heat quantities, the total stream flow is computed [14].

I-9-23 Second Group. The second group, in which the effect is measured of a stream of fluid upon a hot (or cold) body, is represented by the hot-wire anemometer, as described in Chapter I-7. While the hot-wire anemometer is commonly thought of in association with open-air and wind-tunnel velocity measurements, it can be and has been used for velocity measurements in closed conduits. With velocity known, volume rate of flow in a closed conduct is readily determined [15, 16]. However, there is no report of a complete metering unit using the hot-wire principle and indicating volume rate of flow, or a time-integrated quantity of flow.

Tracers

I-9-24 The distinctive feature common to these methods of flow measurement is that a discrete quantity of a foreign substance is injected suddenly into the fluid stream, and the time interval for this substance to be carried to an observation station, or between two observation stations, is measured. From this time the area average velocity is deduced; and, if the average channel area is known, the volume rate of flow may be determined. Normally, the tracer substance will be a liquid when the primary fluid is a liquid and a gas when the primary fluid is gaseous. However, in the case of gas flows the tracer substance may be injected in the liquid state provided it will vaporize instantly on being released into the gas stream.

A second characteristic feature of these procedures is that they give a spot check on the rate of flow rather than a continuous measurement of the flow. While spot measurements may be made in quick succession, it is not usual to continue them for a long time.

I-9-25 While a tracer may be used in checking the rate of flow of liquids in open channels, this is done very seldom for the determination of quantity rate of flow. Hence, the discussion here will treat the use of tracers with flows in closed channels.

It is necessary that the channel from the injection station to the observation station, or to the last observation station if there are more than one, be straight, free from fittings or obstructions, and of uniform cross section or only very slightly converging. This is because the tracer principle depends on the body of the tracer substance being carried along by the primary stream as a slug with a minimum of dispersion.

The important characteristics for a tracer substance are: It must have some property that makes its presence in the primary fluid readily detectable; its presence even in very low concentrations must not be detrimental to the subsequent use of the primary fluid; the relationship between the tracer property to be detected and the equipment and technique for detection must be such that the procedure will be safe and practicable in the location of use. The tracer substances discussed here are: salt (NaCl); discrete heated or chilled quantities of the primary fluid; dyes; radioactive material; carbon dioxide; and nitrous oxide.

I-9-26 Salt. This is commonly called the salt velocity method and sometimes the Allen salt velocity method. It is used only in the measurement of water and is based on the greater electrical conductivity of brine over that of fresh water. A concentrated solution of brine under pressure is suddenly injected into the water stream through a system of injection valves that are so located as to give a uniform distribution of the brine over the cross section of the channel. The pressure for injecting the brine may be supplied by air at about 35 psi above the pressure within the conduit. The injection valves should be spring-loaded to prevent leakage of brine between injections. The operation of the injection system is best controlled by a quick-acting valve as close to the conduit as possible. At all times, there must be no leakage of brine between shots, and the presence of air pockets in the piping must be avoided. In moderate-sized conduits, four pop valves, located at the centroids of the four quadrants of the cross section, may be sufficient. In very large conduits, there should be one pop valve for each 20 sq ft of cross-sectional area of the conduit. Figure I-9-5 shows a typical pop-valve assembly.

If only one set of electrodes is used downstream of the injection section, the salt slug is timed from the pop valves to the electrode section. For this the discharge from the pop valves is sensed by special electrodes, such as a pair of closely spaced wires placed very close to one or more pop valves. Thus, such electrode indicates the exact time the brine enters the conduit with negligible time lag.

I-9-27 The basis for the design of electrodes is to provide equal increments of conductivity for equal segments of conduit cross-sectional area. For both circular and rectangular conduits this may be done by dividing the cross section into strips of equal area or width and locating an electrode at the center of each. For circular conduits it is possible to install an electrode assembly which provides two traverses

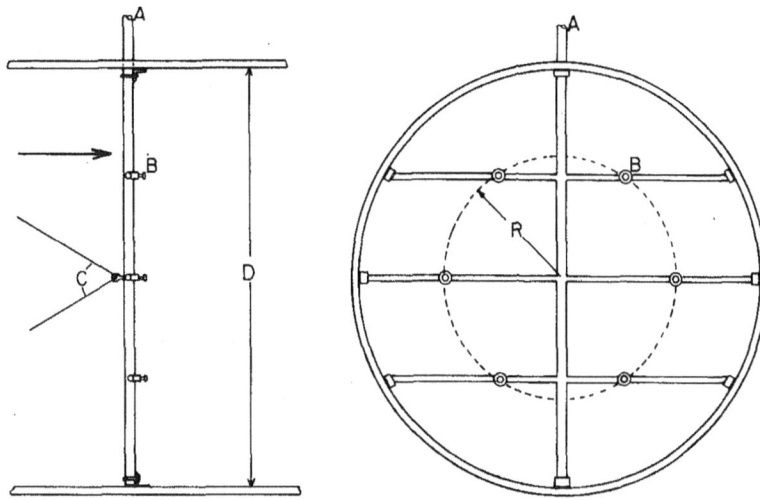

FIG. I-9-5 TYPICAL ARRANGEMENT OF BRINE INJECTION NOZZLES IN
A CONDUIT
A = Supply pipe
B = Injection nozzle
C = Guy wires
R = 0.315 D

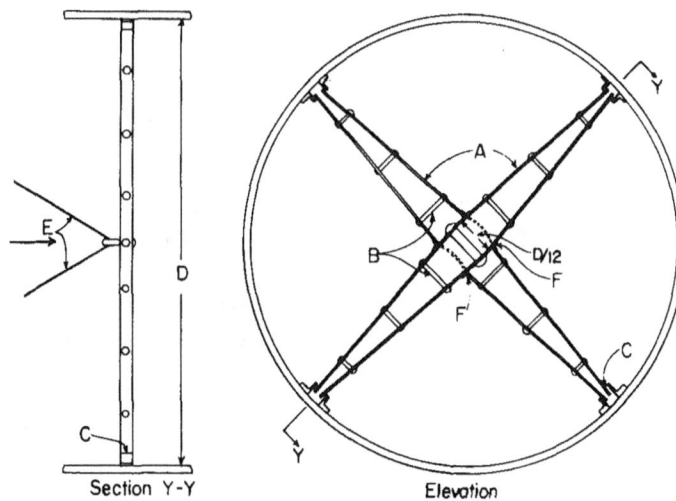

Section Y-Y   Elevation

FIG. I-9-6 A TYPICAL ELECTRODE SYSTEM
IN A LARGE CONDUIT
A = Electrode bars
B = Insulated stiffeners
C = Insulated connection to conduit wall
D = I.D. of conduit to be determined in field
E = Insulated guy wires
F = Insulated joint

133

at right angles across the pipe, as shown in Fig.
I-9-6. An important requirement is for the assembly
to have sufficient strength and stiffness to withstand
the force and induced vibrations from rapidly flowing water. Also, the electrode bars must be insulated
adequately from each other and from the conduit.
The thickness and strength of the insulation must be
sufficient to prevent breakdown from vibrations.
Some plastics meet this requirement. All surfaces
which are not a part of an active electrode area
should be coated with an insulating paint. This
includes the edges and backs of the electrodes, the
end supports, guy wires, and the conduit wall, if
conducting, within 3 ft of the ends of the electrodes.

The portion of the electrode leads within the conduit must be well supported to protect against break-
age from the whipping action of the water.

I-9-28 There should be a minimum distance of at
least 4 diam, or the equivalent between the injection section and the first electrode section, to provide for adequate mixing and consolidation of the
brine slugs before the electrode is reached. Likewise, when two electrodes are used, there must be
at least 4 diam, or equivalent, of conduit between
the electrode sections so that the brine slug will
be entirely past the first section before reaching
the second.

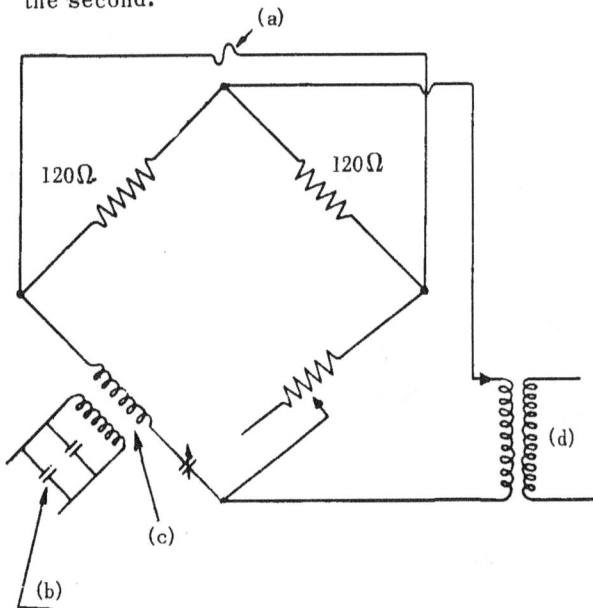

FIG. I-9-7 TYPICAL WIRING DIAGRAM FOR
SALT VELOCITY TEST
(a) = 2500 cycle oscillator
(b) = Electrodes connected in parallel
(c) = Matching transformer from 120 ohms
impedance in primary to variable
secondary, 50, 25 and 10 ohms
(d) = Transformer to amplifier and oscillograph

The volume of the test section of the conduit is to
be determined by careful measurements in the field.
Several determinations of the length should be made.
Where only one downstream electrode section is
used, this length is from the cross-sectional plane of
this electrode section to the plane of the electrode
wires placed at the outlet of the pop valves. Where
two electrode sections are used, the distance is
that between the cross-section planes of the electrodes. At least two and preferably four measurements
of the diameter, or dimensions, of the conduit should
be made in about 20 cross sections uniformly distributed over the length of the test section.

The time of travel of the brine slug between the
electrode stations is recorded on a strip chart meter.
Recording ammeters, milliammeters and oscillo-
graphs are suitable for this purpose. An oscillograph
with an amplifier is preferable because of its greater
sensitivity and stability, shorter natural pen period
and rectilinear recording. A typical wiring diagram
for such a recorder is shown in Fig. I-9-7.

I-9-29 A time record may be marked on the same
chart by a pen actuated by a suitable timer equipped
to give time marks at uniform intervals such as
seconds or half-seconds. Such a timer may be a
precision clock or chronometer fitted with a contact
of zero drag or a shortwave radio receiver tuned to
receive second signals from station WWV.

Note: Station WWV broadcasts on a fundamental frequency of 5 mc. Signals can be received at 10, 15, 20 and 25mc.

The time of travel is the distance, expressed in
seconds, between the abscissa lines which will
divide each of the areas under the conductivity
curves into two equal areas. The positions of these
abscissa lines may be determined by measuring the
areas with a planimeter, by a replot of the curves,
with the time scale on rectilinear coordinate paper
to uniform scales and determining the areas and
corresponding mean ordinates and abscissa, or by
making a cutout of the curves and determining the
desired abscissa (time value) by the knife edge
balancing technique. Figure I-9-8 represents a
typical conductivity diagram.

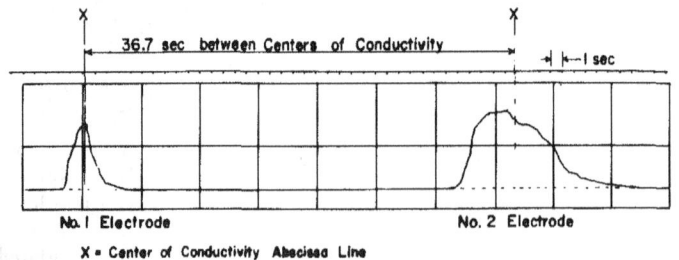

FIG. I-9-8 A SALT CONDUCTIVITY DIAGRAM

I-9-30 The accuracy of the salt velocity flow measurement method is affected by systematic and observational errors. The systematic errors will include those of determining the volume of the test section, those involved in the brine conductivity responsive equipment, and the time-interval measurements. Under favorable conditions and careful measurements, these errors may not exceed ±0.2 to ±0.3 per cent. The observational errors, including those of evaluating the curves, may be lessened by obtaining the travel times for several brine slugs at each rate of flow. For short test sections, with corresponding short time intervals, 10 shots or slugs are recommended, while for very long test sections three to five shots may suffice [17-20].

I-9-31 Hot or Cold Injections. In this case the tracer material may be small quantities of the same fluid as that in the conduit, heated or cooled to a temperature differing appreciably from that of the mainstream. The temperature difference to be used in a particular case will depend on several considerations, such as the type of fluid in the conduit and its physical state, the approximate rate of flow of the mainstream, the volume of fluid injected at each shot, and the length of the test section. A temperature difference of about 25 F may be found to be sufficient.

For detecting the temperature pulses, small fine-filament hot-wire anemometer probes are recommended, with the necessary amplifying circuitry and recording equipment, as described in Chapter I-7. Ordinarily, a single probe at each detection section, placed at approximately the center of the pipe, will be sufficient.

The volume of each injection shot of the fluid should be sufficient to form a "plug" of the fluid having a length equal to between 1/2 and 1 pipe diam. This would be an aspect ratio, ("plug" length/ pipe diam.), of between 0.5 and 1.0. The distance from the injection section to the first temperature probe and between temperature probe stations will depend upon the particular test site conditions, but a distance of at least 10 to 15 ft is suggested as desirable, irrespective of the size of pipe.

The tracer fluid must be injected against the pressure of the mainstream flow and rapidly enough to form a coherent plug. Thus it will temporarily alter the velocity of the mainstream flow at the injection section and will set up a pressure wave which will add to the "noise" picked up by the detector units. The instructions given in the preceding section on determining the dimensions of the test section apply here.

This particular tracer procedure has been tried out to a limited extent in an experimental line. There is no record of it having been used or tried in an actual plant test. In the experimental work the scatter between successive trials ranged up to about 6 per cent with most points in the 3 to 4 per cent range. These figures may be taken as indicative of the probable overall accuracy of the procedure until it has been used sufficiently to provide a basis for more detailed instructions than given here. The outstanding advantage for this particular tracer procedure is that it does not introduce any contaminant into the mainstream [21].

I-9-32 Radioactive Tracers. The use of radioactive tracers may perhaps be used more readily with gases than with liquids. Possibly it would be the best type of tracer for locations where there can be a long distance between detection stations, since increasing the aspect ratio of the tracer plug, by dispersion, by a factor of at least 10 does not seem to affect the accuracy unduly. Another feature is that at the detection stations no probe need extend into the fluid. However, if Geiger counters are used as the detection instruments, some form of opening or "windows" are needed through which the face of the counters may be in actual contact with the fluid at the conduit wall. Scintillation counters to detect the gamma radiation can be used without windows, but the effectiveness of these is increased with windows.

Drawbacks to the use of radioactive tracers are the possible effects of contamination of the fluid stream and the special precautions to be taken in the handling of such material.

For use in a gas stream, argon-41 probably is the most suitable material. It is readily obtainable and has a half-life of only 95 minutes, and within a day its radioactive effects become negligible. In liquid streams it may be possible to use sodium-24, having a half-life of 15 hours, or bromine-82, half-life of 36 hours [6-9].

I-9-33 Carbon Dioxide and Nitrous Oxide. The use of these tracer materials in the measurement of gas streams would be limited to situations in which small traces of these substances would not constitute objectionable contamination. In general, the basis for using these gases as tracers is their high infrared absorption characteristic when passing through an infrared beam. With an infrared detector, concentrations in the order of 10 parts per million can be detected, but naturally higher concentrations will produce sharper signals. A test section of only 5 to 10 ft can be used if space is limited or if it is necessary to use very low concentrations. However, test sections have been used successfully in which the distance from injector to first detector ranged from 2 to 100 ft and that between detectors from 40 to about 300 ft. The possible accuracy by this method may be high, with probable errors in the 1.0 to 0.5 per cent range or lower [21-26].

I-9-34 **Dyes.** Some dyes may be used as a tracer material, especially such as rhodamin and pontasil dyes which will fluoresce. Since fluorescent substances are seldom, if ever, found in natural fluid streams, there would be little or no initial or background trace material to be "zeroed" out in the detection procedure. Thus, the concentration required for detection can be very low, or the detection sensitivity increased manyfold.

The following table lists the tracer materials and detection procedures that have been discussed or referred to.

Table I-9-1   Tracer Materials and Detection Methods

Tracer Substance	Detection Method	Mainstream Fluid
Salt (NaCl) solution	Electrical conductivity or titration	Water
Hot or cold shots mainstream fluid	Thermal (hot-wire) probes	Liquids and gases
Radioactive substances	Geiger or scintillation counters	Liquids and gases
CO_2 and N_2O	Infrared spectrometer	Gases
Rhodamin and pontasil dyes	Fluorimeter	Liquids (water)
Methyl blue and sodium dichromate	Colorimeter or color reagent comparison standard	Liquids (water)

Method of Mixtures

I-9-35 This procedure involves the introduction into the primary fluid stream of a detectable substance at a constant, measured rate. After the added substance has become thoroughly distributed and mixed with the primary stream, a sample of the mixture is drawn off and the proportion of the added substance determined. From this ratio and the known rate of introduction of the detection substance, the rate of flow of the primary stream, is deduced [27].

Some of the important features of this procedure are: It is applicable to the measurement of both liquid and gas flows; it is independent of the size, type and uniformity or lack of uniformity of the conduit; and it can be used with liquid flow in open channels, although the degree of uncertainty will be higher than with flows in closed channels.

Between the section of test substance injection and the section of mixture sample draw-off, thorough mixing must take place. This may be provided by fittings, such as elbows or valves, or special means to promote mixing may be placed within the conduit. Between the injection section and the sample draw-off section there must be no addition to the primary stream. However, after thorough mixing has taken place, a diversion of a part of the stream may be made ahead of the sample draw-off section, since this will not alter the mixture proportion. It is the rate of flow at the injection section that is determined by this procedure. The sample of the mixture for analysis may be a continuous stream or individual samples collected at regular intervals of time.

I-9-36 The detection substance must be such that it will not constitute a detrimental contaminant to the later use of the mainstream. It must be readily detected and such that the proportion of it in the mixture can be determined to a percentage degree at least as good as required for the mainstream determination. The selection of the test substance will be governed, mainly, by the nature of the mainstream fluid and the use or uses that will be made of it after measurement.

The selection of the detection or test substance and the selection of the method of mixture analysis are interdependent. The method of analysis should be one that can be performed rapidly enough that the determination of the mixture proportions will not delay the progress of the test. By several of the methods of analysis it is possible to determine the presence of some test substances from 1 part in 5000 down to 1 part in 10^6.

I-9-37 The method of measuring the quantity rate at which the test substance is introduced must be such as will provide a degree of accuracy at least as good as required for the mainstream. The type of meter used for measuring the quantity rate of flow of the test substance will depend upon the character of the mainstream fluid, whether liquid or gas, the mainstream line pressure, and the estimated rate of flow of the mainstream and that required of the test substance. For metering and maintaining very low injection rates of either a liquid or a gas, a calibrated linear-resistance type of flowmeter might be suitable. Suggested types of displacements meters are: for liquids, a small capacity meter of a type used for dispensing gasoline; for gases, a water-sealed rotating drum type or a small capacity diaphragm type. Either of these last two named meters can be enclosed within a pressure chamber, if necessary, on account of the pressure of the primary stream.

I-9-38 For computing the rate of flow of the mainstream by the mixture method, let

C_m = Fraction by weight

C_v = Fraction by volume

of test substance in mixture at draw-off section

m = Mass rate of flow of the main fluid stream immediately upstream of the section at which the text substance is injected — lb_m/sec

m_X = Mass of test substance injected per sec — lb_m/sec

Then, $m + m_X$ = total mass rate of flow past the draw-off section, lb_m/sec, and

$$C_m = \frac{m_X}{m + m_X} \qquad (I-9-18)$$

so that

$$m = m_X \frac{1 - C_m}{C_m} \qquad (I-9-19)$$

If it is preferred to express the rate of flow in volume units, then using q and q_X to represent the volume rates of flow of the mainstream and test substance, respectively,

$$q = q_X \frac{1 - C_v}{C_v} \qquad (I-9-20)$$

If the mainstream and test substance are liquids, the units for q and q_X could be either cfs or gpm. The densities of the two streams should be as nearly the same as possible. If the streams are both gases, the unit would be cfs—referred to the same pressure, temperature and relative humidity, usually those of the mainstream at the injection section.

I-9-39 Some possible combinations of test substances and analytical methods for use with some primary fluid streams are listed in Table I-9-2.

Table I-9-2 Mixture Method Test Substances and Methods of Analysis

Main Fluid Stream	Test Substance	Analytical Method
Water	Sodium thiosulphate	Titration with iodine
Water or steam	Carbon dioxide	Separation and absorption with KOH
Fuel oil, gasoline and similar liquids	D-monobromonaphalene	Refractive index
Air	Hydrogen, ammonia or carbon dioxide	Thermal conductivity or flame ionization
Air of fuel gas	Benzene or carbon dioxide	Refractive index

Pressure-Time or "Gibson" Method

I-9-40 The following letter symbols are used in describing this method:

A = Cross-sectional area of pipe — ft^2

D = Pipe diameter — ft

E = Modulus of elasticity of pipe — lb_f/ft^2

f = Darcy-Weisback friction factor

g = Local acceleration of gravity — ft/sec^2

g_c = Proportionality factor in the force-mass-acceleration relation

h = Liquid pressure, general — ft

h_c = Liquid pressure, total, at zero flow — ft

h_f = Liquid pressure at gate under flowing conditions, gate open — ft

L = Length of conduit from entrance to point of measurement — ft

m = Mass rate of fluid flow at steady state before start of gate closure — lb_m/sec

p = Pressure — lb_f/ft^2

q = Volume rate of fluid flow at steady state before start of gate closure — cfs

s = Space coordinate along pipe line — ft

t = Time interval — sec

V = Velocity of fluid — ft/sec

V_c = Velocity of fluid at gate closure — ft/sec

V_f = Velocity of fluid under steady-state conditions before start of gate closing   ft/sec

Z = Height above the datum   ft

d/dt = Total time derivative

α (alpha)	= Angle pipe center line makes with a horizontal line	radian
κ (kappa)	= Bulk modulus of elasticity of the fluid	$\mathrm{lb}_f/\mathrm{ft}^2$
λ (lambda)	= Lagrange multiplier	
ρ (rho)	= Density of the fluid	$\mathrm{lb}_m/\mathrm{ft}^3$
τ_w (tau)	= Shear stress of fluid on pipe wall	$\mathrm{lb}_f/\mathrm{ft}^2$
υ (upsilon)	= Velocity of the pressure wave created by closing of valve	ft/sec
ϕ (phi)	= Pipe stress factor	

I-9-41 The pressure-time on Gibson method was developed initially as a means of determining the rate of flow in a penstock to a hydraulic turbine, as represented schematically by Fig. I-9-9. While it is an inherently accurate method of determining a steady flow, it involves stopping the flow. However, this does not require the installation of a valve or gate other than those installed for normal operation.

I-9-42 To develop the physical principles involved, consider the instantaneous closing of a valve in a horizontal pipe flowing full. The fluid next to the valve is brought to rest by the impulse of a higher pressure at the valve. As soon as this incremental volume is brought to rest, it is at the higher pressure; and the same impulse is applied to the next incremental volume upstream bringing it to rest, and so on. Thus a high-pressure wave is set up which travels upstream at a wave velocity, υ, which depends on the fluid in the pipe and the elasticity of the pipe walls. For steel pipes υ is nearly the acoustic velocity, but υ is markedly reduced if the conduit is flexible tubing. In addition, υ is severely reduced (to less than the acoustic velocity in air alone) if the slightest amount of gas is trapped in the line.

I-9-43 The momentum flux equation states that the resultant axial force on the fluid in a typical volume is equal to the net efflux of axial momentum plus the rate of increase of axial momentum inside the typifying volume. In Fig. I-9-10, the typical volume is shown at the instant the pressure wave is passing through it. On the downstream end (at right), the fluid is stopped; there is no efflux of momentum, but an unbalanced pressure force exists. On the upstream end there is an influx of momentum equal to $\rho A V^2$, since this part of the fluid is not affected yet by the fluid stoppage downstream. Inside of the typical volume, momentum is being destroyed at the rate of $\rho A V \upsilon$. The momentum flux equation for the volume is:

$$-\rho g (\Delta h) A = -\rho A V^2 - \rho A V \upsilon$$

or

$$(\Delta h) = \frac{\upsilon}{g} \, V \left(1 + \frac{V}{\upsilon} \right) \qquad \text{(I-9-21)}$$

where (Δh) is the pressure rise due to the instantaneous closure of the valve downstream. Differentiating this equation gives the increase in pressure due to a decrease in the fluid velocity

$$d(\Delta h) = \frac{\upsilon}{g} \, dV \left(1 + \frac{2V}{\upsilon} \right) \qquad \text{(I-9-22)}$$

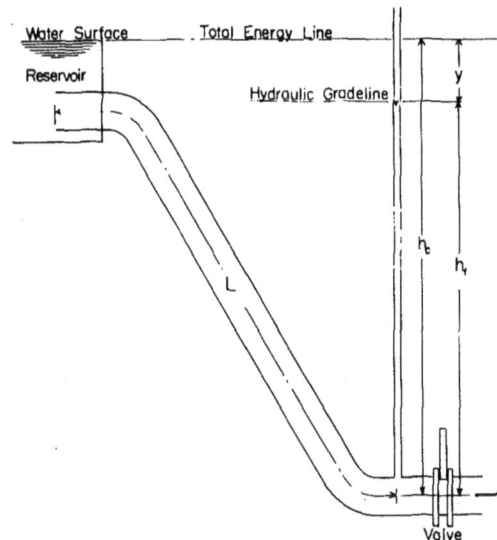

FIG. I-9-9 SCHEMATIC REPRESENTATION OF A RESERVOIR-PENSTOCK SYSTEM

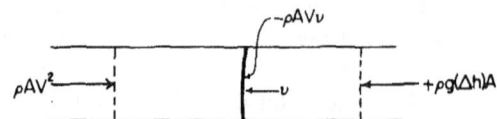

FIG. I-9-10 PRESSURE WAVE PASSING THROUGH CONTROL VOLUME

Since in most cases v is very much greater than V, these equations can be approximated by neglecting terms of V/v. This is the basic relation used by Gibson to convert pressure data into a flow measurement. Strictly, these equations apply only when no pressure waves are reflected from the upstream end of the pipe.

After the valve closure, a pressure wave travels upstream, bringing the fluid to rest. At the instant the wave reaches the reservoir, $t = L/v$, all the kinetic energy of flowing fluid has been stored as elastic compression energy; and the fluid is under a new pressure, higher than the reservoir pressure. The fluid begins a return flow into the reservoir starting at the upstream end and progressing downstream. This decompression returns the pressure to what was normal before the valve closure and the fluid to normal velocity, but in the reverse direction. At $t = 2L/v$, all the fluid is flowing upstream from the valve; however, since the valve is closed, no fluid is available to maintain the flow. A low-pressure wave begins at the valve and progresses upstream, again bringing the fluid to rest. This pressure is as far below the normal pressure as the first wave was above it, assuming no losses. It is during this part of the cycle that the pressure may decrease to that of the vapor pressure of the fluid. If this happens, liquid column separation will occur with possible disastrous effect. At $t = 3L/v$, this negative wave reaches the reservoir, and all the fluid is at rest. An unbalanced pressure exists, and flow enters the pipe, starting at the reservoir and traveling downstream. Again, pressures return to normal and the velocity to normal in the downstream direction. At $t = 4L/v$, this wave reaches the valve, and the flow conditions are identical with those existing when the valve was closed $4L/v$ seconds before. This cycle then repeats; only frictional and

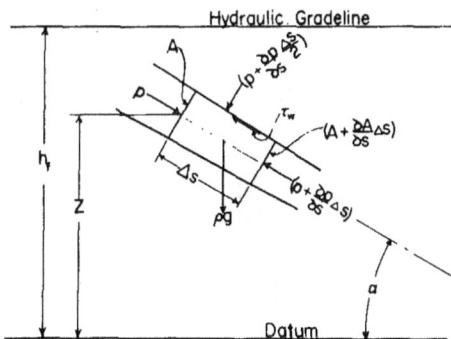

FIG. I-9-11 FORCES ACTING ON A FREE BODY OF FLUID

inelastic losses slowly dissipate the energy until the flow comes to rest permanently.

I-9-44 Referring to Fig. I-9-11 and using Newton's second law, the resultant forces in the direction of flow on the free body of fluid are: (end pressure forces) + (sidewall pressure component) + (gravitational forces) − (wall shear force) = (fluid inertia). In mathematical symbols,

$$\left[pA - \left(p + \frac{\partial p}{\partial s}\, \Delta s \right)\left(A + \frac{\partial A}{\partial s}\, \Delta s \right) \right]$$
$$+ \left[\left(p + \frac{\partial p}{\partial s}\, \frac{\Delta s}{2} \right) \frac{\partial A}{\partial s}\, \Delta s \right]$$
$$+ \left[\rho g \Delta s \left(A + \frac{\partial A}{\partial s}\, \frac{\Delta s}{2} \right) \sin \alpha \right]$$
$$- (\tau_w D \Delta s) = \rho \left(A + \frac{\partial A}{\partial s}\, \frac{\Delta s}{2} \right) \Delta s\, \frac{dV}{dt} \quad \text{(I-9-23)}$$

Terms of the order Δs^2 are neglected, and all terms are divided by the mass of the element of fluid. The friction can be represented as the Darcy-Weisbach type, so that the following equations are used:

$$- A \frac{\partial p}{\partial s}\, \Delta s + A \rho g \Delta s\, \sin \alpha$$
$$- \tau_w \pi D \Delta s = \rho A\, \Delta s\, \frac{dV}{dt} \quad \text{(I-9-24)}$$

$$\tau_w = \frac{\rho f V^2}{8} \quad \left(\text{Darcy-Weisbach friction} \right) \quad \text{(I-9-25)}$$

$$p = \rho g (h_f - Z) \quad \text{(I-9-26)}$$

$$\frac{\partial p}{\partial s} = \rho g \left(\frac{\partial h_f}{\partial s} - \frac{\partial Z}{\partial s} \right) \quad \text{(I-9-27)}$$

Variations in density are small compared to variations in pressure:

$$\frac{\partial Z}{\partial s} = - \sin \alpha \quad \text{(I-9-28)}$$

This is the simplified equation expressing Newton's second law

$$g \frac{\partial h}{\partial s} + \frac{dV}{dt} + \frac{fV}{2D}\, |V| = 0 \quad \text{(I-9-29)}$$

The absolute value sign denotes that the friction force always opposes the velocity. In using this method to determine rate of flow, the flow is reduced slowly enough that flow reversal should never take place. In equation (I-9-29) $dV/dt \cong \partial V/\partial t$.

I-9-45 In considering the conservation of mass, or flow continuity, the compressibility of the fluid cannot be neglected. The pressure wave velocity in pipes, v, is always less than the acoustic velocity in an infinite volume of the same fluid. This is due to the work done by the pressure wave on the pipe as it passes through. The flexibility of the pipe, linear or nonlinear, has a great deal to do with v as well as with the dissipation of its intensity. Different methods of pipe support also affect v by determining the Poisson-ratio effect on the pipe stresses caused by the pressure wave. However, it is again noted that 2 per cent of entrapped air will have a larger effect on v than any of these; and, whenever possible, it is recommended that v be measured.

For deriving an equation for the conservation of mass, it will be assumed the pipe is supported in such a manner that its length remains constant. Then, in an elemental section of the pipe, as represented by Fig. I-9-12, (the flow in) − (the flow out) = (the fluid stored), or

or

$$\rho A V - \left[\rho A V + \frac{\partial}{\partial s}(\rho A V)\Delta s\right] = \frac{\partial}{\partial t}(\rho A \Delta s) \quad \text{(I-9-30)}$$

Divide each term by the mass of the element, $\rho A \Delta s$,

$$-\frac{V}{\partial s} - \frac{V}{A}\frac{\partial A}{\partial s} - \frac{V}{\rho}\frac{\partial \rho}{\partial s}$$

$$= \frac{1}{\Delta s}\frac{\partial \Delta s}{\partial t} + \frac{1}{A}\frac{\partial A}{\partial t} + \frac{1}{\rho}\frac{\partial \rho}{\partial t} \quad \text{(I-9-31)}$$

The terms $V\,\partial A/A\,\partial s$, $V\,\partial \rho/\rho\,\partial s$, and $1\,\partial s/\Delta s\partial t$ are assumed to be negligible compared to the partial derivatives with respect to time, so that the conservation of mass equation becomes

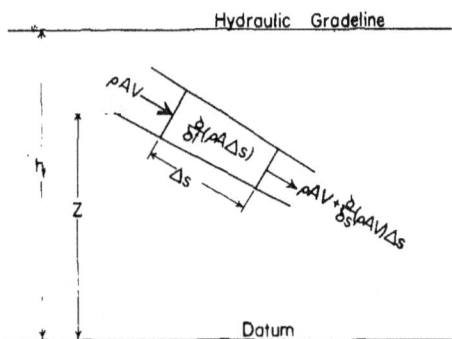

FIG. I-9-12 MASS FLOW THROUGH AN ELE-
MENTAL SECTION OF PIPE

$$-\frac{\partial A}{\partial s} = \frac{1}{A}\frac{\partial A}{\partial t} + \frac{1}{\rho}\frac{\partial \rho}{\partial t} \quad \text{(I-9-32)}$$

The two elastic storage terms can be related to the fluid pressure as follows:

$$\frac{1}{\rho}\frac{\partial \rho}{\partial t} = \frac{1}{\kappa}\frac{\partial p}{\partial t} \quad \text{(I-9-33)}$$

$$\frac{1}{A}\frac{\partial A}{\partial t} = \frac{\phi}{E}\frac{\partial p}{\partial t} \quad \text{(I-9-34)}$$

$$\frac{1}{\rho}\frac{\partial \rho}{\partial t} + \frac{1}{A}\frac{\partial A}{\partial t} = \frac{1}{\kappa}\left(1 + \frac{\phi\kappa}{E}\right)\frac{\partial p}{\partial t} \quad \text{(I-9-35)}$$

Let

$$v = \pm \frac{\sqrt{\kappa/\rho}}{\sqrt{1 + \frac{\phi\kappa}{E}}} \quad \text{(I-9-36)}$$

which is the acoustic velocity through unconfined fluid reduced by a factor describing the effect of elastic pipe walls.

$$\phi = D/e \qquad \text{for thin-walled pipes} \quad \text{(I-9-37)}$$

$$\phi = \frac{2(D_o^2 + D_i^2)}{(D_o^2 - D_i^2)} \quad \text{for thick-walled pipes}$$

where e = thickness of pipe wall (ft) and subscripts i and o refer to the inside and outside diameters of the pipe.

The mass conservation equation may now be written

$$-\frac{\partial V}{\partial s} = \frac{1}{\rho v}\frac{\partial p}{\partial t} \quad \text{(I-9-38)}$$

Using equation (I-9-26) relating pressure to the hydrostatic column and the fact that Z does not vary with time,

$$-\frac{\partial V}{\partial s} = \frac{g}{v^2}\frac{\partial h}{\partial t} \quad \text{(I-9-39)}$$

Combining equation (I-9-39) with (I-9-29), Newton's second law, by the Lagrange multiplier technique, two values of the multiplier will turn up giving two equivalent equations in symmetrical form.

$$g\frac{\partial h}{\partial s} + \frac{\partial V}{\partial t} + \frac{fV^2}{2D} + \lambda\frac{\partial V}{\partial s} + \frac{g}{v^2}\frac{\partial h}{\partial t} = 0 \quad \text{(I-9-40)}$$

The total derivatives are defined

$$\frac{dh}{dt} = \frac{\partial h}{\partial s}\frac{ds}{dt} + \frac{\partial h}{\partial t} \quad \text{(I-9-41)}$$

$$\frac{dV}{dt} = \frac{\partial V}{\partial s}\frac{ds}{dt} + \frac{\partial V}{\partial t} \qquad (I\text{-}9\text{-}42)$$

The Lagrange multiplier is chosen to use equations (I-9-41) and (I-9-42) and reduce equation (I-9-40) to a pair of symmetric ordinary differential equations.

$$\frac{ds}{dt} = \frac{v^2}{\lambda} = \lambda \qquad (I\text{-}9\text{-}43)$$

$$\lambda = \pm v \qquad (I\text{-}9\text{-}44)$$

The result is the equation which is the connecting link of the time-pressure method of flow measurement. The two (positive and negative) values of lambda are associated with the upstream and downstream wave directions.

$$\frac{+g\lambda}{a^2}\frac{dh}{dt} + \frac{dV}{dt} + \frac{fV^2}{2D} = 0 \qquad (I\text{-}9\text{-}45)$$

$$\frac{ds}{dt} = \pm v \qquad (I\text{-}9\text{-}46)$$

The positive value of λ is equation (I-9-45) is valid only for the positive direction of v, and vice versa.

I-9-46 The time-pressure (Gibson) method is really a summation over time of the pressure waves generated by a valve closure, with a correction term added to account for the dissipating effect of pipe friction. The technique of making a test, as described in Pars. I-9-48 through 51, is basically a graphical integration of the pressure measurement recording with an iteration procedure to establish the value of pipe-friction effect in that time interval. Then the pressure recording is reduced by the factor, g/v, to produce the velocity measurement.

$$\int_0^{t_c}\frac{g}{u}\frac{dh}{dt}\,dt + \int_0^{t_c}\frac{fV^2}{2D}\,dt = \int_{V_f}^{V_c} -\,dV \qquad (I\text{-}9\text{-}47)$$

Time, o, is the instant of starting the closing of the valve, and V_f is the velocity of the fluid, at the valve section, up to time 0. Subscript c refers to conditions when the valve has been closed. Thus, t_c is the time interval, in seconds, required to close the valve. Ideally the velocity V_c should be zero, but often there is a leakage which should be measured.

Equation (I-9-47) is the general expression of the time-pressure method. Note that the wave velocity bears directly on the pressure interpretation of velocity, and better precision is obtained if v is measured before each test. Equation (I-9-47) can be simplified to

$$\int_{h_f}^{h_c}\frac{g}{v}\,dh + \int_0^{t_c}\frac{fV^2}{2D}\,dt = \int_{V_f}^{V_c} -\,dV \qquad (I\text{-}9\text{-}48)$$

I-9-47 This method has a peculiar inherent precision in that the pressure generated by the closure of a valve depends on the speed of closing the valve. The only restriction is that the time of the entire test, i.e., t_c, be long compared to $4L/v$. The pressure term can be made much greater than the friction loss term and can be made as much as 20 times larger than the velocity term, and yet it is always proportional to the velocity it is desired to know. The accuracy of the measurement depends also on the mass of fluid in the pipe. Generally, the velocities being measured are not large, and it is the destruction of momentum which generates the pressure signal. For the case when V_c can be taken equal to zero, equation (I-9-48) simplifies to:

$$\frac{g}{v}(h_c - h_f) + \int_0^{t_c}\frac{fV^2}{2D}\,dt = V_f \qquad (I\text{-}9\text{-}49)$$

and

$$q = AV_f \qquad (I\text{-}9\text{-}50)$$

I-9-48 To apply the relations developed above to the determination of the rate of flow in a conduit, it is necessary to record changes of pressure with time by some convenient means. One method has been to photograph the movement of the top of a mercury column which responds to the changes of pressure. A second has been by recording the movement of a piston in a cylinder to which water pressure from the conduit is admitted, the piston working against a calibrated spring. A third method would be with a pressure transducer of the strain gage or piezoelectric crystal type, the output of which is fed to one pen of a two-pen chronograph, the second pen of this chronograph giving second or half-second signals. Regardless of the method, the diagram should show, in addition to the pressure variations as the gate is closed:

1. The average (steady-state) static pressure, p_f, under flowing conditions before starting to close the

gate or valve (p_f being the pressure of the fluid column, h_f (Fig. I-9-9).)

2. The pressure changes as the valve is closed.

3. The decaying pressure waves after the valve is closed.

4. The static pressure, p_c, after the valve is closed and the pressure oscillations have ceased (p_c being the pressure of the fluid column h_c).

5. Second or half-second signals by which the time sequence of valve closing and pressure changes can be evaluated.

When a mercury manometer is used to measure the pressure, a diagram, such as that shown in Fig. I-9-13, is produced. The fairly regular decaying waves to the right of the line B-C are due to the damped oscillations of the mercury column; and, because of the high inertia of the mercury, such a

gage shows almost none of the pressure waves set up in the water column. On the other hand, when a pressure pickup unit that has almost zero inertia is used, e.g., a piezoelectric crystal, the resulting graph will show the pressure waves, harmonics as well as primary, which are set up in the water column by the closure of the valve. These pressure waves will not have a uniform period. Also, due to the harmonic pressure waves, the resulting graph will be far from a smooth curve.

I-9-49 For a record produced by photographing a mercury manometer or column, the action of the mercury after the valve is closed is one of free damped vibration. It can be demonstrated that such a vibration can be expressed mathematically as

$$y = e^{-ax} \sin x \qquad \text{(I-9-51)}$$

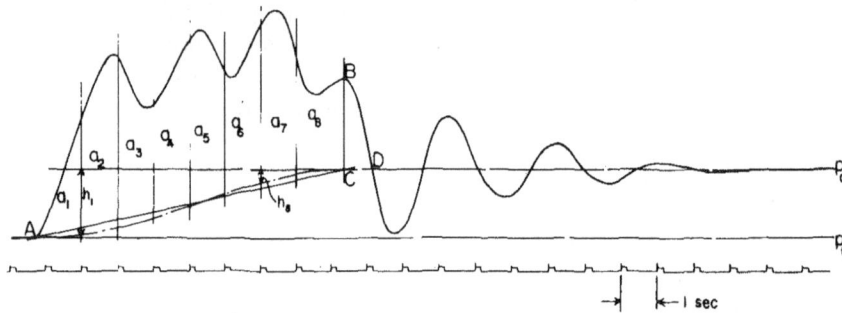

FIG. I-9-13 REPRESENTATION OF A PRESSURE-TIME RECORD OB-
TAINED WITH A MERCURY MANOMETER

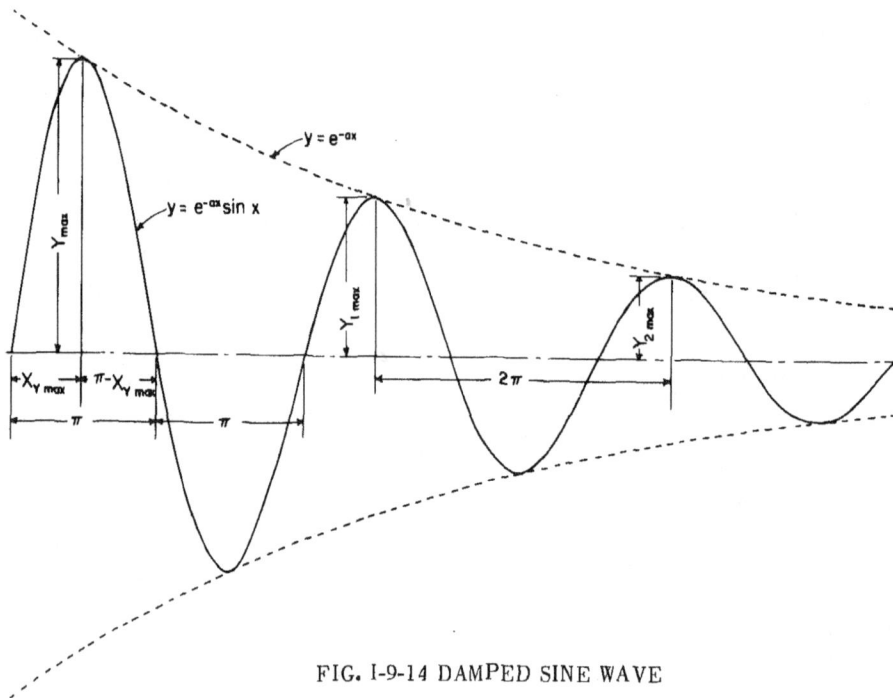

FIG. I-9-14 DAMPED SINE WAVE

The graph of equation (I-9-51) is shown in Fig. I-9-14. It will be noted that this curve is enveloped by the mirror images of the function, $y = e^{-ax}$, and that these two functions must coincide when $\sin x = 1$, or every half-cycle. The location of the value of x for the maximum value of y may be determined by differentiating equation (I-9-51) and equating to zero; thus,

$$\frac{dy}{dx} (e^{-ax} \sin x) = e^{-ax} \cos x + \sin x(-ae^{-ax}) = 0$$

$$= \cos x - a \sin x = 0 \qquad \text{(I-9-52)}$$

From equation (I-9-52)

$$a = \frac{\cos x}{\sin x} = \cot x \qquad \text{(I-9-53)}$$

or the maximum value of y occurs when $x = \cot^{-1} a$.

The value of a may be obtained by considering any two adjacent maximum values of y, such as Y_{1max} and Y_{2max} (Fig. I-9-14). Taking their ratios

$$\frac{Y_{1max}}{Y_{2max}} = \frac{e^{-a(x'' - 2\pi)} \sin (x'' - 2\pi)}{e^{-ax''} \sin x''} \qquad \text{(I-9-54)}$$

Since x'' and $x'' - 2\pi$ are separated by exactly one cycle $\sin (x'' - 2\pi) = \sin x''$ and equation (I-9-54) reduces to

$$\frac{Y_{1max}}{Y_{2max}} = \frac{e^{-a(x'' - 2\pi)}}{e^{-ax''}} = e^{2\pi a} = e^{\theta} \qquad \text{(I-9-55)}$$

where θ is the logarithmic decrement, $\theta = \log_e (Y_{1max}/Y_{2max})$. Substituting this value in equation (I-9-53)

$$\frac{\theta}{2\pi} = \cot x = \frac{1}{2\pi} \log_e (Y_{1max}/Y_{2max}) \qquad \text{(I-9-56a)}$$

or

$$x_{Ymax} = \cot^{-1}\left(\frac{\theta}{2\pi}\right) \qquad \text{(I-9-56b)}$$

The relationship between x_{Ymax} and $\pi - x_{Ymax}/\pi$ is shown in Table I-9-3 and Fig. I-9-15.

Table I-9-3   Computation of $(\pi - x_{Ymax})/\pi$

$\dfrac{Y_{1max}}{Y_{2max}}$	$v = \log_e \dfrac{Y_{1max}}{Y_{2max}}$	$x_{Ymax} = \cot^{-1}\dfrac{v}{2\pi}$	$\dfrac{\pi - x_{Ymax}}{\pi}$
1.0	0	1.57080	0.5000
1.5	0.40547	1.50622	0.5206
2.0	0.69315	1.46084	0.5350
2.5	0.91629	1.42594	0.5461
3.0	1.09861	1.39771	0.5551
3.5	1.25276	1.37387	0.5627
4.0	1.38630	1.35364	0.5691

I-9-50 To complete the determination of the rate of flow in the conduit before starting to close the valve, by graphical integration of the impulse-momentum diagram represented by Fig. I-9-13, proceed as follows:

1. Draw the horizontal lines representing p_f and p_c across the diagram, that for p_f from left to right, that for p_c from right to left.

2. Determine and mark point, A, near the start of the diagram, where the pressure just begins to depart from p_f. This represents the start of the valve closing.

3. Designate as D the point where the graph of decreasing pressure crosses the p_c line.

4. To locate the vertical line B-C, which represents the complete closure of the valve, or end of the diagram:

   (a) Measure the values of Y_{1max} and Y_{2max} in the first two after waves. Using the ratio, Y_{1max}/Y_{2max}, determine from Fig. I-9-15 the value of $(\pi - x_{Ymax})/\pi$.

   (b) Measure the half-cycle displacement of the first after wave, F, using any convenient scale. Multiply this value of F by the value of

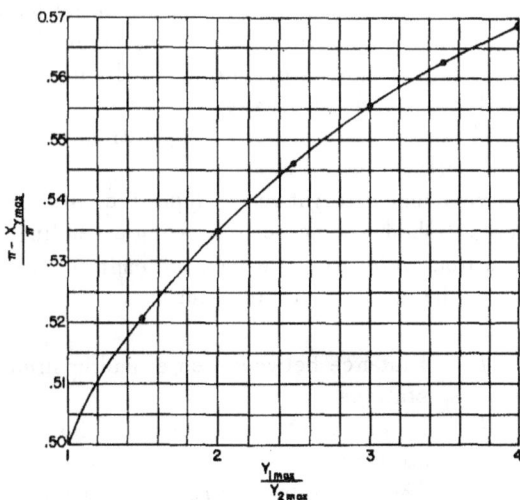

FIG. I-9-15 LOCATION OF X_{Ymax}

$(\pi - x_{Y \max})/\pi$ obtained in (a). This product is the distance D to C.

(c) Locate point C by measuring off the distance C-D on the p_c line. Draw the vertical line through C, locating B, the point on the pressure curve when the valve became closed.

5. Extend the vertical line B-C to intersect the time signal line. Draw a vertical line through A to intersect the time signal line. Connect points A and C with a straight line.

6. Draw a number of vertical lines which will intersect both the pressure curve A-B and the line A-C. (While it may be convenient that these coincide with second signal marks, this is not essential, nor is it essential they be equally spaced.) Designate as a_1, a_2, — the small areas enclosed by these lines — the pressure curve and line A-C.

7. Measure the vertical distance, $y = p_c - p_f$.

8. Measure the areas, a_1, $a_2 \cdots a_n$, using a planimeter or by counting the squares on cross-section paper. Determine the total area, $a' = a_1 + a_2 + \cdots + a_n$. (The units for a as well as for y above may be sq in. and in., sq cm and cm, or other convenient units.)

9. Compute the values of $h_1 = y(1 - a_1/a')^2$, $h_2 = y [1 - (a_1 + a_2)/a']^2$, and so on to $h_n = y [1 - (a_1 + a_a + \cdots a_n)/a']^2$.

10. On the line separating areas a_1 and a_2 measure down from the p_c line a distance equal to h_1. Similarly, on the line between a_2 and a_3 the distance h_2 is measured, and so on. Draw a smooth curve from A through the points thus located to C.

11. Using this curved line as a boundary instead of the straight line A-C, repeat steps (8), (9) and (10) until a curve is located that will coincide with the points located by h_1, h_2, etc.

Note: If the pressure record is made by photographing a mercury column, the theoretical width of the aperture should be zero. Actually, a finite width must be used, and there is a maximum width of slot that may be used and still obtain a true record. Let w represent this maximum width; it must be found experimentally. If a wider slot is used, the diagram obtained will be too large, and a correction for slot width, i.e., a "slot correction" should be made to the areas. This correction is made by deducting from the areas a_1, a_2, — a_n, an area equal to w times the vertical height of the rising (pressure increasing) side, if any, of each of the areas. This of course should be done before proceding with steps (9), (10) and (11).

12. Let $\bar{a} = a_1 + a_2 + \cdots a_n$, as determined by the final curve between A and C, the pressure curve between A and B and the close-off line, B-C. It is expressed in sq in. or other convenient units.

r = Ordinate scale factor — the height in inches on the diagram, representing a change of pressure in the conduit of 1 ft of water

s = Abscissa scale factor — the length in inches on the diagram representing one second of time

L = Length of conduit test section, ft

q = Rate of flow in conduit immediately before the start of valve closing, cfs

Then

$$q = \frac{g_c \, \bar{a} \, A}{r \, s \, L} \qquad \text{(I-9-57)}$$

I-9-51 In the development of equation (I-9-57) and in the preceding discussion of the theory and basis for the equation, it is assumed the valve closed completely, there is no leakage, or, if there is a slight leakage, it is altogether too small to be significant. However, if there is significant leakage, it must be measured and added to the flow rate computed by equation (I-9-42). If the leakage is large compared to the total rate of flow (e.g., 20 per cent thereof), the effect of residual velocity with the valve closed must be taken into account in the integration of equation (I-9-48).

Laboratory studies of the pressure-time method have indicated there may be a Reynolds number effect which would cause the indicated rate to be low. This effect decreases as the Reynolds number is increased and becomes negligible at Reynolds numbers over 100,000.

Sound and Light Velocity Methods

I-9-52 For the most part these methods make use of the difference in the time required for sound or light to travel a given distance in the direction of flow and against the flow. Sound may be used with either liquids or gases but the use of light is limited, practically, to gases.

I-9-53 A procedure and the computations involved may be outlined as follows.

In a very long straight run of pipe three stations are established, the center one being a sending station from which sound waves are emitted, the other two being the receiving stations.

Let

L_1 = Distance between center and upstream stations

L_2 = Distance between center and downstream stations

t_1 and t_2 = The time intervals required for the sound to travel the distances, L_1 and L_2, respectively

V = The average velocity of the fluid which is sought

V_s = The velocity of sound in the fluid when the fluid is not in motion but at the same density as when under the flowing conditions in the line (This must be determined by experiment if not already known.)

Now

$$L_1 = t_1(V_s - V) \qquad (I\text{-}9\text{-}58)$$

and

$$L_2 = t_2(V_s + V) \qquad (I\text{-}9\text{-}59)$$

from which

$$V = \frac{V_s(t_1 - t_2) + (L_2 - L_1)}{(t_1 + t_2)}$$
$$= \frac{L_2 t_1 - L_1 t_2}{2 t_1 t_3} \qquad (I\text{-}9\text{-}60)$$

The units for the distances and velocities may be ft and ft/sec or meters and meters/sec.

For use with sound, piezoelectric crystals have been used both as a source of sound and as receivers [35-37].

I-9-54 The procedure for using a light beam would be somewhat similar to that outlined for sound. The use of a laser beam has received some attention and experimentation.

Floats and Moving Screens

I-9-55 These are applicable to observing the rate of flow of liquids only, water primarily.

Floats are usable in open channels of more or less uniform cross section. In general, one or more floats will be used when a knowledge of the rate of flow to within 10 to 20 per cent will be adequate.

With most floats it is desirable, but not essential, for the channel section used to be straight. However, departures from straightness should be gradual, with no abrupt changes of direction. The channel section should be free of obstructions against which the float might lodge, even momentarily. Also, the stream should be free of ice cover.

The simplest kind of float would be a block of wood, or other floatable material, carried along on the surface of the stream. From the time determined possibly with a stop watch, for the float to be carried over a known length of channel, the apparent

velocity of the stream is obtained. This apparent velocity will be between 5 and 30 per cent higher than the area-average velocity. This large difference between the apparent and area-average velocities may be reduced by using a submerged float. Such a float may be a watertight container which can be filled with ballast, such as shot, to which is attached a flag rod of wood, or tubing, long enough to extend well above the surface of the stream. For use, the quantity of ballast should be adjusted so that, with the increasing buoyancy on the flag rod as the float sinks, the main body of the float will be at about the lower one-third of the stream depth. The apparent velocity is determined by timing the travel of the flag rod over the measured length of channel. Even with this type of float, the apparent velocity will be higher than the area-average velocity by a probable 4 to 10 per cent and possibly as much as 20 per cent.

From the area-average velocity and the average cross-sectional area of the stream, the quantity rate of flow can be computed.

I-9-56 Moving Screens are restricted to use in straight channels of very regular cross section and relatively smooth surface.

In practice a very light, varnished canvas or plastic screen to insure imperviousness is stretched over a stiff frame which is suspended from a wheeled carriage mounted on tracks along the edges of the channel. The carriage is kept light to help keep the frictional drag to a minimum. The screen itself must nearly fill the channel, but a clearance of not more than 1/2 in. may be left between the screen and the channel walls to provide freedom of movement. If the frictional drag of the carriage is kept very low, the leakage area around the screen may be as much as 5 per cent of the total channel cross-sectional area with very little influence on the accuracy of measurement.

The test and inlet sections of the channel must be of a regular and uniform cross section throughout. The length of the inlet section must be sufficient for uniform flow to be established and to allow sufficient distance for the movement of the screen and carriage to become uniform before entering the test section. Beyond the test section, a shorter and possibly slightly larger section must be provided in which the screen can be retracted.

For accurate results, timing of the travel of the carriage through the test section is done by electrical contacts recorded on a chronograph. By providing a number of contacts at equal distances along the test section, the uniformity of the carriage motion can be determined.

I-9-57 From the time of passage and the measured length of the test section the mean velocity of the screen may be determined. In order to obtain the value of the rate of discharge, it is necessary to multiply this mean velocity by the average cross-sectional area of the stream. The width of the stream remains fixed and may be determined accurately by a suitable number of measurements before and after the test. The depth of the stream varies from test to test. In addition, the depth of the stream varies during the test by reason of the difference in head created by the frictional drag of the carriage. For this reason it is necessary to observe that the level of the water has a direct bearing on the discharge measurements. If the discharge measurement, as determined through the test section, is being applied to the flow downstream from the test section, then the average water level on the downstream side of the screen should be used. Conversely, if the test measurement is being used to check the rate of flow from a source upstream from the test and inlet sections, then the average water level on the upstream side of the screen should be used in computing the final discharge.

It is possible to measure the rate of flow of water by this method to within 1 per cent under favorable conditions. In some laboratories special attention is given to maintaining the dimensions and straightness of the channel surfaces to very close tolerances, such that throughout the length of the channel the differences are less than 0.5 per cent. Clearances between the screen and channel surfaces are kept below 0.2 in., and usually the rate of screen travel is kept below 3 ft/sec. Under such conditions it may be possible that rates of flow can be determined to an absolute accuracy to between ± 0.2 and 0.5 per cent.

Miscellaneous Meters and Methods

I-9-58 The fluid meters and methods of fluid measurement described in this report are not all inclusive. There have been developed or proposed numerous meters and instruments for the measurement of fluids and fluid flows, as indicated by the many patents issued year after year. Some of these apply to a special situation, while others are developed to utilize some phenomenon associated with fluid flow. As implied by the reference to patents, almost all of these are proprietary. In the past, a few have be-

come of sufficient use or importance to be included in this and other similar publications. Similarly, future editions of this report will cover the metering equipment that becomes of interest from time to time.

References

[1] "An Alternating Field Induction Flow Meter of High Sensitivity," A. Kolin; *Journal of Applied Physics*, vol. 15, Feb. 1944, p. 150. Also, *Review of Scientific Instruments*, vol. 16, May 1945, p. 109.

[2] "The Theory of Electromagnetic Flow-Measurement," J. A. Shercliff; Cambridge University Press, London, 1962.

[3] "An Investigation of Electromagnetic Flowmeters," H. G. Elrod and R. R. Fouse; *Trans. ASME*, vol. 74, May 1952, p. 589.

[4] "Liquid Metal Magnetic Flowmeters, W. C. Gray and E. R. Astley; *Journal Instrument Society of America*, vol. 1, June 1954, p. 15.

[5] "An Induction Flowmeter Design Suitable for Radioactive Liquids," W. G. James; *Review of Scientific Instruments*, vol. 22, Dec. 1951, p. 989.

[6] "Standard Handbook for Electrical Engineers," A. E. Knalton; McGraw-Hill Book Co., New York, 1949, 8th ed., p. 36.

[7] "Experimental Engineering and Manual, for Testing," R. C. Carpenter and H. Diederichs; John Wiley & Sons, New York, 1915, 7th ed., p. 400.

[8] "The Commercial Metering of Air, Gas and Steam," J. L. Hodgson; *Proceedings of the Institute of Civil Engineers* (London), vol. 204, 1916-1917, p. 108. Also (reprinted), *Instrument Engineering*, Apr. 1954, p. 110; to Oct. 1955, p. 174; to Oct. 1956, p. 41; concl. Apr. 1957, p. 61.

[9] "Flowmeter," J. M. Pearson; U.S. Patent No. 2,624,198, Jan. 6, 1953.

[10] "A Fast Responsive True Mass-Rate Flowmeter," V. T. Li and S. Y. Lee; *Trans. ASME*, vol. 75, July 1953, p. 835.

[11] "Momentum Principle Measures Mass Rate of Flow," V. A. Olando and F. B. Jennings; *Trans. ASME*, vol. 76, Aug. 1954, p. 961.

[12] "Mass Flow Meter," R. B. Dowdell; U. S. Patent No. 3,063,295, Nov. 13, 1962.

[13] "The Measurement of Gases," C. C. Thomas; *Journal of the Franklin Institute*, vol. 172, Nov. 1911, p. 411.

[14] "The Joliet Reference Meter," H. S. Bean, M. E. Benesh and F. C. Witting; *Journal of Research*, National Bureau of Standards, vol. 17, Aug. 1936, R.P. 908.

[15] "The Hot Wire Anemometer: Its Application to the Investigation of the Velocity of Gases in Pipes," J. S. G. Thomas; *Philosophical Magazine*, 6th series, vol. 39, May 1920, p. 505.

[16] "On the Precision Measurement of Air Velocity by Means of the Linear Hot Wire Anemometer," L. V. King; *Philosophical Magazine*, 6th series, vol. 29, Apr. 1915, p. 556.

[17] "Contribution to the Study of Allen Salt-Velocity Method of Water Measurement," M. A. Mason; *Journal of the Boston Society of Civil Engineers*, vol. 27, July 1940, p. 207.

[18] "Modern Equipment for Application of Salt Velocity Method of Discharge Measurement for Performance Test," C. W. Thomas and R. B. Dexter; *Proceedings of the International Association of Hydraulic Research*, The Hague, 1955.

[19] "Discharge Measurements by the Allen Salt Velocity Method," L. J. Hooper; Paper E-1, Symposium of Flow Measurement in Closed Channels, East Kilbride, Glasgow, Sept. 1960.

[20] "Effects of Brine Dispersion by the Allen Salt Velocity Method," L. J. Hooper; *Trans. ASME, Journal Engineering for Power*, vol. 83, Apr. 1961, p. 194.

[21] "The Use of Short Run Tracer Detection Techniques for Pulsative Flow Measurement," C. R. Sparks, G. E. Buss and J. C. Wachel; American Gas Association, Project NQ-15, Orifice Metering of Pulsating Gas Flow, Report of Dec. 31, 1961 (not published).

[22] "Gas Flow Rate Determination by Radioactive Tracer," R. T. Ellington, W. R. Staats and D. V. Kniebes; Institute of Gas Technology, Apr. 1959.

[23] "Radioactive Tracer for Flow Rate Measurement in Natural Gas Pipelines," D. V. Kniebes, P. V. Burket and W. R. Staats; Institute of Gas Technology, July 1959.

[24] "Accurate Measurement of Turbulent Flow in Pipes Using the Isotope Velocity Method and the Effects of Some Restrictions on Optimum Operation," C. G. Clayton, W. E. Clark and A. M. Ball; Paper E-4, Sumposium on Flow Measurement in Closed Channels, East Kilbride, Glasgow, Sept. 1960.

[25] "The Accurate Measurement of Turbulent Flow in Pipes Using Radioactive Isotopes — the Isotope Dilution Method," C. G. Clayton, W. E. Clark and E. A. Spencer; Paper E-3, Symposium on Flow in Closed Channels, East Kilbride, Glasgow, Sept. 1960.

[26] "Tracer Techniques Used in Flow Measurement,"

T. D. Watkins; ASME Flow Measurement Symposium, Sept. 1966, p. 139.

[27] "Steam Flow Measurements," E. G. Bailey; *Journal ASME*, vol. 38, Oct. 1916, p. 775.

[28] "New Method of Water Measurement in Efficiency Tests of 37,500 HP Turbines," N. R. Gibson; *Power*, vol. 53, No. 12, 1921, p. 452.

[29] "Gibson Method and Apparatus for Measuring Flow of Water in Closed Conduits," N. R. Gibson; *Trans. ASME*, vol. 45, 1923, p. 285.

[30] "A Discussion of the Suitability of the Gibson Method and Apparatus for Measuring the Quantity of Water Supplied to the Boulder (Hoover) Dam Turbines during the Efficiency Tests," W. D. Dickinson and L. C. Weathers; U. S. Bureau of Reclamation Technical Memorandum No. 431, Jan. 1935.

[31] "Hydraulics," King, Wisler and Woodburn; J. Wiley & Sons, New York, 5th ed.

[32] "Handbook of Fluid Dynamics," V. Streeter; McGraw-Hill Book Co., New York, 1961, 1st ed., Ch. 20.

[33] "Some Problems Related to an Experience Gained from the Gibson Method of Flow Measurement in Norway, K. Alming; Paper F-1, Symposium on Flow Measurement in Closed Channels, East Kilbride, Glasgow, Sept. 1960.

[34] "Small Scale Experiments on the Gibson Method," N. G. Calvert and J. M. Drabble; Paper F-2, Symposium on Flow Measurement in Closed Channels, East Kilbride, Glasgow, Sept. 1960.

[35] "Ultrasonic Flowmeter," J. Kritz; *Instruments and Automation*, vol. 28, Nov. 1955, p. 1912.

[36] "The Ultrasonic Measurement of Hydraulic Turbine Discharge," R. C. Swengel, W. B. Hess and S. K. Waldorf; *Trans. ASME*, vol. 75, 1953.

[37] "Demonstration of Principles of Ultrasonic Flowmeters," R. C. Swengel, W. B. Hess and S. K. Waldorf; *Electrical Engineering*, vol. 73, Dec. 1954, p. 1082.

Part Two

APPLICATION OF FLUID METERS–
ESPECIALLY DIFFERENTIAL
PRESSURE TYPES

INTRODUCTION

$\mathbf{P}$art II of this edition of *Fluid Meters* presents the recommended conditions, procedures and data for measuring the flow of fluids, particularly with the three principal differential pressure meters: the orifice, the flow nozzle, and the Venturi tube. Included here are the factors and other data that may be needed in computing the flows, using the equations developed in Part I. These data and procedures should provide a degree of accuracy suitable for most fluid measurements, whether it be a commercial transfer measurement or associated with a performance guarantee.

The accuracy of flow measurements made with orifices, flow nozzles, or Venturi tubes depends, in part, upon the values of the coefficients of discharge used in computing the flow, and these are affected by the design and quality of construction of the primary elements. Inasmuch as the coefficients presented here are the results of many thousands of tests made in many different laboratories, both in the U.S. and abroad, under many different operating conditions, it follows that certain tolerances must be applied to cover the spread of these data.

Also, other tolerances must be assigned for other variables, such as the measurements of diameters, pressures, and temperatures. The combination of all of these tolerances is then a measure of the final accuracy one can expect to achieve for a flow measurement made under a particular set of operating conditions. If this tolerance is larger than acceptable for a particular application, then the primary element together with the flow section in which it is mounted, should be individually calibrated.

151

Chapter II-I

Conversion Factors, Constants and Data on Fluids and Materials

II-I-I Pursuant to statute authority, the following *exact* conversion factors between the international metric units (S.I. units) and the foot-pound units have been adopted and published by the National Bureau of Standards [1, 2]:

$$1 \text{ foot} = 0.3048 \text{ meter}$$
$$1 \text{ inch} = 0.0254 \text{ meter} = 2.54 \text{ centimeter}$$
$$1 \text{ pound} = 0.453\ 592\ 37 \text{ kilogram}$$
$$1 \text{ cubic foot} = 28.316\ 846\ 592 \text{ cubic decimeters}$$
$$= 28.316\ 846\ 592 \text{ liters}$$

In addition, the standard atmospheric pressure and acceleration due to gravity are [3], respectively,

$$p_o = 101\ 325 \text{ Newtons/m}^2 = 14.695\ 95 \text{ lb}_f/\text{in.}^2$$
$$g_o = 980.665 \text{ cm/sec}^2 \quad = 32.174\ 06 \text{ ft/sec}^2$$

From the exact relations given above the following conversion factors are derived:

1. For density:

$$\rho \ (\text{gram}_m/\text{cc}) \times 62.427\ 96 = \rho \ (\text{lb}_m/\text{ft}^3) \qquad [\text{II-1-1}]$$

2. For absolute viscosity:

$$\mu \ (\text{gram}_m/\text{sec-cm}) \times 0.067\ 197 = \mu \ (\text{lb}_m/\text{sec-ft})$$
$$(\text{II-I-2})$$

3. For kinematic viscosity:

$$\nu \ (\text{cm}^2/\text{sec}) \times 0.001\ 076\ 39 = \nu \ (\text{ft}^2/\text{sec}) \quad (\text{II-I-3})$$

II-I-2 A graphical comparison of temperature scales is given by Fig. II-I-1. There are tables of equivalents in many handbooks, and conversions of temperatures between Celsius and Fahrenheit scales can be made by the relation:

$$\text{degrees C} = 5/9 \ (\text{degrees Fahrenheit} - 32) \quad (\text{II-I-4}).$$

II-I-3 Frequently pressures measured with liquid manometers must be converted into pounds per square inch. In many such cases it may be sufficient to use a factor based upon some average temperature for the manometer fluid and to take no further account of the effects of temperature. Assuming a manometer temperature of 68 F (20 C), such factors are

$$\text{In. of mercury} \times 0.4893$$
$$= \text{Lb}_f/\text{per sq in} \qquad (\text{II-I-5})$$
$$\text{In. of mercury under water} \times 0.4532$$
$$= \text{Lb}_f/\text{per sq in} \qquad (\text{II-I-6})$$
$$\text{In. of water} \times 0.0361$$
$$= \text{Lb}_f/\text{per sq in} \qquad (\text{II-I-7})$$
$$\text{In. of mercury} \times 13.57$$
$$= \text{In. of water at the}$$
$$\text{same temperature}$$
$$\text{between 45 F and}$$
$$\text{81F} \qquad (\text{II-I-8})$$

In other cases, the conditions may justify the use of a more exact value to be obtained by taking account of the manometer temperature. Such values may be obtained from Fig. II-I-2.

Note: The factors in Fig. II-I-2 give pressures in pounds-force per sq in. at local gravity. If, as a further refinement, it is desired to obtain pounds-force at standard gravity, it will be necessary to multiply the values by g/g_o where g is the value of the local acceleration of gravity and $g_o = 32.174$ ft/sec^2, the standard value of gravitational acceleration.

II-I-4 When differential pressure meters are used to meter fluids at temperatures considerably above or below ordinary room temperatures, the thermal expansion of the primary element must be taken into account. Figure II-I-3 gives the area expansion factors, F_a, to be used when the material of the primary element is steel, steel alloy, bronze or monel. The factor is to be used as a multiplier of a or d^2 in such equations as (I-5-33) and the working equations given later in Par. II-III-39.

II-I-5   Values of the velocity of approach factor, $E = 1/\sqrt{1 - \beta^4}$, used with many differential pressure meters, are given in Table II-I-1 and Figs. II-I-4(a) and (b).

Density, Viscosity and Compressibility Data

II-I-6   The density of mercury, Table II-I-2, is from data published in 1964 by tne National Physical Laboratory, England [4]. The data for Table II-I-3 on the density of dry air are from National Bureau of Standards Circular 564 [5].

II-I-7   The density of water Table II-I-4 is from equations given in the *1967 ASME Steam Tables* [6]. At temperatures and pressures intermediate beween those of the table, linear interpolation may be used.

For all data on the density, specific volume, compressibility and other thermodynamic properties of steam, the *1967 ASME Steam Tables* should be used, and such data are not repeated here.

II-I-8   Values of the viscosity of water are given in Figs. II-I-5 and II-I-6, and those for the viscosity of steam in Fig. II-I-7 are from the ASME Steam Tables. The data on the viscosities of nonhydrocarbon gases, Figs. II-I-8 and II-I-9, are from Circular 564, while those for methane and other hydrocarbon gases are from a report by N.L. Carr [5, 7]. The viscosity of most gases increases slightly with increasing pressure, but the rate of increase is not the same for all gases, as shown by the upper corner insert of Fig. II-I-8.

The most common method of determining the viscosity of liquids is to observe the time in seconds required for a definite volume to pass through a small aperture or a short capillary tube. From this observed

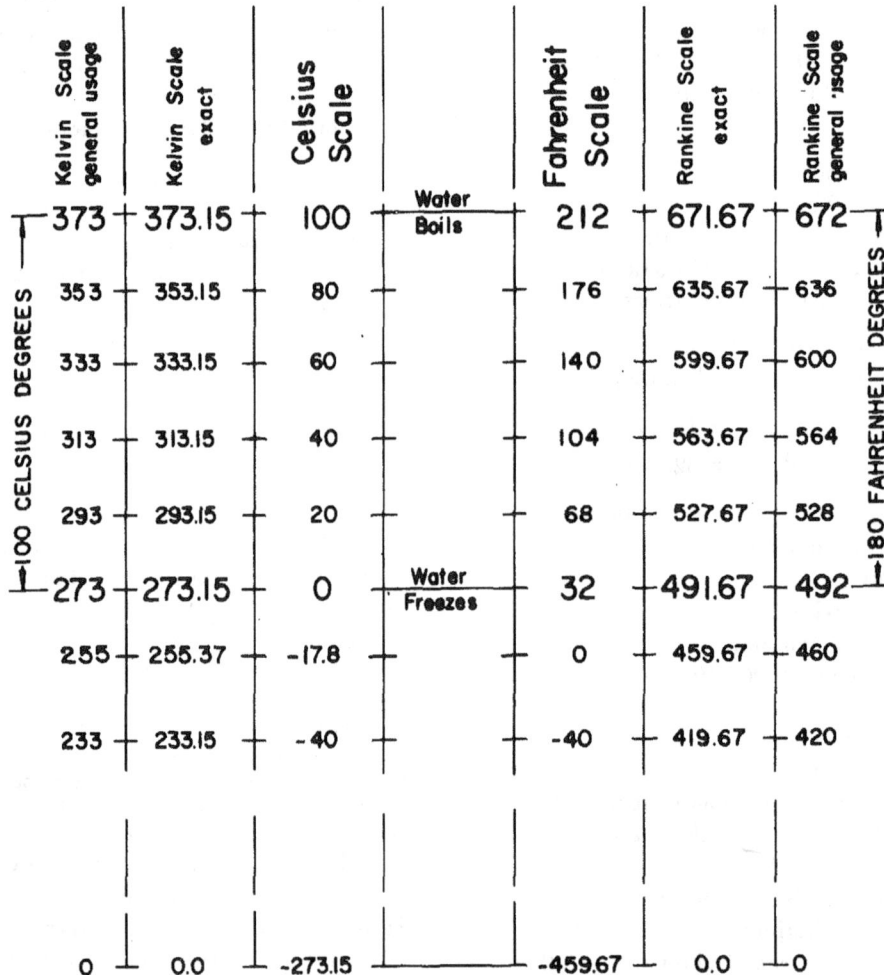

FIG. II-I-1   TEMPERATURE SCALES

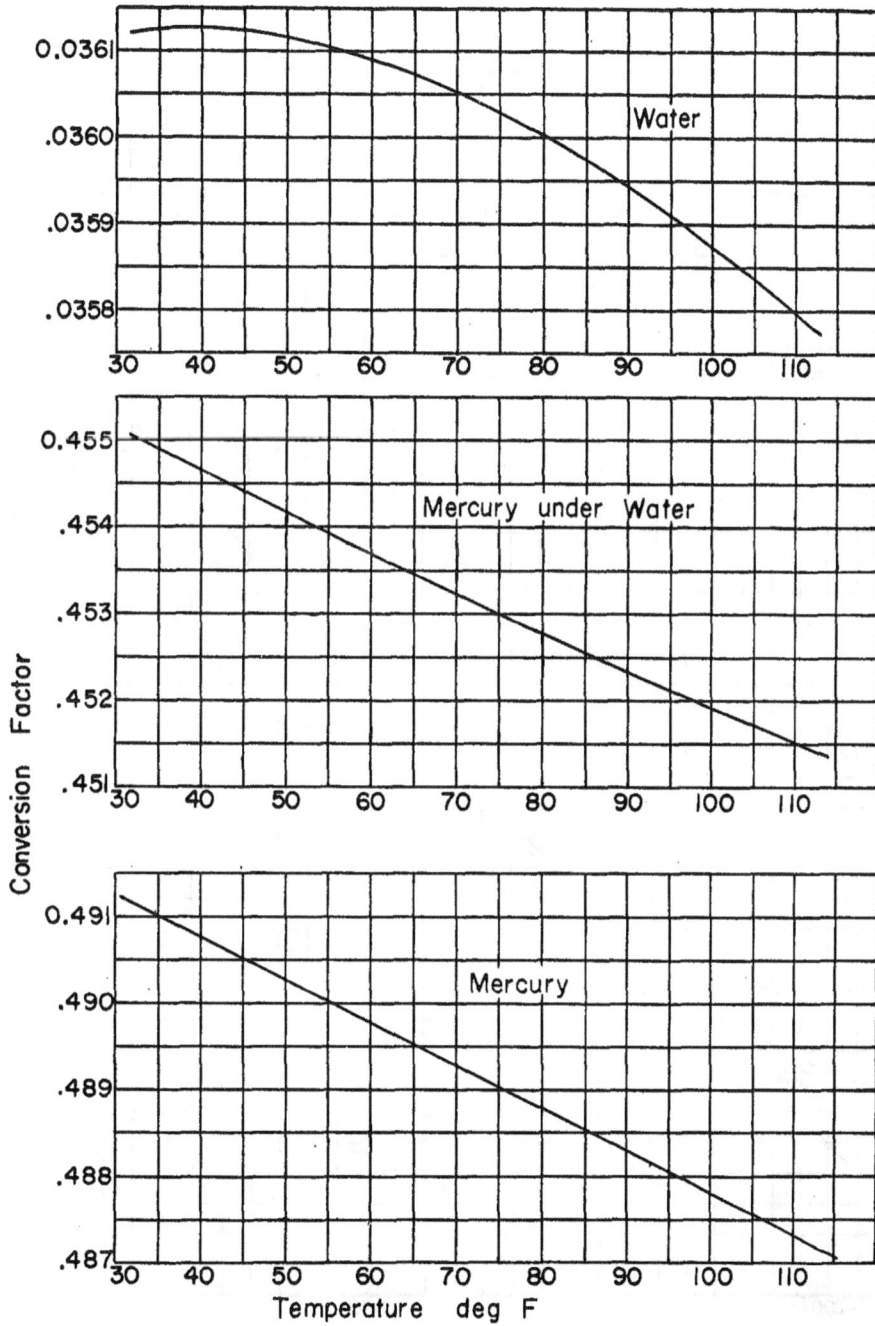

FIG. II-I-2 FACTORS FOR CONVERTING INCHES OF
FLUID TO PSI AT LOCAL GRAVITY

155

time, t, in seconds, the kinematic viscosity is computed by an empirical equation. The empirical equations applying to the four most commonly used viscosimeters give values in terms of *stokes*, i.e., cm²/sec. These equations are as follows:

1. For Saybolt Universal [8, 9]:

$$\nu \ (\text{cm}^2/\text{sec}) = 0.00226t - 1.95/t \qquad \text{(II-I-9)}$$

when $32 < t < 100$

$$\nu \ (\text{cm}^2/\text{sec}) = 0.00220t - 1.35/t \qquad \text{(II-I-10)}$$

when $t > 100$

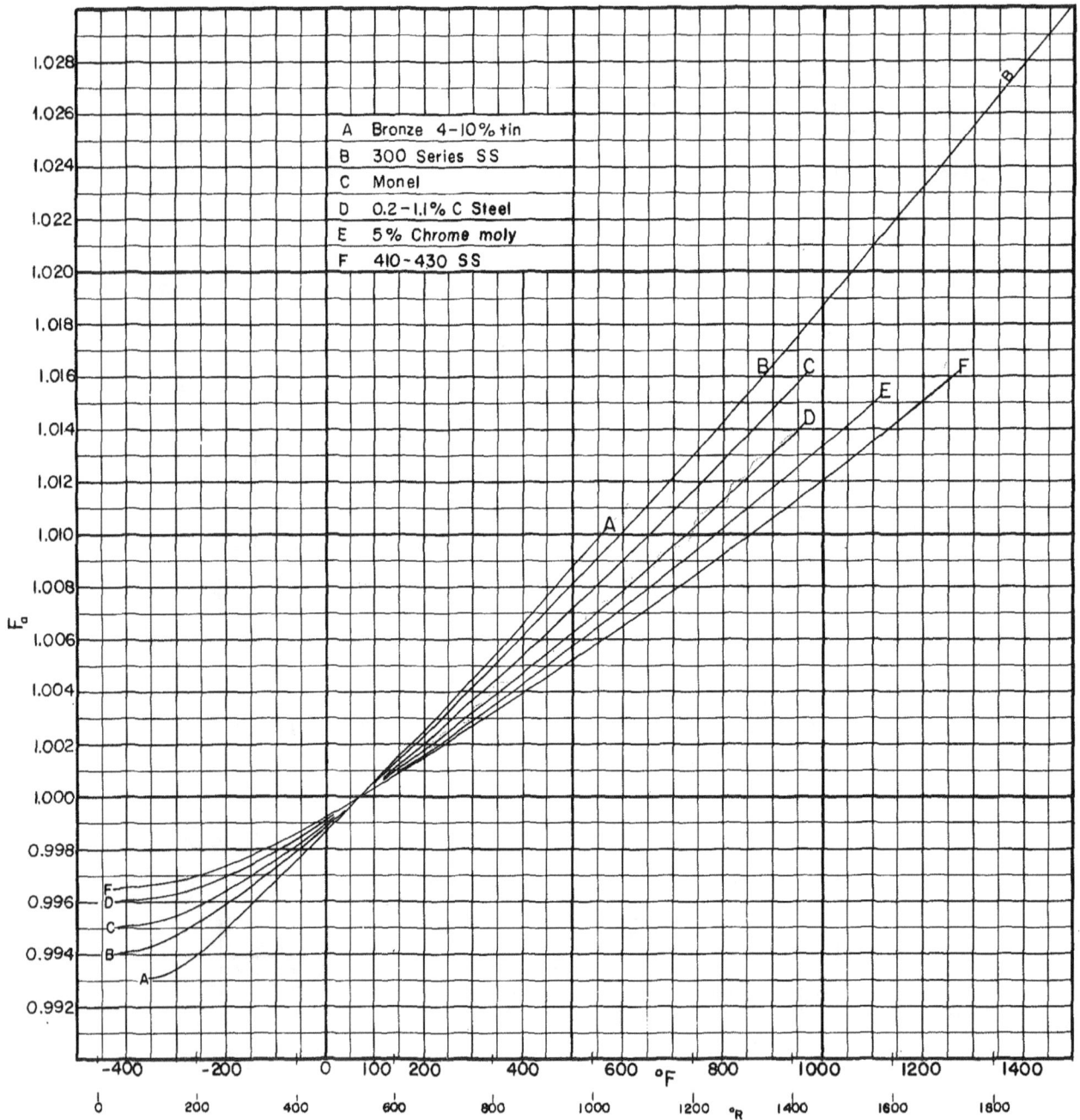

FIG. II-I-3 AREA FACTORS, F_a, FOR THE
THERMAL EXPANSION OF PRIMARY
ELEMENTS

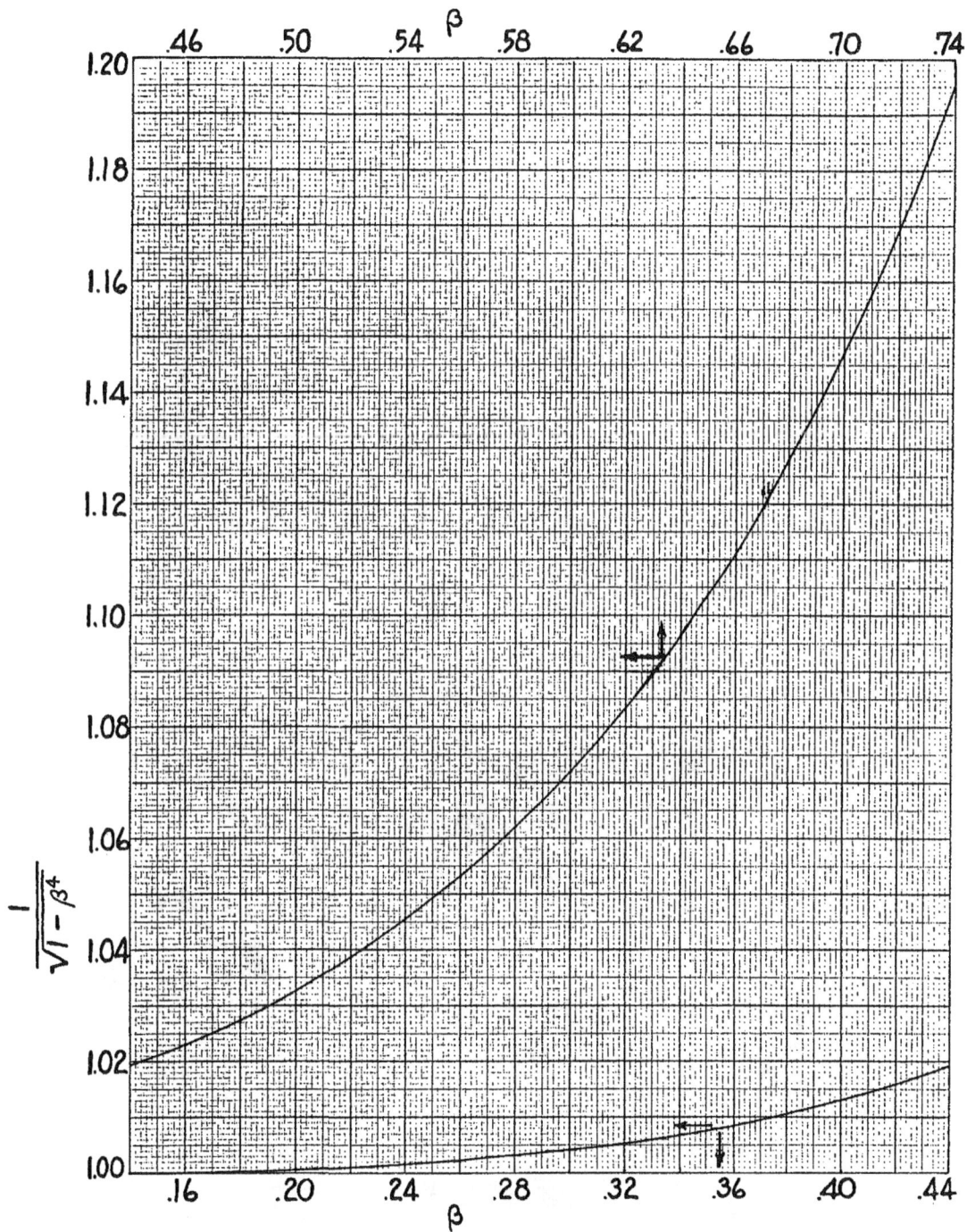

FIG. II-I-4(a) VALUES OF VELOCITY OF
APPROACH FACTOR, $E = 1/\sqrt{1-\beta^4}$

157

2. For Saybolt Furol [10, 11]:

$$\nu \ (\mathrm{cm^2/sec}) = 0.0224t - 1.84/t \qquad \text{(II-I-11)}$$

when $25 < t < 40$

$$\nu \ (\mathrm{cm^2/sec}) = 0.0216t - 0.60/t \qquad \text{(II-I-12)}$$

when $t > 40$

3. For Redwood Standard No. 1 [11]:

$$\nu \ (\mathrm{cm^2/sec}) = 0.00260t - 1.79/t \qquad \text{(II-I-13)}$$

when $34 < t < 100$

$$\nu \ (\mathrm{cm^2/sec}) = 0.00247t - 0.50/t \qquad \text{(II-I-14)}$$

when $t > 100$

4. For Engler [9]:

$$\nu \ (\text{·   ·sec}) = 0.00147t - 3.74/t \qquad \text{(II-I-15)}$$

ʒad of using the equations, values of ν may be
ʒed from the curves of Fig. II-I-10.

ʒedwood and Engler viscosimeters are used very
little in this country.

Attention is called to the fact that the values of ν
obtained by any of the above equations are at best
only close approximations and sometimes may be in

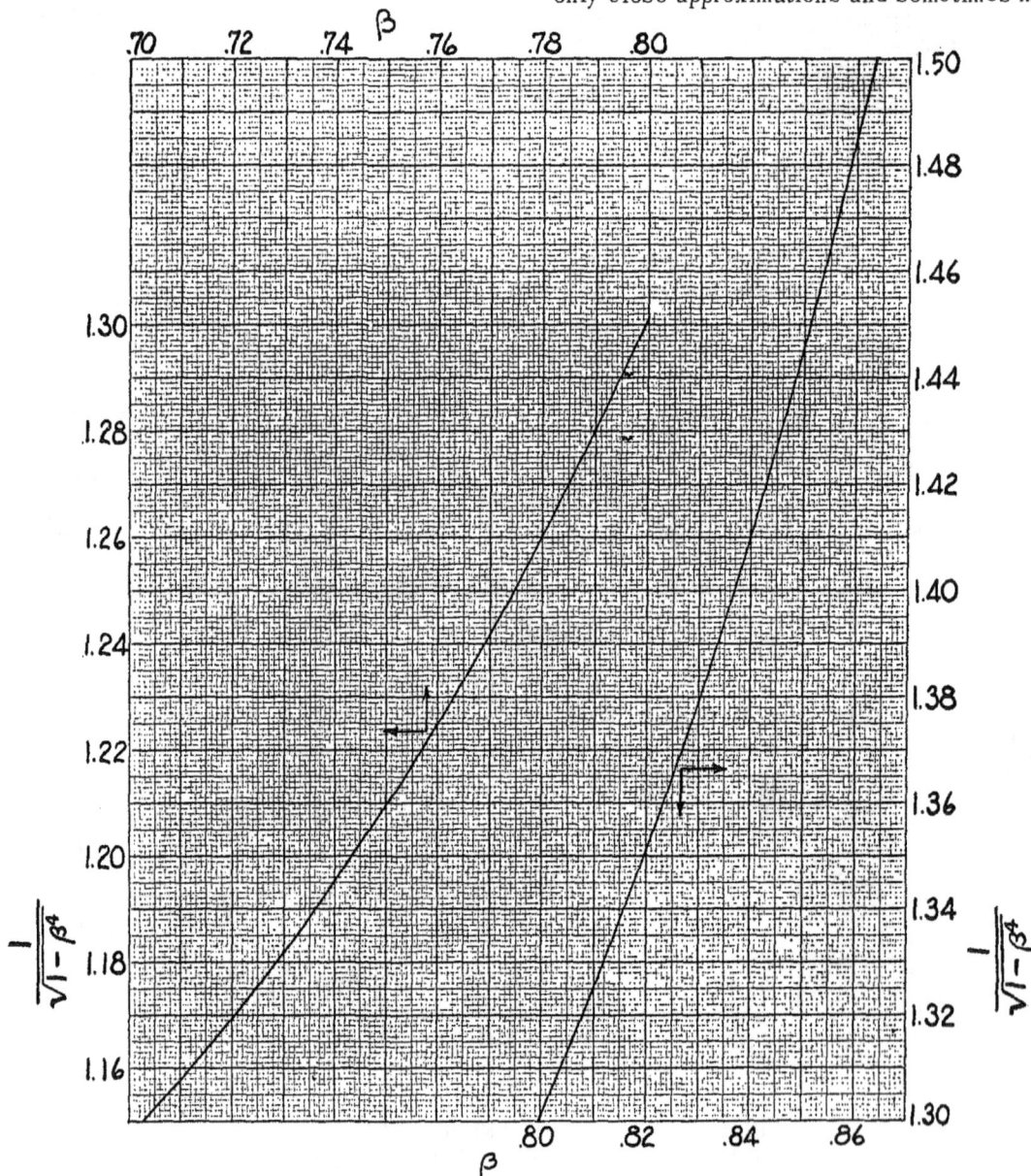

FIG. II-I-4(b) VALUES OF VELOCITY OF
APPROACH FACTOR, $E = 1/\sqrt{1-\beta^4}$

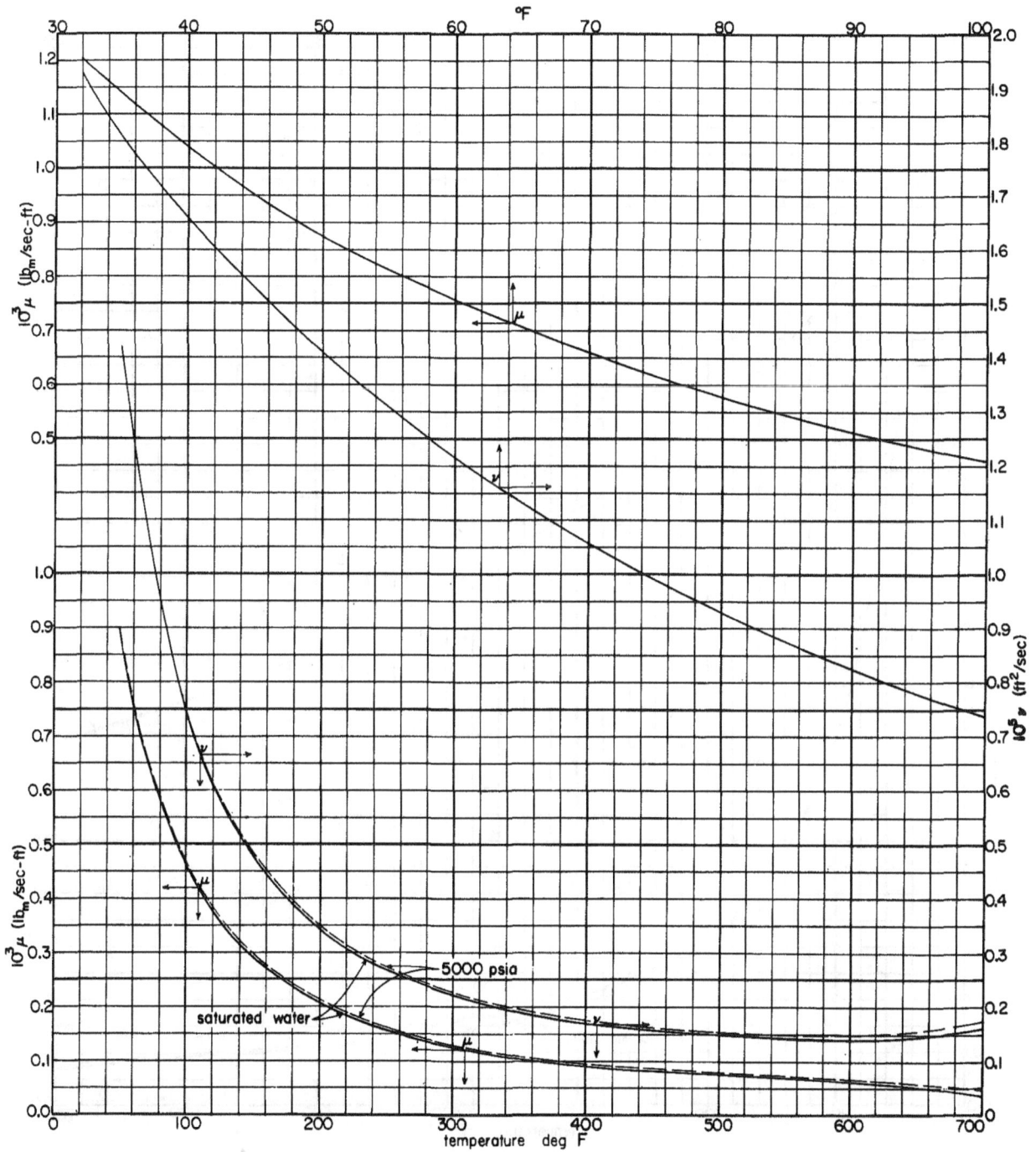

FIG. II-I-5 VISCOSITY OF WATER, μ, IN LB_m/
SEC-FT AND KINEMATIC VISCOS-
ITY, ν, IN FT^2/SEC

159

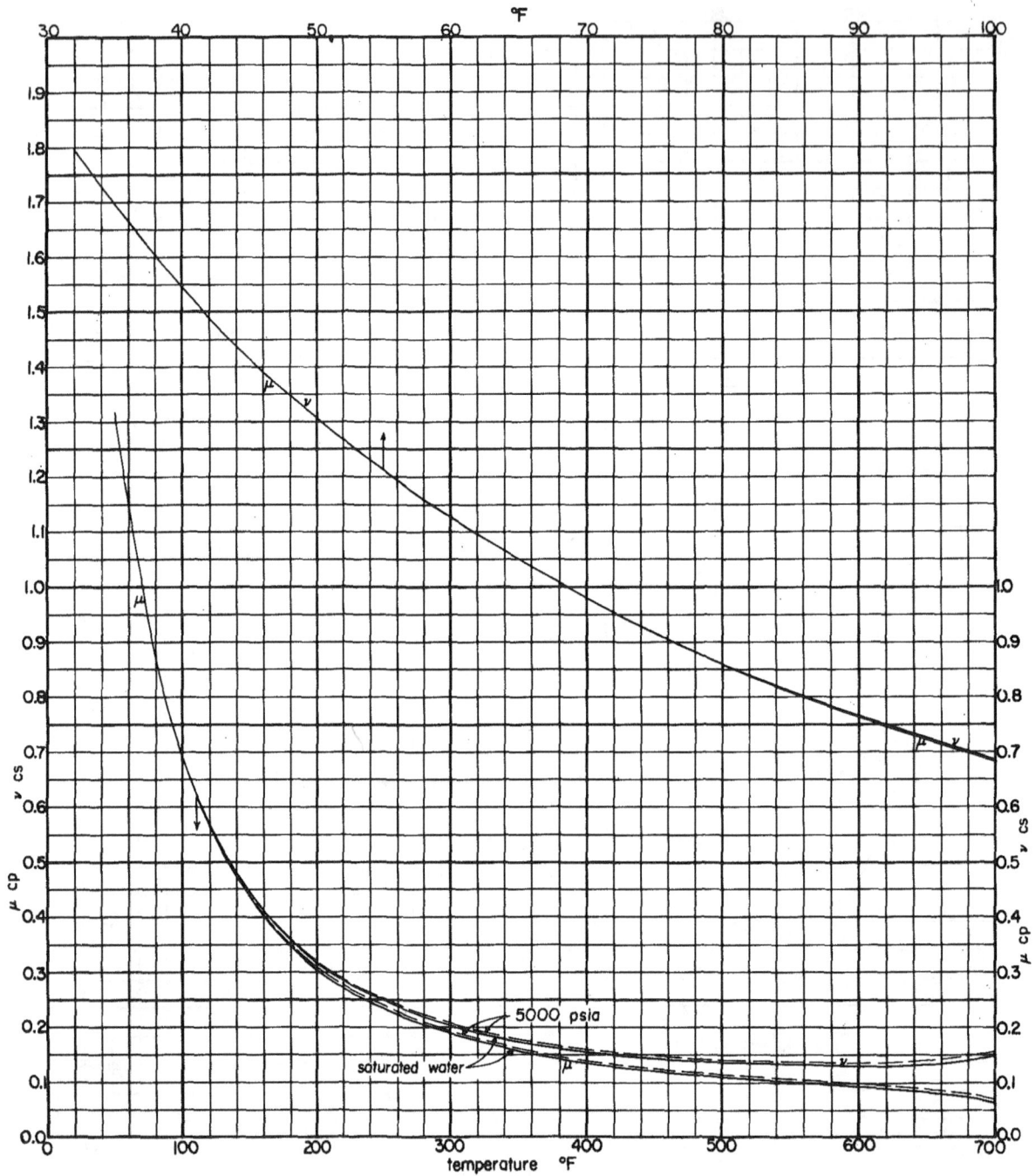

FIG. II-I-6 VISCOSITY OF WATER, μ, IN
CENTIPOISE AND KINEMATIC
VISCOSITY, ν, IN CENTISTOKE

160

error by 2 to 3 per cent. The principal reason for this is that the manner in which the viscosimeters are used does not provide for a sufficiently accurate measurement of the temperature and, therefore, also of the volume of the liquid as it flows through the metering passage. The fact that calculations at the time breaks of 100 sec or 40 sec may not give identical results with the two equations for a particular instrument will cause no significant error. There are tables that may be used in place of the equations [8, 9]. Should it be necessary to determine more exact values of ν, reference should be made to some of the literature on viscosity determinations [12–15].

As stated in Chapter I-3, the name of the metric unit for the coefficient of absolute viscosity, or dynamic viscosity, is "poise" and that for kinematic viscosity is "stoke." The corresponding quantities in ft-lb units have no generally accepted names. In industrial practice in this country as well as in technical laboratories, the use of viscosity values in poise (or centipoise) and stokes (or centistokes) in place of the ft-lb units has become common. For this reason most of the viscosity data given here are in both systems of units.

II-I-9 The curves giving the compressibilities of air, hydrogen and carbon dioxide, Figs. II-I-11, II-I-12 and II-I-13, are from data in Circular 564. Figure II-I-14, on the compressibility of methane, is based on data correlated by Beitler, Darrow and Zimmerman, combined with data given by Din [16, 17, 18, 19].

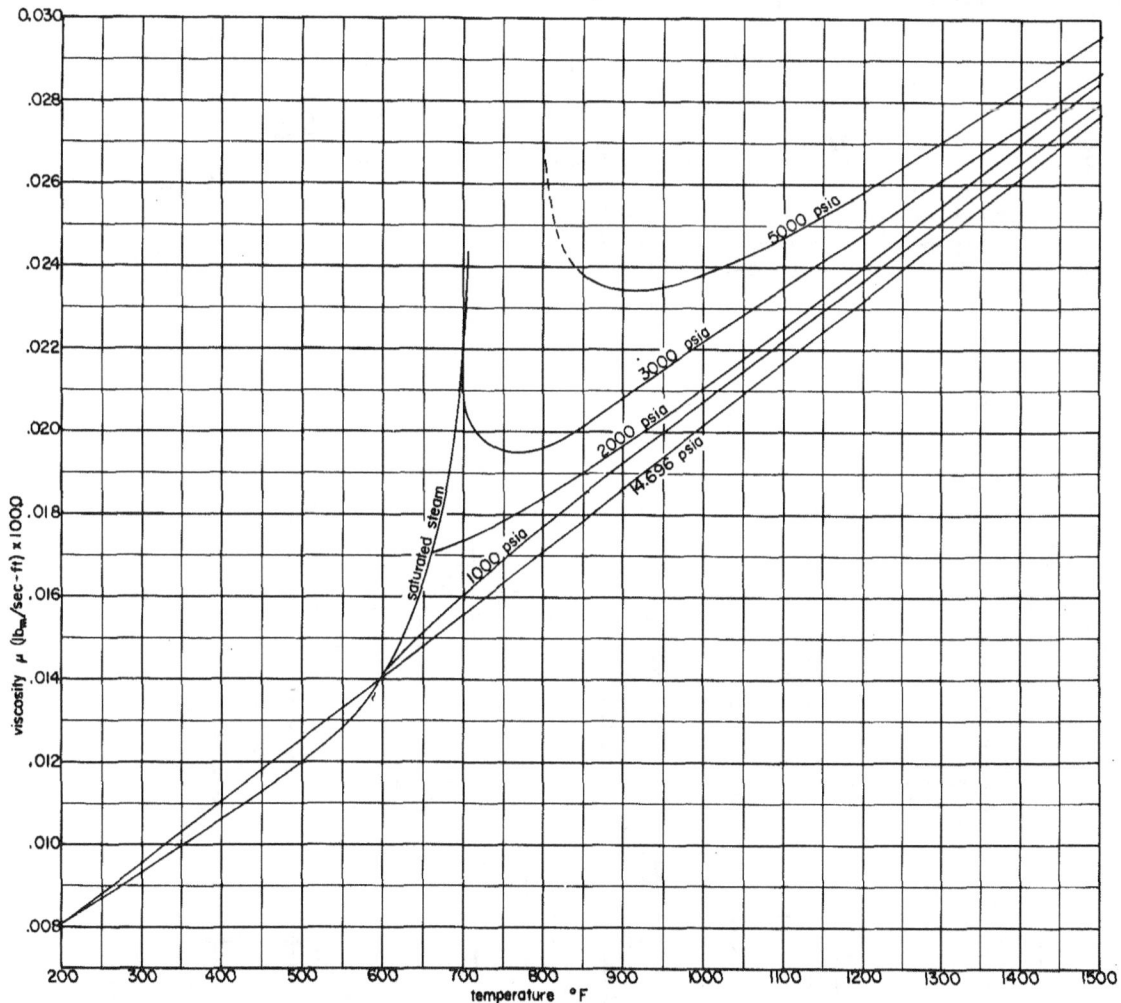

FIG. II-I-7 VISCOSITY OF STREAM, μ, IN
LB$_m$/SEC-FT

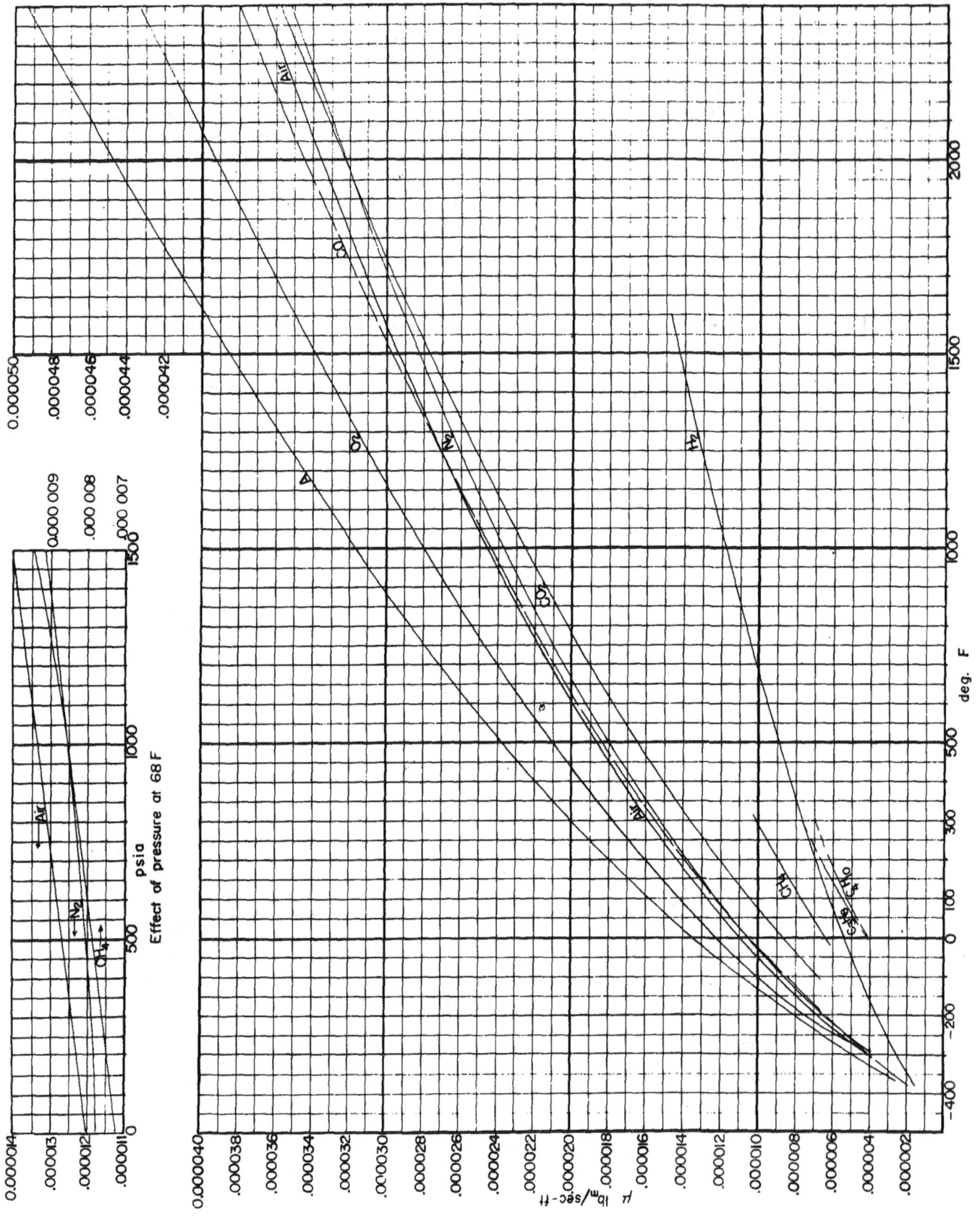

FIG. II-I-8 VISCOSITY OF GASES, μ, IN LB$_m$/SEC-FT

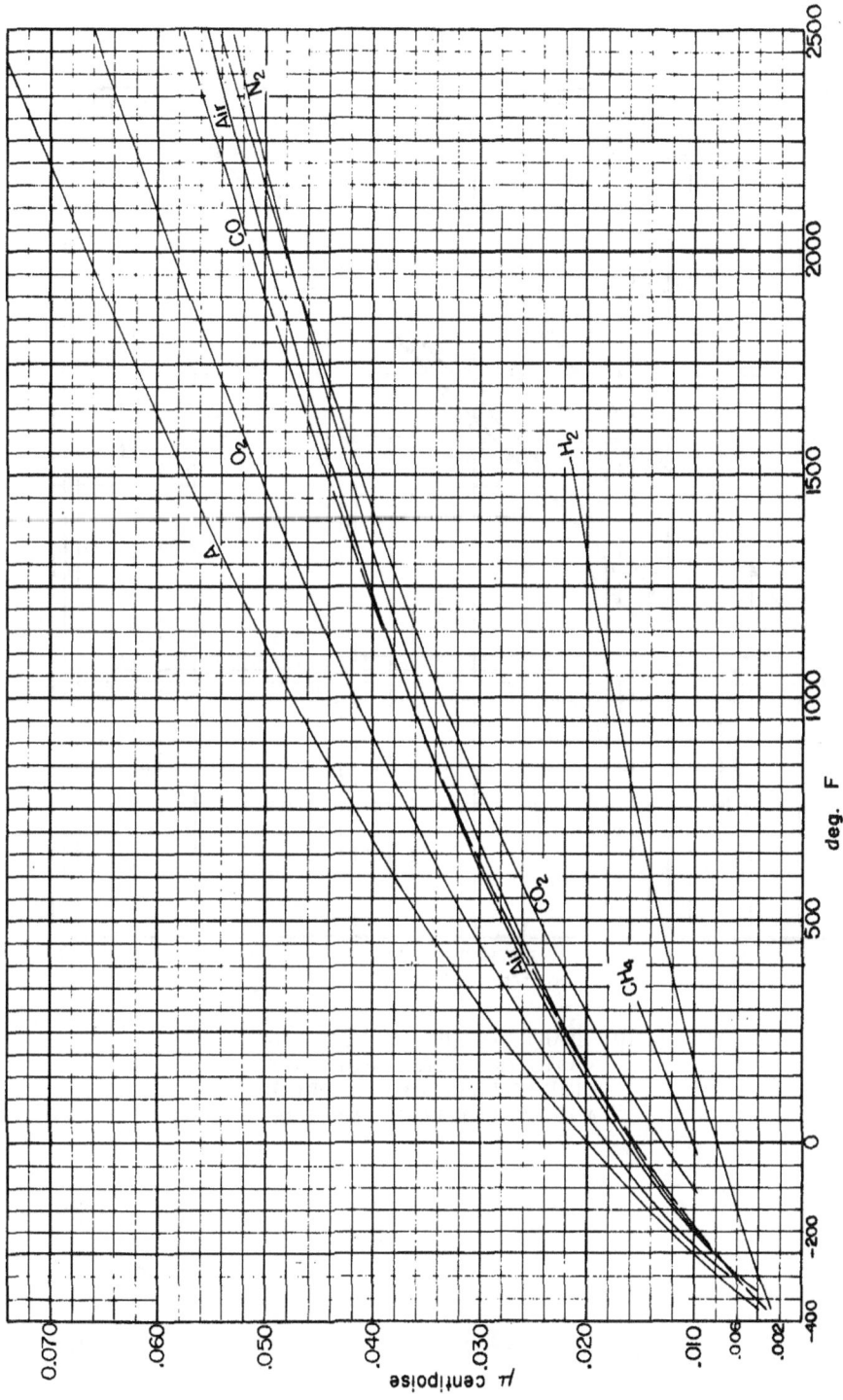

FIG. II-I-9 VISCOSITY OF GASES AT 1
ATMOSPHERE, μ IN CENTIPOISE

163

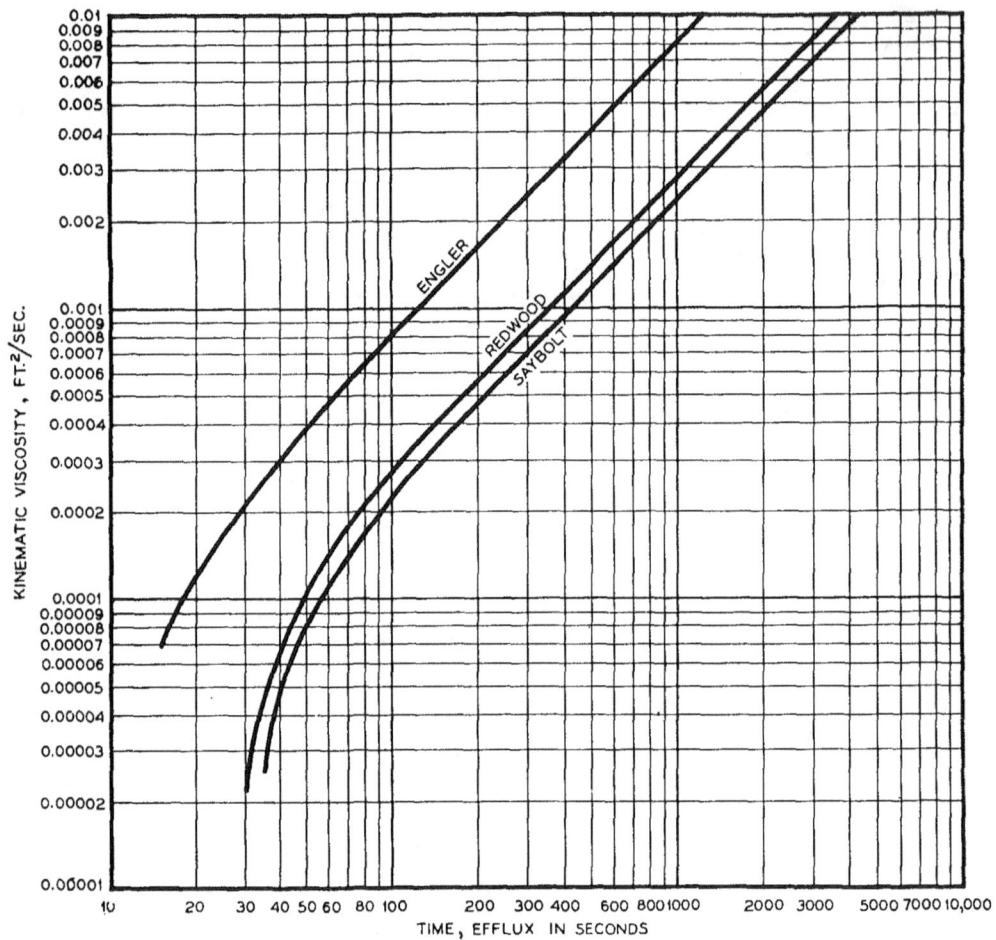

FIG. II-I-10 CHART FOR OBTAINING KINE-
MATIC VISCOSITY FROM FLOW
TIME FOR SAYBOLT, REDWOOD
AND ENGLER VISCOSIMETERS

164

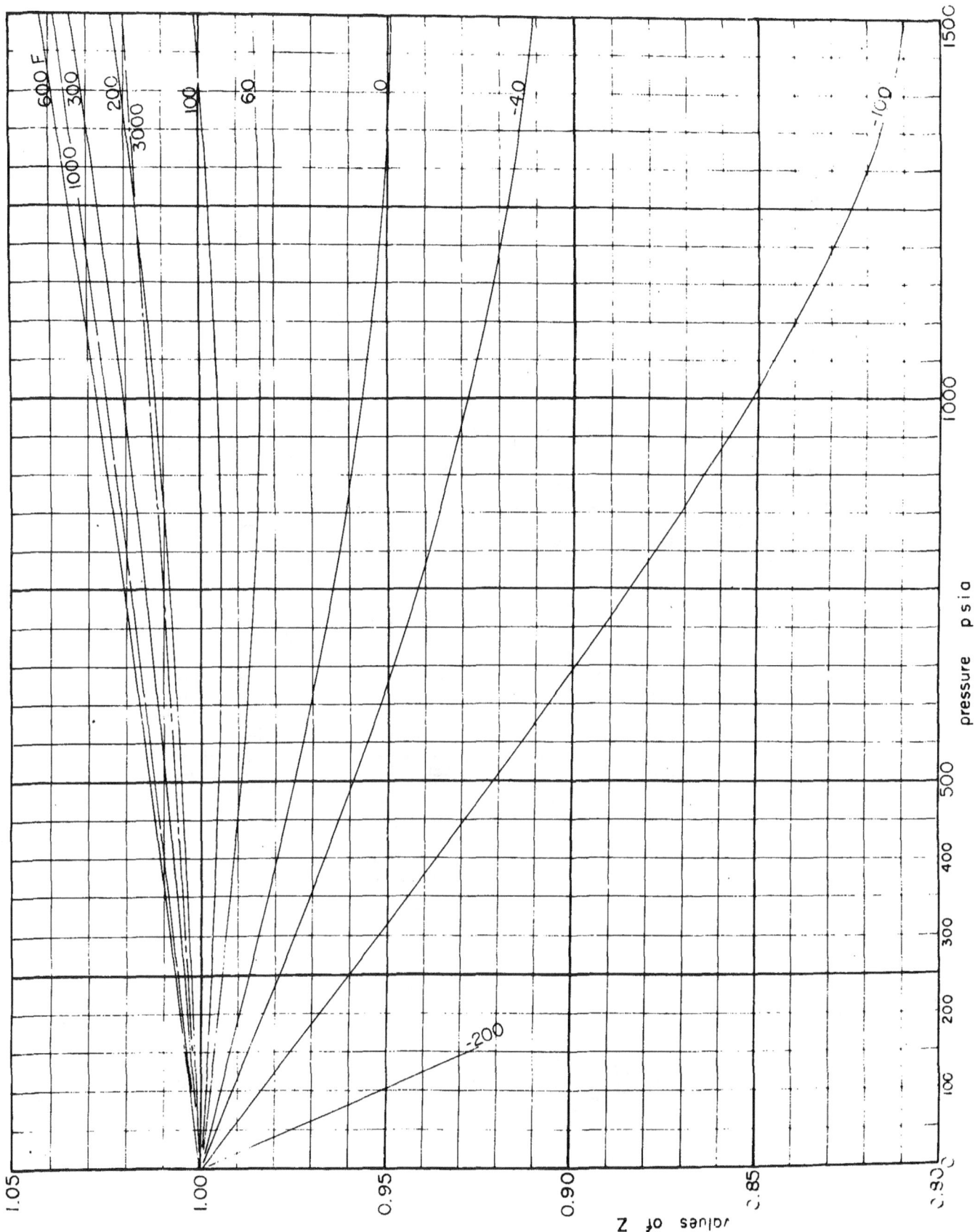

FIG. II-I-11 COMPRESSIBILITY FACTOR, Z,
OF DRY AIR

165

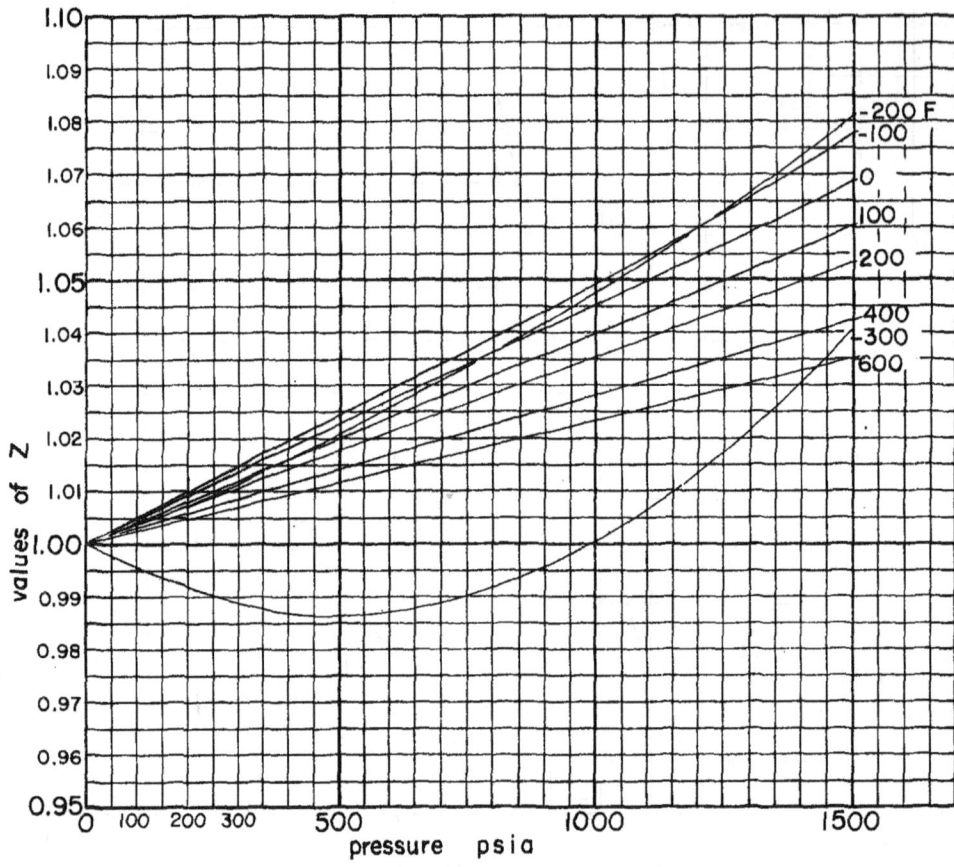

FIG. II-I-12 COMPRESSIBILITY FACTOR, Z,
OF NORMAL HYDROGEN

166

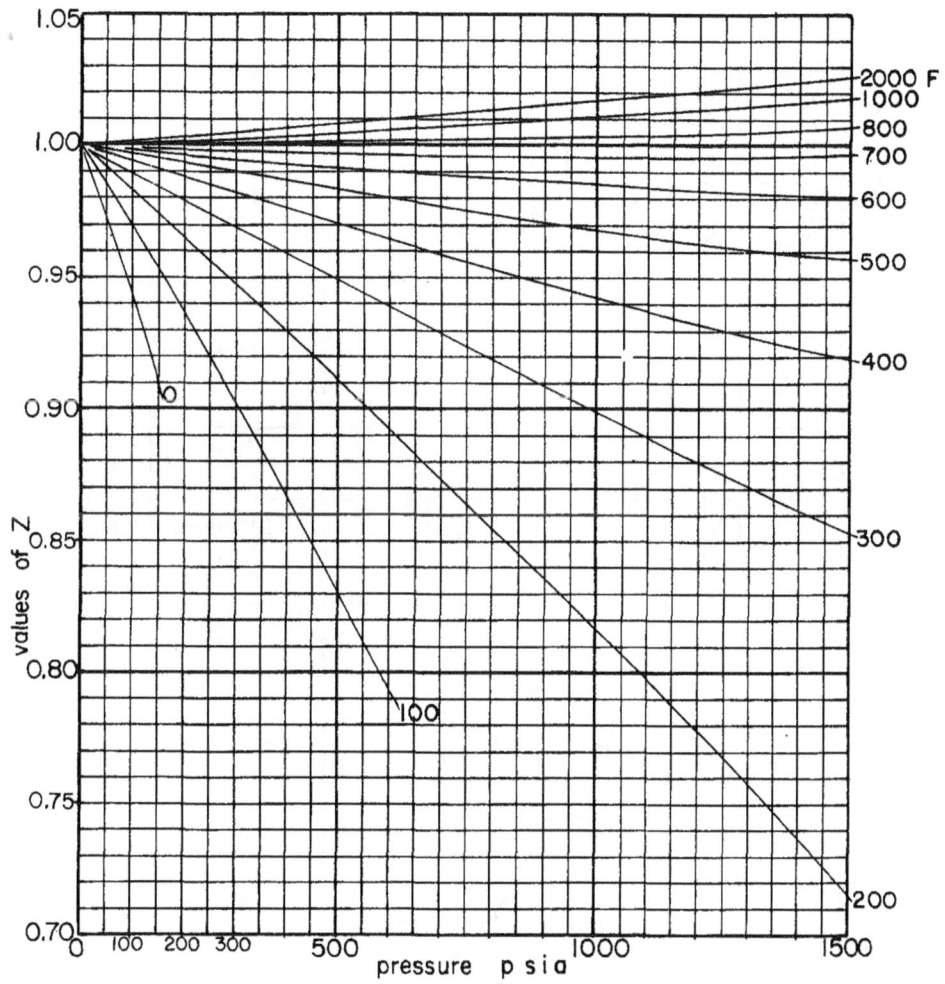

FIG. II-I-13 COMPRESSIBILITY FACTOR, Z,
OF CARBON DIOXIDE

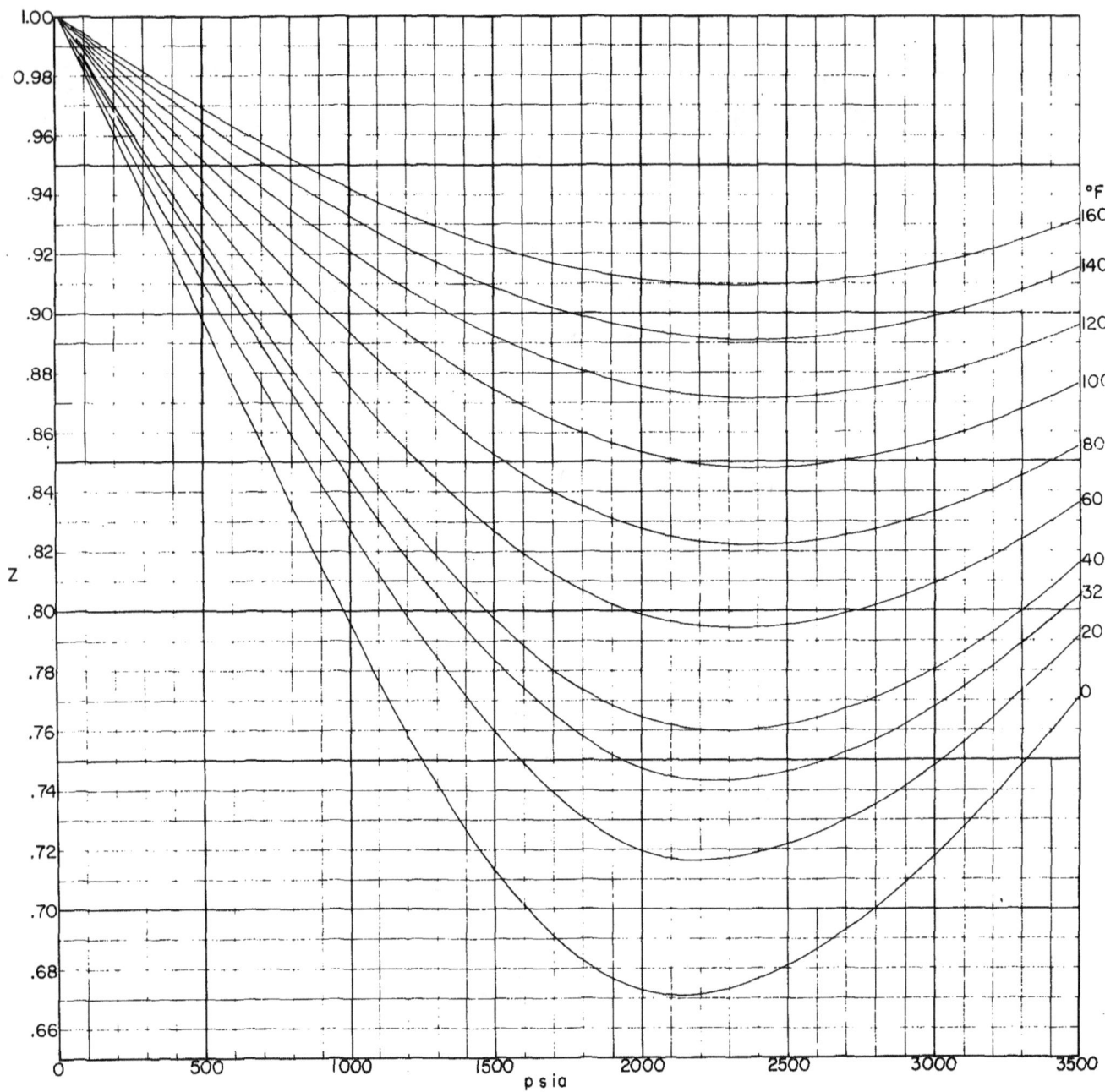

FIG. II-I-14 COMPRESSIBILITY FACTOR, Z,
OF METHANE

Table II-I-1 Velocity of Approach Factors

β	$\dfrac{1}{\sqrt{1-\beta^4}}$	β	$\dfrac{1}{\sqrt{1-\beta^4}}$	β	$\dfrac{1}{\sqrt{1-\beta^4}}$	β	$\dfrac{1}{\sqrt{1-\beta^4}}$
0.100	1.000 04	0.625	1.086 30	0.725	1.175 47	0.770	1.241 81
.150	1.000 25	.630	1.089 48	.726	1.176 72	.771	1.243 57
.200	1.000 80	.635	1.092 77	.727	1.177 97	.772	1.245 34
.220	1.001 17	.640	1.096 17	.728	1.179 23	.773	1.247 12
.240	1.001 66	.645	1.099 68	.729	1.180 50	.774	1.248 92
.260	1.002 29	.650	1.103 31	.730	1.181 78	.775	1.250 73
.280	1.003 08	.652	1.104 79	.731	1.183 07	.776	1.252 56
.300	1.004 07	.654	1.106 30	.732	1.184 37	.777	1.254 41
.320	1.005 28	.656	1.107 82	.733	1.185 67	.778	1.256 27
.340	1.006 74	.658	1.109 37	.734	1.186 99	.779	1.258 14
.350	1.007 58	.660	1.110 93	.735	1.188 32	.780	1.260 03
.360	1.008 50	.662	1.112 52	.736	1.189 66	.781	1.261 94
.370	1.009 50	.664	1.114 13	.737	1.191 01	.782	1.263 86
.380	1.010 59	.666	1.115 76	.738	1.192 36	.783	1.265 80
.390	1.011 77	.668	1.117 41	.739	1.193 73	.784	1.267 76
.400	1.013 05	.670	1.119 09	.740	1.195 11	.785	1.269 73
.410	1.014 43	.672	1.120 78	.741	1.196 50	.786	1.271 72
.420	1.015 93	.674	1.122 50	.742	1.197 90	.787	1.273 72
.430	1.017 54	.676	1.124 25	.743	1.199 31	.788	1.275 75
.440	1.019 28	.678	1.126 02	.744	1.200 73	.789	1.277 79
.450	1.021 15	.680	1.127 81	.745	1.202 16	.790	1.279 85
.460	1.023 16	.682	1.129 63	.746	1.203 60	.791	1.281 92
.470	1.025 32	.684	1.131 47	.747	1.205 05	.792	1.284 02
.480	1.027 64	.686	1.133 33	.748	1.206 52	.793	1.286 13
.490	1.030 13	.688	1.135 23	.749	1.207 99	.794	1.288 26
.500	1.032 79	.690	1.137 15	.750	1.209 48	.795	1.290 41
.510	1.035 64	.692	1.139 09	.751	1.210 98	.796	1.292 58
.520	1.038 69	.694	1.141 06	.752	1.212 50	.797	1.294 77
.530	1.041 95	.696	1.143 06	.753	1.214 02	.798	1.296 97
.540	1.045 43	.698	1.145 09	.754	1.215 55	.799	1.299 20
.550	1.049 15	.700	1.147 15	.755	1.217 10	.800	1.301 45
.555	1.051 10	.702	1.149 23	.756	1.218 65	.801	1.303 72
.560	1.053 12	.704	1.151 35	.757	1.220 22	.802	1.306 00
.565	1.055 20	.706	1.153 50	.758	1.221 81	.803	1.308 31
.570	1.057 36	.708	1.155 67	.759	1.223 40	.804	1.310 64
.575	1.059 58	.710	1.157 88	.760	1.225 01	.805	1.312 99
.580	1.061 88	.712	1.160 12	.761	1.226 63	.806	1.315 36
.585	1.064 26	.714	1.162 39	.762	1.228 26	.807	1.317 76
.590	1.066 71	.716	1.164 69	.763	1.229 91	.808	1.320 18
.595	1.069 24	.718	1.167 03	.764	1.231 57	.809	1.322 61
.600	1.071 86	.720	1.169 40	.765	1.233 24	.810	1.325 08
.605	1.074 56	.721	1.170 59	.766	1.234 93	.811	1.327 56
.610	1.077 36	.722	1.171 80	.767	1.236 63	.812	1.330 07
.615	1.080 24	.723	1.173 02	.768	1.238 34	.813	1.332 60
.620	1.083 22	.724	1.174 24	.769	1.240 07	.814	1.335 15

Table II-I-2 Density of Mercury

Temperature (°F)	ρ (lb_m/ft³)	Temperature (°F)	ρ (lb_m/ft³)	Temperature (°F)	ρ (lb_m/ft³)	Temperature (°F)	ρ (lb_m/ft³)
- 5	851.888 14	35	848.456 54	75	845.047 35	115	841.656 89
- 4	.802 05	36	.371 01	76	844.962 45	116	.571 98
- 3	.715 98	37	.286 11	77	.876 92	117	.487 71
- 2	.629 92	38	.200 59	78	.792 02	118	.403 43
- 1	.543 88	39	.115 06	79	.707 12	119	.318 53
0	.457 85	40	.029 53	80	.622 21	120	.234 25
1	.371 84	41	847.944 01	81	.537 31	121	.149 97
2	.285 84	42	.858 48	82	.452 41	122	.065 07
3	.199 63	43	.772 95	83	.367 51	123	840.980 79
4	.113 90	44	.688 05	84	.282 61	124	.896 51
5	.027 95	45	.602 53	85	.197 70	125	.811 61
6	850.942 01	46	.517 00	86	.112 80	126	.727 33
7	.856 09	47	.432 10	87	.027 90	127	.643 06
8	.770 19	48	.346 57	88	843.943 00	128	.558 15
9	.684 30	49	.261 04	89	.858 10	129	.473 88
10	.598 43	50	.175 52	90	.773 19	130	.389 60
11	.512 57	51	.089 99	91	.688 29	131	.305 32
12	.426 73	52	.005 09	92	.604 01	132	.221 04
13	.340 90	53	846.919 56	93	.519 11	133	.136 77
14	.255 09	54	.834 66	94	.434 21	134	.052 49
15	.169 56	55	.749 13	95	.349 31	135	839.968 21
16	.084 04	56	.664 23	96	.264 41	136	.883 93
17	849.998 51	57	.578 71	97	.179 50	137	.799 65
18	.912 36	58	.493 80	98	.095 23	138	.715 38
19	.826 83	59	.408 28	99	.010 32	139	.631 10
20	.740 68	60	.323 38	100	842.925 42	140	.546 82
21	.655 16	61	.238 47	101	.841 14	145	.125 43
22	.569 63	62	.152 95	102	.756 24	150	838.704 67
23	.483 48	63	.068 05	103	.671 96	155	.283 90
24	.397 95	64	845.983 14	104	.587 06	160	837.863 76
25	.312 43	65	.897 62	105	.502 16	165	.443 62
26	.226 90	66	.812 72	106	.417 88	170	.023 48
27	.141 37	67	.727 19	107	.332 98	175	836.603 97
28	.055 22	68	.642 29	108	.248 70	180	.184 45
29	848.969 70	69	.557 39	109	.163 80	185	835.765 56
30	.884 17	70	.472 48	110	.079 52	190	.346 67
31	.798 64	71	.387 58	111	841.994 62	195	834.927 78
32	.713 12	72	.302 68	112	.910 34	200	.508 88
33	.627 59	73	.217 15	113	.825 44	205	.090 62
34	.542 07	74	.132 25	114	.741 16	210	833.672 97
						212	833.505 67

Table II-I-3 Density, ρ, of Dry Air at a Pressure of 1 Atm (14.696 psia)

Temperature ($^\circ$R)	($^\circ$F)	ρ (lb$_m$/ft^3)	Temperature ($^\circ$R)	($^\circ$F)	ρ (lb$_m$/ft^3)
189.67	-270	0.212 663	559.67	100	0.070 890
199.67	-260	.201 515	569.67	110	.069 639
209.67	-250	.191 546	579.67	120	.068 436
			589.67	130	.067 274
219.67	-240	0.182 545	599.67	140	.066 144
229.67	-230	.174 376			
239.67	-220	.166 909	609.67	150	0.065 062
249.67	-210	.160 064	619.67	160	.064 013
259.67	-200	.156 141	629.67	170	.062 996
			639.67	180	.062 011
269.67	-190	0.147 972	649.67	190	.061 058
279.67	-180	.142 612			
289.67	-170	.137 583	659.67	200	0.060 122
299.67	-160	.132 933	669.67	210	.059 226
309.67	-150	.128 583	679.67	220	.058 354
			689.67	230	.057 507
319.67	-140	0.124 522	699.67	240	.056 683
329.67	-130	.120 623			
339.67	-120	.117 112	709.67	250	0.055 884
349.67	-110	.113 722	719.67	260	.055 101
359.67	-100	.110 541	729.67	270	.054 350
			739.67	280	.053 616
369.67	- 90	0.107 530	749.67	290	.052 897
379.67	- 80	.104 681			
389.67	- 70	.101 976	759.67	300	0.052 203
399.67	- 60	.099 401	769.67	310	.051 525
409.67	- 50	.096 963	779.67	320	.050 863
			789.67	330	.050 217
419.67	- 40	0.094 639	799.67	340	.049 596
429.67	- 30	.092 427			
439.67	- 20	.090 328	809.67	350	0.048 966
449.67	- 10	.088 294	819.67	360	.048 377
459.67	0	.086 365	829.67	370	.047 796
			839.67	380	.047 231
469.67	10	0.084 524	849.67	390	.046 674
479.67	20	.082 757			
489.67	30	.081 045			
491.67	32	.080 722 3			
499.67	40	.079 431			
509.67	50	0.077 865			
519.67	60	.076 355			
529.67	70	.074 918			
539.67	80	.073 522			
549.67	90	.072 182	859.67	400	0.046 125

171

Table II-I-4 Density of Saturated and Compressed Liquid Water (lb_m/ft^3)

Temperature (°F)	Pressure (psia)						
	Saturated	500	1000	1500	2000	2500	3000
32	62.4140	62.5217	62.6288	62.7355	62.8415	62.9470	63.0519
33	.4167	.5240	.6308	.7370	.8426	.9477	.0522
34	.4191	.5260	.6324	.7382	.8434	.9480	.0521
35	62.4212	62.5277	62.6336	62.7390	62.8438	62.9481	63.0517
36	.4229	.5289	.6345	.7395	.8439	.9478	.0510
37	.4242	.5299	.6351	.7397	.8437	.9471	.0500
38	.4252	.5305	.6353	.7395	.8432	.9462	.0487
39	.4258	.5308	.6352	.7390	.8423	.9450	.0471
40	62.4261	62.5307	62.6348	62.7383	62.8412	62.9435	63.0453
41	.4261	.5304	.6341	.7372	.8397	.9417	.0431
42	.4257	.5297	.6330	.7358	.8380	.9396	.0407
43	.4251	.5287	.6317	.7342	.8360	.9373	.0380
44	.4241	.5274	.6301	.7322	.8337	.9347	.0350
45	62.4229	62.5258	62.6282	62.7300	62.8312	62.9318	63.0318
46	.4213	.5239	.6260	.7275	.8283	.9286	.0284
47	.4194	.5218	.6235	.7247	.8252	.9252	.0246
48	.4173	.5193	.6208	.7216	.8219	.9216	.0207
49	.4149	.5166	.6178	.7183	.8183	.9177	.0165
50	62.4122	62.5136	62.6145	62.7148	62.8144	62.9135	63.0120
51	.4092	.5104	.6110	.7109	.8103	.9091	.0073
52	.4059	.5068	.6072	.7069	.8060	.9045	.0024
53	.4024	.5031	.6031	.7026	.8014	.8996	62.9973
54	.3986	.4990	.5988	.6980	.7966	.8946	.9919
55	62.3946	62.4947	62.5943	62.6932	62.7915	62.8892	62.9864
56	.3903	.4902	.5895	.6882	.7863	.8837	.9806
57	.3858	.4854	.5845	.6829	.7808	.8780	.9746
58	.3810	.4804	.5793	.6775	.7750	.8720	.9684
59	.3760	.4752	.5738	.6718	.7691	.8658	.9620
60	62.3707	62.4697	62.5681	62.6658	62.7630	62.8595	62.9554
61	.3652	.4640	.5622	.6597	.7566	.8529	.9485
62	.3595	.4581	.5560	.6533	.7500	.8461	.9415
63	.3535	.4519	.5497	.6468	.7432	.8391	.9343
64	.3474	.4455	.5431	.6400	.7363	.8319	.9269
65	62.3410	62.4390	62.5363	62.6330	62.7291	62.8245	62.9194
66	.3344	.4322	.5293	.6258	.7217	.8170	.9116
67	.3275	.4251	.5221	.6185	.7142	.8092	.9036
68	.3205	.4179	.5147	.6109	.7064	.8013	.8955
69	.3132	.4105	.5071	.6031	.6984	.7931	.8872

Table II-I-4 (Continued)

Temper- ature (°F)	Pressure (psia)						
	Saturated	500	1000	1500	2000	2500	3000
70	62.3058	62.4029	62.4993	62.5952	62.6903	62.7848	62.8787
71	.2981	.3950	.4914	.5870	.6820	.7763	.8700
72	.2902	.3870	.4832	.5787	.6735	.7677	.8612
73	.2822	.3788	.4748	.5701	.6648	.7588	.8522
74	.2739	.3704	.4663	.5614	.6559	.7498	.8430
75	62.2654	62.3618	62.4575	62.5525	62.6469	62.7406	62.8337
76	.2568	.3530	.4486	.5435	.6377	.7313	.8242
77	.2479	.3440	.4395	.5342	.6283	.7217	.8145
78	.2389	.3349	.4302	.5248	.6188	.7121	.8047
79	.2297	.3255	.4207	.5152	.6090	.7022	.7947
80	62.2203	62.3160	62.4111	62.5055	62.5992	62.6922	62.7846
81	.2107	.3063	.4013	.4955	.5891	.6820	.7743
82	.2009	.2964	.3913	.4854	.5789	.6717	.7638
83	.1910	.2864	.3811	.4752	.5685	.6612	.7532
84	.1809	.2762	.3708	.4647	.5580	.6505	.7424
85	62.1706	62.2658	62.3603	62.4542	62.5473	62.6397	62.7315
90	.1166	.2113	.3055	.3988	.4915	.5835	.6748
95	.0585	.1529	.2467	.3397	.4320	.5236	.6145
100	61.9964	.0906	.1841	.2769	.3689	.4602	.5508
105	.9307	.0246	.1180	.2105	.3023	.3934	.4838
110	61.8612	61.9551	62.0483	62.1408	62.2325	62.3234	62.4136
115	.7884	.8821	61.9754	.0678	.1594	.2502	.3404
120	.7121	.8059	.8992	61.9916	.0832	.1740	.2641
125	.6326	.7265	.8198	.9123	.0040	.0949	.1850
130	.5500	.6440	.7375	.8301	61.9219	.0129	.1031
135	61.4643	61.5584	61.6521	61.7450	61.8369	61.9281	62.0184
140	.3757	.4700	.5640	.6570	.7492	.8406	61.9311
145	.2842	.3787	.4730	.5663	.6588	.7504	.8412
150	.1899	.2847	.3793	.4730	.5658	.6577	.7488
155	.0928	.1880	.2830	.3770	.4702	.5624	.6538
160	60.9932	61.0887	61.1841	61.2786	61.3721	61.4647	61.5565
165	.8909	60.9868	.0827	.1776	.2716	.3647	.4568
170	.7862	.8824	60.9789	.0743	.1687	.2622	.3549
175	.6789	.7756	.8726	60.9686	.0635	.1575	.2506
180	.5693	.6665	.7640	.8605	60.9560	.0506	.1442
185	60.4573	60.5549	60.6531	60.7502	60.8463	60.9414	61.0356
190	.3430	.4411	.5400	.6377	.7344	.8301	60.9249
195	.2265	.3250	.4246	.5230	.6204	.7167	.8121
200	.1076	.2068	.3070	.4062	.5042	.6012	.6972
205	59.9866	.0863	.1873	.2872	.3860	.4837	.5803

Table II-I-4 (Continued)

Temperature (°F)	Pressure (psia)						
	Saturated	500	1000	1500	2000	2500	3000
210	59.8635	59.9636	60.0655	60.1662	60.2657	60.3641	60.4615
215	.7382	.8389	59.9416	.0430	.1433	.2425	.3406
220	.6108	.7120	.8156	59.9179	.0190	.1190	.2178
225	.4813	.5830	.6875	.7907	59.8927	59.9935	.0932
230	.3497	.4520	.5574	.6615	.7644	.8661	59.9666
235	59.2161	59.3189	59.4253	59.5304	59.6342	59.7367	59.8381
240	.0804	.1838	.2912	.3973	.5020	.6055	.7077
245	58.9428	.0467	.1551	.2622	.3679	.4723	.5755
250	.8031	58.9075	.0171	.1252	.2319	.3373	.4415
255	.6614	.7663	58.8770	58.9862	.0940	.2005	.3056
260	58.5177	58.6231	58.7350	58.8453	58.9542	59.0617	59.1678
265	.3720	.4779	.5910	.7025	.8125	58.9211	.0283
270	.2244	.3306	.4450	.5577	.6689	.7786	58.8869
275	.0747	.1814	.2970	.4110	.5234	.6343	.7437
280	57.9231	.0301	.1471	.2624	.3761	.4881	.5987
285	57.7695	57.8768	57.9952	58.1118	58.2268	58.3401	58.4519
290	.6139	.7215	.8413	57.9593	.0756	.1902	.3032
295	.4563	.5641	.6854	.8048	57.9225	.0385	.1528
300	.2966	.4046	.5275	.6484	.7675	57.8848	.0005
305	.1350	.2431	.3675	.4900	.6106	.7293	57.8463
310	56.9713	57.0795	57.2056	57.3296	57.4517	57.5719	57.6903
315	.8056	56.9137	.0415	.1672	.2908	.4126	.5324
320	.6378	.7459	56.8754	.0028	.1281	.2513	.3727
325	.4680	.5758	.7072	56.8363	56.9633	.0882	.2111
330	.2960	.4036	.5369	.6678	.7965	56.9230	.0476
335	56.1220	56.2291	56.3644	56.4972	56.6277	56.7560	56.8821
340	55.9458	.0524	.1897	.3245	.4568	.5869	.7148
345	.7674	55.8735	.0128	.1496	.2839	.4158	.5454
350	.5869	.6922	55.8337	55.9726	.1088	.2427	.3742
355	.4042	.5085	.6523	.7934	55.9317	.0675	.2009
360	55.2192	55.3225	55.4687	55.6119	55.7524	55.8902	56.0255
365	.0320	.1340	.2826	.4282	.5709	.7108	55.8482
370	54.8424	54.9430	.0942	.2422	.3872	.5293	.6688
375	.6506	.7495	54.9033	.0538	.2012	.3456	.4872
380	.4563	.5534	.7099	54.8630	.0129	.1597	.3035
385	54.2597	54.3546	54.5140	54.6698	54.8223	54.9715	55.1177
390	.0606	.1531	.3155	.4742	.6293	.7810	54.9296
395	53.8590	53.9489	.1144	.2760	.4338	.5882	.7393
400	.6548	.7418	53.9105	.0751	.2359	.3930	.5467
405	.4481	.5318	.7039	53.8717	.0354	.1954	.3518

Table II-I-4 (Continued)

Temperature (°F)	Pressure (psia)						
	Saturated	500	1000	1500	2000	2500	3000
410	53.2387	53.3187	53.4944	53.6655	53.8324	53.9953	54.1545
415	.0267	.1026	.2819	.4565	.6267	.7927	53.9548
420	52.8119	52.8833	.0665	.2447	.4183	.5875	.7527
425	.5942	.6607	52.8480	.0300	.2071	.3796	.5480
430	.3737	.4348	.6262	52.8122	52.9930	.1691	.3407
435	52.1503	52.2053	52.4012	52.5913	52.7760	52.9557	53.1307
440	51.9238	51.9723	.1728	.3673	.5560	.7395	52.9181
445	.6942	.7354	51.9409	.1399	.3329	.5204	.7027
450	.4615	.4948	.7054	51.9092	.1066	.2983	.4844
455	.2255	.2501	.4661	.6749	51.8770	.0730	.2633
460	50.9862	51.0012	51.2229	51.4370	51.6441	51.8446	52.0391
465	.7434	50.7479	50.9757	.1954	.4076	.6129	51.8118
470	.4971	.4971	.7243	50.9499	.1675	.3778	.5814
475	.2472	.2472	.4686	.7003	50.9236	.1392	.3477
480	49.9935	49.9935	.2082	.4465	.6758	50.8970	.1106
485	49.7359	49.7359	49.9431	50.1884	50.4240	50.6510	50.8700
490	.4744	.4744	.6731	49.9257	.1680	.4011	.6258
495	.2087	.2087	.3978	.6582	49.9077	.1473	.3779
500	48.9387	48.9387	.1170	.3857	.6427	49.8892	.1261

Table II-I-5 Physical Data on Some Common Commercial Gases

Gas Name	Formula	Molecular Weight (basis C^{12})	Density at 32 F and 14.696 psia (lb_m/ft^3) ρ_o	Specific Gravity (see par. I-3-23) G	Ratio Specific Heats (see Note (e)) γ	Boiling Point at 14.696 psia (°R) T_b	Critical Temperature (°R) T_c	Critical Pressure (psia) p_c	Critical Volume (ft^3/lb_m) v_c
Air		28.9644(a)	0.0807223(b)	1.00000	1.41	142.0	238.4	547	0.0517
Argon	Ar	39.948		1.3792	1.67	157.4	272.08	705.4	.0301
Acetylene	C_2H_2	26.0382	.06860	0.89897	1.24	340.7	557.1	905	.0661
Ammonia	NH_3	17.0306	.0452	.58798	1.31	431.6	731.1	1657	.0684
Benzene	C_6H_6	78.11	.0548	2.6967		635.9	1010.9	700.9	.0527
Butane-n	C_4H_{10}	58.1243	.15805	2.0068	1.09	490.8	765.3	550.7	.0704
Butane-iso	C_4H_{10}	58.1243	.15788	2.0068	1.10	470.6	734.6	529.1	.0725
Carbon dioxide	CO_2	44.00995	.12342 (b)	1.5194	1.30	350.4	547.7	1073	.0348
Carbon monoxide	CO	28.01055	.078065(b)	0.96707	1.40	143.0	241.7	510	.0515
Ethane	C_2H_6	30.0701	.07987	1.0382	1.19	332.2	549.8	708.3	.0787
Ethylene	C_2H_4	28.0542	.07391	0.96858	1.24	305.0	509.5	742.1	.0705
Ethyl alcohol	C_2H_5OH	46.07	49.2759 (liq)	1.5905	1.13	632.75	929.3	927.3	.0581
Helium	He	4.0026	.011143(c)	.13819	1.66	7.669	672.41	1306.4	
Hydrogen	H_2	2.0159	.0056114(b)	.069599	1.41	36.8	59.9	188	.5168
Methyl alcohol	CH_3OH	32.04	49.6942 (liq)	1.1061	1.203	608.06	923.7	1156.6	.0588
Hydrogen sulphide	H_2S	34.0799	.09050	1.1766	1.32	383.2	672.4	1306	
Methane	CH_4	16.0430	.042355	0.55389	1.31	201.0	343.2	673.1	.0993
n-Octane	C_8H_{18}	114.23	43.9257 (liq)	3.9438		715.968	1024.5	361.5	.0684
Nitrogen	N_2	28.0134	.078064 (b)	.96717	1.40	139.3	226.9	492	.0515
Oxygen	O_2	31.9988	.0892102(b)	1.1047	1.40	162.3	277.9	730	.0373
Propane	C_3H_8	44.0972	.11806	1.5225	1.33	416.0	666.	617.4	.0730
Sulphur dioxide	SO_2	64.07	.1826	2.212		473.7	774.6	1141.9	.0308
Water (steam, dry)	H_2O	18.0153		0.62198	1.30	671.7	1165.1(d)	3208.2(d)	.05078(d)

(a) U.S. Standard Atmosphere, U.S. Government Printing Office, 1962, p. 9.
(b) National Bureau of Standards Circular 564, Tables of Thermodynamic Properties of Cases, 1960.
   (Interpolation within these tables should be by a method of second differences [20].
(c) U.S. Bureau of Mines Journal of Chemistry and Engineering Data, vol. 5, Jan. 1960, p. 51.
(d) *1967 ASME Steam Tables*.
(e) Values for an ideal gas and a reversible system.

References

[1]   Federal Register, July 1, 1959, Doc. 59-5442.

[2]   Units of Weights and Measures, Definitions and Tables of Equivalents; U.S. Dept. of Comm., Natl. Bu. of Standards, Miscl. Publication 286, May 1967.

[3]   U.S. Standard Atmosphere 1962, U.S. Government Printing Office, Washington D.C. 20025.

[4]   1967 ASME Steam Tables, Thermodynamic and Transport Properties of Steam.

[5]   The Density of Mercury, P. H. Bigg; *British Jol. Applied Physics*, vol. 15, 1964, p. 1111.

[6]   Absolute Viscosity of Water at 20 C; J. F. Swindells, J. R. Coe and T. B. Gadfrey; *Journal of Resh., Natl. Bu. of Stds.*, vol. 48, Jan. 1952, p. 1, RP 2279.

[7]   Thermodynamic Functions of Gases, F. Din; vol. 3, 1961, p. 24.

[8]   Standard Method of Conversion of Kinematic Viscosity to Saybolt Universal Viscosity; ASTM D 446-53.

[9]   ASTM Viscosity Tables; *ASTM Special Technical Publication No. 43A.*

[10]   Standard Method of Conversion of Kinematic Viscosity to Saybolt Furol Viscosity; ASTM D666-53.

[11]   Kinematic Viscosity and Times of Outflow from Commercial Viscometers, F. H. Garner and C. 1. Kelly; *Physics*, vol. 4, 1933, p. 97.

[12]   The Viscosity, Thermal Conductivity and Prandtl Number of Air, O_2, N_2, NO, H_2, CO, CO_2, H_2O, He and A; J. Hilsenrath and Y. S. Touloukian; *Trans. ASME*, vol. 76, Aug. 1954, p. 967.

[13]   Measurement of the Viscosity of Five Gases at Elevated Pressures by the Oscillating Disk Method; J. Kestin and K. Pilarczyk; *Trans. ASME*, vol. 76, Aug. 1954, p. 987.

[14]   Viscosities of Natural Gas Components and Mixtures; N. L. Carr; Inst. of Gas Tech., *Research Bul. 23*, June 1953.

[15]   Determination of Viscosity of Exhaust-Gas Mixtures at Elevated Temperatures; J. C. Westmoreland; *NACA Tech. Note 3180*, June 1954.

[16]   Tables of Thermal Properties of Gases, National Bu. of Stds. *Circular 564*, 1955. (Also reprinted as Tables of Thermodynamic and Transport Properties of Air, A, CO_2, CO, H_2, N_2, O_2 and Steam; Pergamon Press, London.

[17]   Supercompressibility Factors for Natural Gas; R. H. Zimmerman and S. R. Beitler; *Trans. ASME*, vol. 74, Aug. 1952, p. 945.

[18]   Evaluation fo Compressibility Factors for Natural Gas Mixtures; R. G. Darrow; MS Thesis, Ohio State University, 1953.

[19]   A Method of Predicting Supercompressibility Factors of Natural Gases; R. H. Zimmerman, S. R. Beitler and R. G. Darrow; *Trans. ASME*. vol. 80 No. 7, Oct. 1958, p. 1358.

Chapter II-II

General Requirements for
Fluid Metering: Installation

II-II-1 The fluid to be measured may be incompressible or compressible, i.e., liquid or gaseous. In locations where there may be a choice of measuring the flow of a fluid in either the liquid or gaseous state, the measurement should be made with the liquid if at all possible. This is because in the present state-of-the-art measurements made of a liquid are more reliable.

The fluid must remain in a single phase. For example, superheated steam must remain superheated. In the case of liquids, the pressure throughout the entire metering unit must be sufficiently high to prevent evaporation or the escape of any dissolved gases. Also, the fluid stream must be free of pulsations. This applies for almost all types of meters, but particularly those of the differential-pressure type.

The fluid should not contain suspended particles, as, for example, sand. Colloidal solutions with an index of dispersion not materially different than that of a homogeneous liquid, for example, milk, may be measured. Solutions of coarse dispersion should be excluded. Fluids having high viscosities such that the Reynolds numbers are below the limits shown by the tables or curves of coefficients of discharge should be avoided if a differential-pressure-type meter is to be used.

II-II-2 Installations. The conditions under which meters such as the orifice, flow nozzle and Venturi tube are installed may have as much effect on the accuracy of the flow measurement as the degree of perfection of manufacture or the characteristics of the elements themselves. The rate of flow computed from the differential pressure produced by these elements may be in error to an unacceptable degree if the piping arrangements are such that distorted flow conditions result. Distortions of the velocity traverse, helical swirls or vortices can affect the accuracy of the flow measurement. A projecting gasket, misalignment or a burr on a pressure tap can cause considerable error. Therefore, the following instructions should be followed carefully.

Whenever possible, it is preferable to locate the primary element in a horizontal line. If the unit is in a vertical pipe, the pressure tubing to the secondary element, including reservoirs, if used, should be adjusted to the same elevation as described in Pars. II-II-9 and II-II-11 below.

II-II-3 To insure accurate flow measurement, the fluid should enter the primary element with a fully developed velocity profile, free from swirls or vortices. Such a condition is best achieved by the use of adequate lengths of straight pipe, both preceding and following the primary element. The *minimum* desirable lengths of such piping are shown in the eight diagrams of Fig. II-II-1. Each diagram shows a somewhat different arrangement of piping. That diagram which corresponds the closest to the actual piping arrangement for the meter location should be used to determine the required lengths of straight pipe on inlet and outlet. These lengths have been determined as necessary to hold errors due to piping configurations to less than ±0.5 per cent.

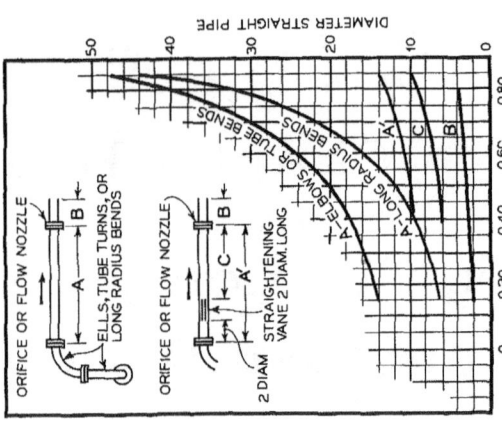

(A) FOR ORIFICES AND FLOW NOZZLES
ALL FITTINGS IN SAME PLANE

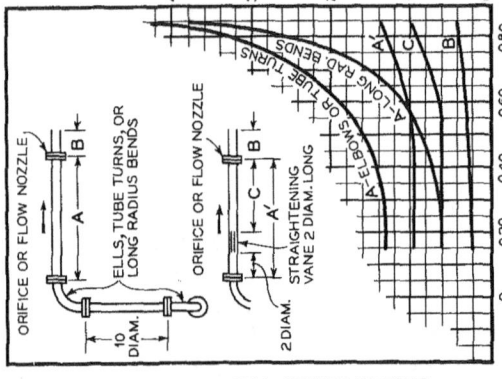

(B) FOR ORIFICES AND FLOW NOZZLES
ALL FITTINGS IN SAME PLANE

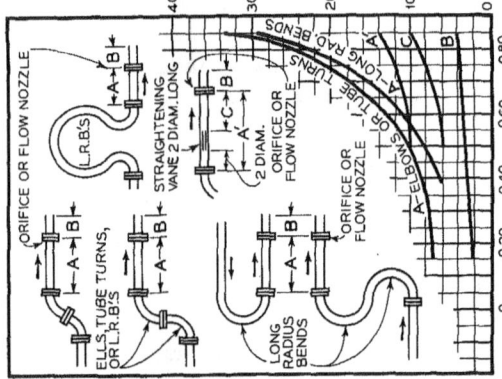

(C) FOR ORIFICES AND FLOW NOZZLES
FITTINGS IN DIFFERENT PLANES

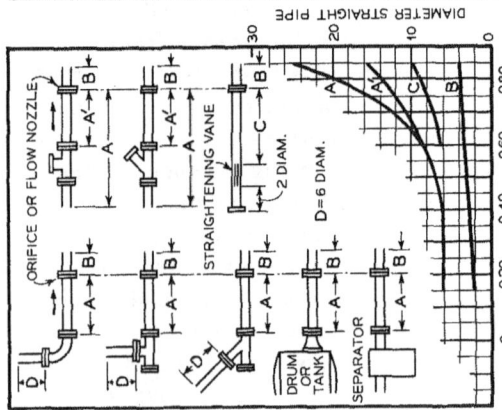

(D) FOR ORIFICES AND FLOW NOZZLES
FITTINGS IN DIFFERENT PLANES

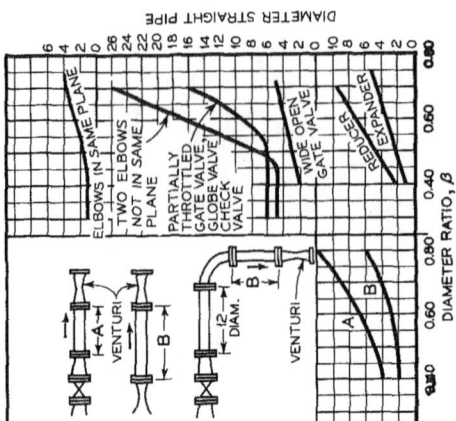

(E) FOR ORIFICES AND FLOW NOZZLES
WITH REDUCERS AND EXPANDERS

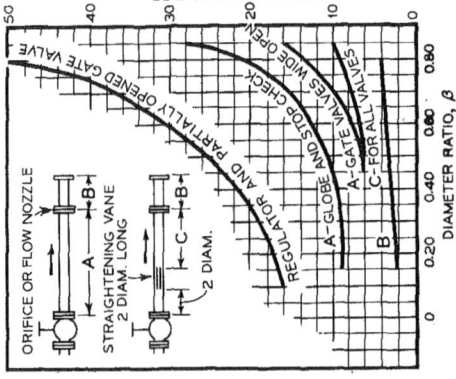

(F) FOR ORIFICES AND FLOW NOZZLES
IN ATMOSPHERIC INTAKE

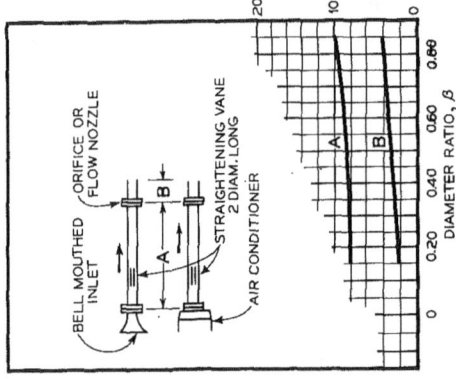

(G) VALVES AND REGULATORS

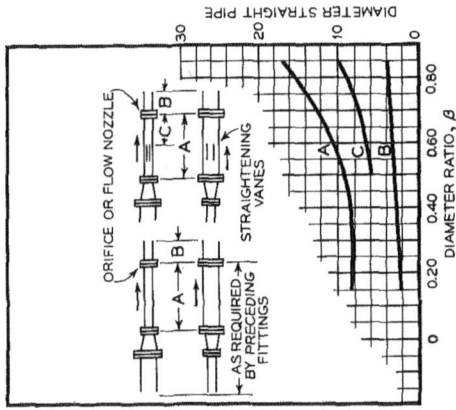

(H) FOR VENTURI TUBES

FIG. II-II-1 RECOMMENDED MINIMUM LENGTHS OF PIPE PRECEDING
AND FOLLOWING ORIFICES, FLOW NOZZLES AND
VENTURI TUBES (ALL CONTROL VALVES, INCLUDING
REGULATORS, SHOULD BE LOCATED ON OUTLET SIDE
OF PRIMARY ELEMENT.)

180

If it is impossible to arrange the piping so as to provide the recommended lengths of straight pipe or should there be uncertainty as to the diagram to follow, straightening vanes may be used. Diagrams (A) through (H) (Fig. II-II-1) show the minimum straight pipe that may be used with straightening vanes. The vanes should be preceded by two or more diameters of straight pipe. Greater lengths of straight pipe, both with and without vanes, than the minimums shown in Fig. II-II-1 should be provided whenever possible.

If it is impossible to provide the recommended minimum lengths, even with the use of straightening vanes, an additional tolerance of ±0.5 per cent should be applied to the flow measurement. This additional tolerance should be treated in the same way tolerances for other elements are treated as described in Par. II-V-4.

II-II-4 The use of any type of differential pressure meter in a pipe line that includes a reciprocating pump should be avoided. However, with liquids, as explained in Par. I-3-46 (Chapter I-3), a measurement corrected for pulsation effects may be possible if a time average of the square root of the instantaneous differential pressures at the meter can be obtained by the recorder. On the other hand, with compressible fluids no reliable measurement is possible; the pulsations from a reciprocating pump must be reduced or completely suppressed by some means between the source and the primary element.

If the pump is a centrifugal or turbine type, the preferred location of the primary element is on the inlet side and as far from the pump as possible. If the meter is located on the discharge side of the pump, straightening vanes will be needed to eliminate swirl. There should be a minimum of eight pipe diameters between the vanes and primary element. If there are fittings between the pump and primary element, the vanes should be located according to the appropriate arrangement for the fittings as shown in Fig. II-II-1.

II-II-5 Cross sections of recommended designs of straightening vanes are shown in Fig. II-II-2. In the tubular and cross-plate designs, the maximum distance between tube centers or passage centers should not exceed one-fourth the pipe diameter, D; and the overall length should be at least eight times this dimension. These vanes may be constructed of thin-walled tubes, or plates, welded together. The perforated-plate design has plates held one pipe diameter apart by spacers; each has a large number of small holes, and the total area of these holes should be equal to at least 50 per cent of the cross-sectional area of the pipe. Regardless of the design of straightening vane used, they must be secured firmly in place within the pipe.

The pressure drop through the tubular and cross-plate designs is about the same as that for 20 diameters of the pipe. For the perforated-plate design, the loss is about the same as the differential pressure across a sharp-edged orifice of 0.75-diameter ratio.

II-II-6 Internal Pipe Surface and Diameter. The internal surface of the pipe immediately preceding and following an orifice or flow nozzle should be straight, free from mill scale, pits or holes, reamer scores or rifling, bumps or other irregularities. The surface roughness should not be greater than 350 microinch. The pipe should be near enough to a cylindrical shape that no diameter departs from the average diameter, D, by more than 0.33 per cent. If, to secure this degree of surface roughness and pipe roundness, boring is necessary, such boring should extend for at least $4D$ preceding and $2D$ following the inlet face of the orifice or nozzle. The bored portion should be faired into the unbored portion at an included angle of less than 30 deg. The depth of boring should be the minimum required to obtain the desired condition, and the boring should be done after any necessary welding of flanges and pressure connections has been done.

II-II-7 The internal pipe diameter, D, should be measured on four or more diameters in the plane of the inlet pressure tap hole, p_1. Check measurements should be made on three or more diameters in two additional cross sections so as to cover at least two pipe diameters from the inlet face of the orifice plate of flow nozzle, or past the weld, whichever is the greater distance. The values of all such inlet-section diameters should agree within 4 per cent when the diameter ratio, β, of the orifice or flow nozzle to be used will be 0.2 and within 0.5 per cent when the diameter ratio, β, is to be 0.75. For intermediate values of β, a linear relation can be used. The average of all diameters near the plane of the inlet pressure tap should be used in computing the diameter ratio, β, of the primary element.

Measurements of the diameter of the outlet section should be made in the plane of the outlet pressure tap to insure that the diameter of the outlet section agrees with that of the inlet section within twice the percentage spreads given above for the diameters of the inlet section.

For use with a Venturi tube, the internal surface of the pipe, attaching to the tube inlet, should be free of such imperfections as mentioned above.

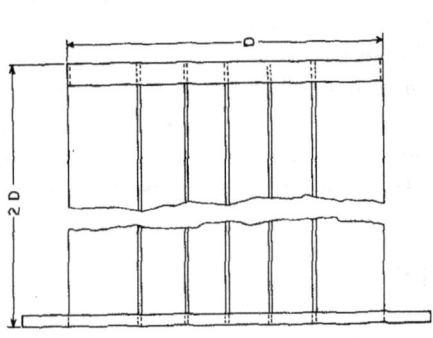

Cross Plates

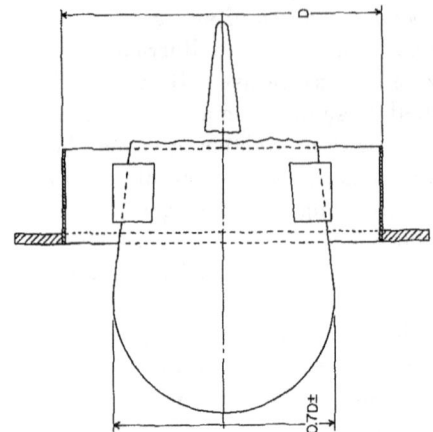

Centered Teardrop

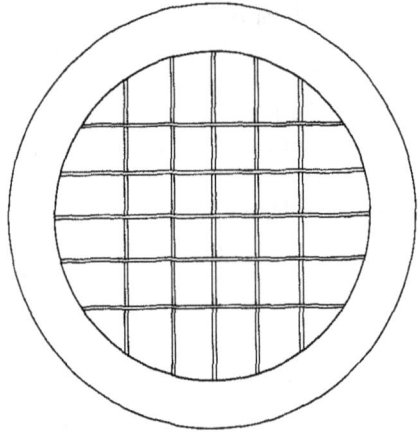

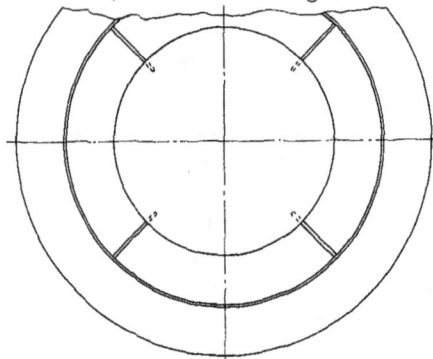

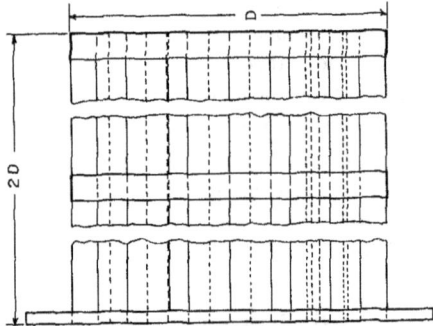

Tubular

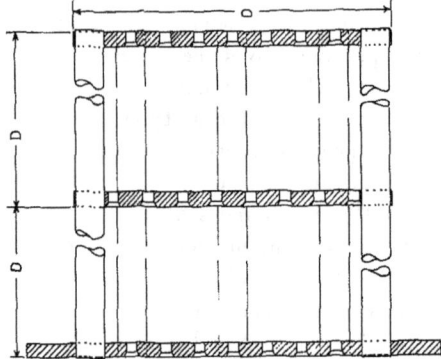

Perforated Plates

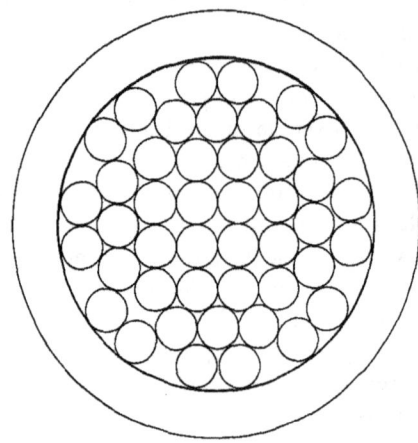

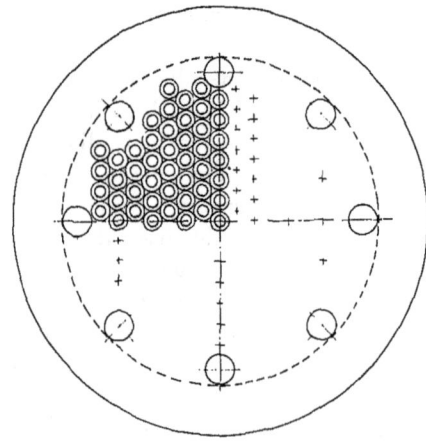

FIG. II-II-2 RECOMMENDED DESIGNS OF STRAIGHTENING VANES

182

NAMEPLATE TO BE IN LINE
WITH PRESSURE CONNECTION
NIPPLES AND WITH STAMPING
ON NAMEPLATE FACING THE
ONCOMING FLOW

PRESSURE CONNECTION
NIPPLE AND WELDING
ADAPTER

CENTERING PINS SOMETIMES
USED WITH ORIFICES

FLOW NOZZLE MUST BE CENTERED AND MUST
POINT IN DIRECTION OF FLOW IN PIPE

ANY BEVELLING OF ORIFICE
MUST BE ON OUTLET SIDE

ORIFICE OR FLOW NOZZLE FLANGE
CENTERED INSIDE FLANGE BOLTS

p_2

ϕp_1

B

A

SEE NOTE 1

GASKETS MAY BE INSTALLED ON EACH SIDE OF
ORIFICE OR FLOW NOZZLE FLANGE. HOWEVER, THE
GASKETS MUST NOT EXTEND INSIDE THE PIPE

CENTERING PINS MAY BE USED WITH
ORIFICES TO AID IN CENTERING IN PIPE

① ② ③ ④

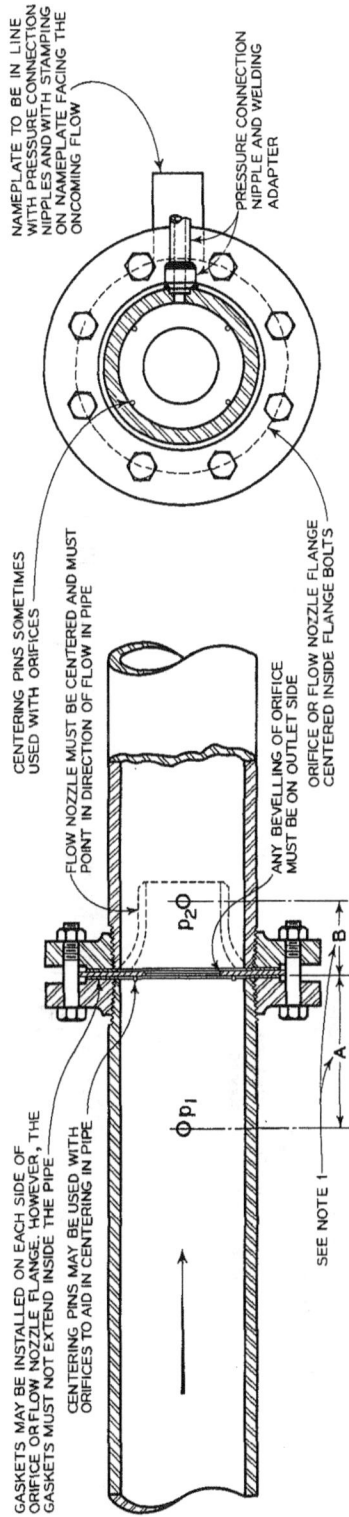

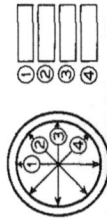

POINTS OF MEASUREMENT IN SAME PLANE FOR
DETERMINING ROUNDNESS OF PIPE

NOTES:

1. IN HORIZONTAL PIPES, WITH STEAM OR GAS FLOW,
PRESSURE TAPS SHOULD BE ON SIDE OR TOP OF
PIPE.; WITH LIQUIDS PRESSURE TAPS SHOULD BE
ON SIDE.

2. IF PROVIDED, DRAIN HOLE IN ORIFICE OR FLOW
NOZZLE SHOULD BE LOCATED AT BOTTOM OF PIPE
FOR STEAM OR GAS FLOW, OR AT TOP OF PIPE
FOR WATER OR OTHER LIQUID FLOW.

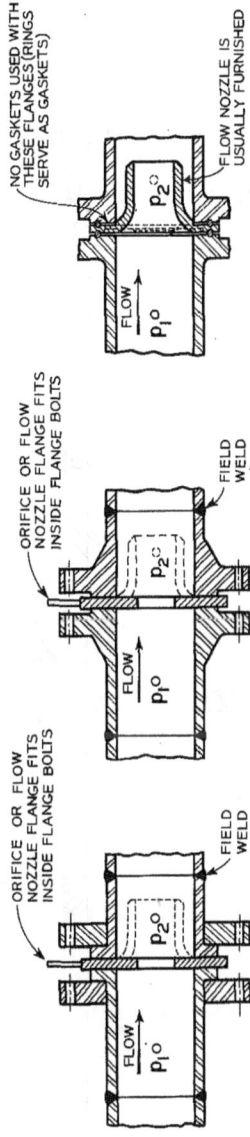

NO GASKETS USED WITH
THESE FLANGES (RINGS
SERVE AS GASKETS)

FLOW NOZZLE IS
USUALLY FURNISHED

FLOW
p_1O
p_2O

RING JOINT FLANGES

ORIFICE OR FLOW
NOZZLE FLANGE FITS
INSIDE FLANGE BOLTS

FIELD
WELD

FLOW
p_1O
p_2O

WELD NECK FLANGES

ORIFICE OR FLOW
NOZZLE FLANGE FITS
INSIDE FLANGE BOLTS

FIELD
WELD

FLOW
p_1O
p_2O

VAN STONE OR
LAP JOINT FLANGES

TONGUE FLANGE TO BE RE-
CESSED WIDTH OF ORIFICE OR
FLOW NOZZLE FLANGE.

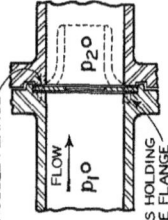

FLOW
p_1O
p_2O

FLAT HEAD SCREWS HOLDING
ORIFICE TO TONGUE FLANGE

TONGUE AND
GROOVE FLANGES

ORIFICE OR FLOW
NOZZLE FLANGES FITS
INSIDE FLANGE BOLTS

FLOW
p_1O
p_2O

FIELD WELD
INSIDE & OUT

SLIP-ON FLANGES

FIG. II-II-3 METHODS OF INSTALLING ORIFICE PLATES AND FLOW
NOZZLES BETWEEN FLANGES FOR LOW-PRESSURE,
LOW-TEMPERATURE SERVICE

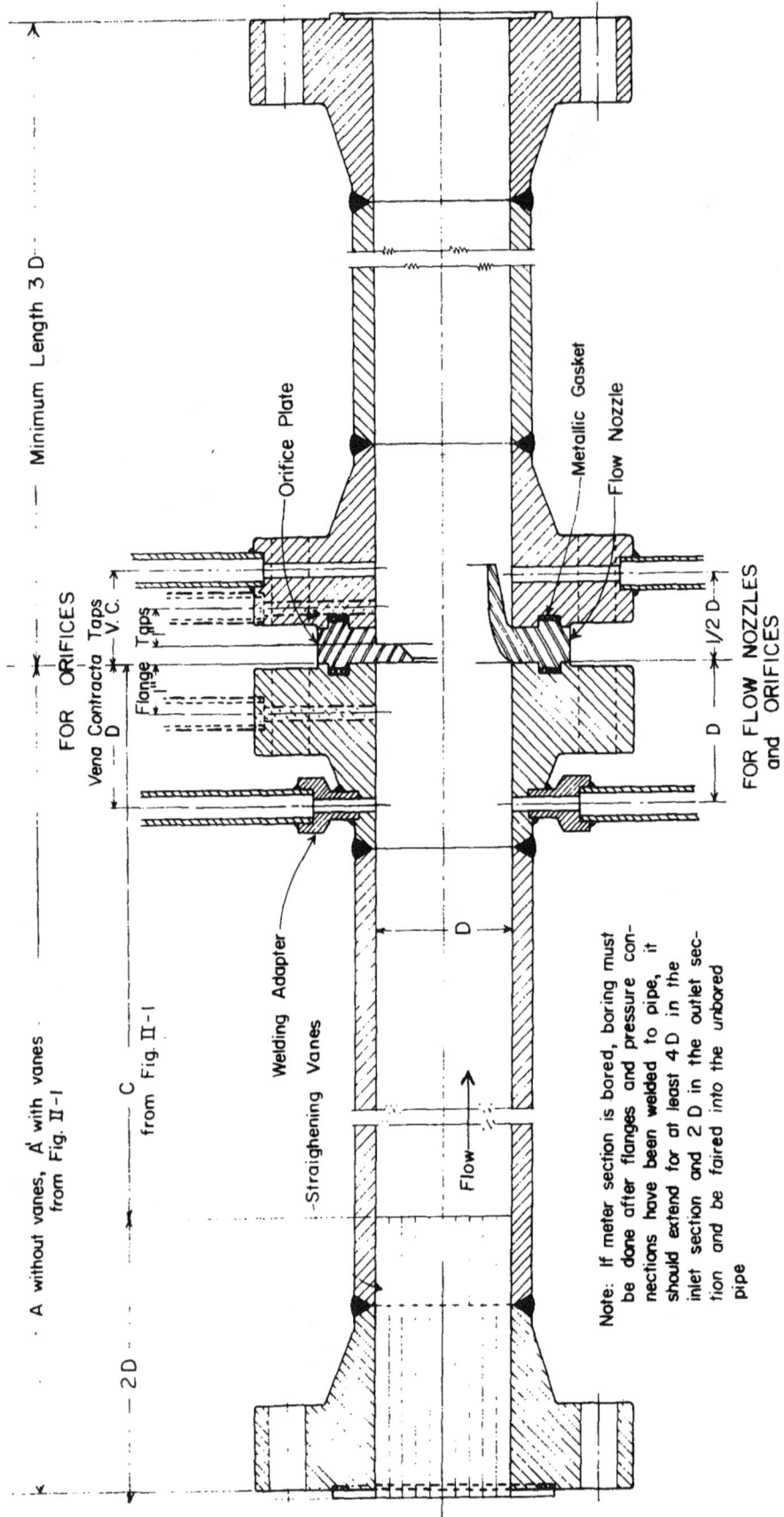

FIG. II-II-4 A RECOMMENDED DESIGN OF ORIFICE OR FLOW-NOZZLE METER SECTION FOR HIGH-TEMPERATURE, HIGH-PRESSURE SERVICE

Minimum Length 3 D

FOR ORIFICES

Vena Contracta Taps V.C. — D

Flange Taps

A without vanes, A' with vanes
from Fig. II-I

C
from Fig. II-I

2D

Orifice Plate

Metallic Gasket

Flow Nozzle

1/2 D

D

D

FOR FLOW NOZZLES
and ORIFICES

Welding Adapter

Straightening Vanes

Flow

Note: If meter section is bored, boring must be done after flanges and pressure connections have been welded to pipe, it should extend for at least 4D in the inlet section and 2 D in the outlet section and be faired into the unbored pipe

The average diameter of the pipe, D, where it joins the Venturi tube shall be within ±1 per cent of the diameter of the tube inlet, and the out-of-roundness should not exceed 2 per cent. Desirably, the pipe following a Venturi tube should be of the same nominal size.

II-II-8 Installation of an Orifice or Flow Nozzle. When installed between pipe-line flanges, the center of a concentric orifice or of a nozzle throat should be within ± 1/32 in. of the axis of the pipe. Several optional methods of installing and centering an orifice plate or flow nozzle for low-pressure low-temperature service are shown in Fig. II-II-3. The centering of an eccentric or segmental orifice should be such that no portion of the round hole will be closer than 1 per cent of the pipe diameter to the pipe wall. A method of centering and holding an orifice plate or flow nozzle between flanges for high-pressure high-temperature service is shown in Fig. II-II-4.

When male and female or tongue and groove flanges are used, the outside diameter of the orifice plate or the nozzle flange should be made to fit inside the female or groove flange.

For use in ring joint flanges, a flow-nozzle flange can be made thick enough so that a ring groove can be cut into each side of this flange, thereby allowing for the usual ring on each side. For an orifice plate a ring which will fit in the grooves of the flanges may be fitted to the outer circumference of the plate, or the face of the inlet flange may be recessed by an amount equal to the thickness of the orifice plate and the plate attached to the flange. This procedure may be used with a flow nozzle also, provided that it does not require an undesirably thin nozzle flange.

II-II-9 Gaskets. In all cases the inside diameter of the gasket must be made large enough and the gasket so positioned that, when in service, it will not protrude beyond the inner surface of the pipe at any point.

II-II-10 Pressure Connections. The locations of pressure tap holes used with orifices and flow nozzles are referred to the inlet face of the orifice plate or flow-nozzle flange as the datum plane, except for flange taps used with orifices. For orifices, the locations for which data are given are flange, D and $1/2D$ and vena contracta taps. For flow nozzles, data are given for pipe-wall taps at D and $1/2D$ and for D and the nozzle throat. The locations of these several pairs of pressure taps are specified in the sections dealing with each differential pressure producer in Ch. II-III.

The recommended maximum diameter, δ, of pressure tap holes through the pipe wall or flange are given in Table II-II-1. With clean fluids smaller diameters may be desirable.

Table II-II-1 Recommended Maximum Diameters of Pressure Tap Holes

Nominal Inside Pipe Diam. D	Max. Diam. δ (Fig. II-II-5)
Under 2	1/4
2,3	3/8
4 to 8	1/2
10 and over	3/4

All dimensions are given in inches.

There must be no burrs, wire edges or other irregularities on the inside of the pipe at the nipple connections or along the edge of the hole through the pipe wall. The diameter of the hole should not *decrease* within a distance of 2δ from the inner surface of the pipe but may be increased within a lesser distance.

Where the pressure hole breaks through the inner surface of the pipe there *must* be no roughness, burrs nor wire edge. The edge (corner) of the hole may be left truly square or it may be dulled (rounded) very slightly.

Connections to the pressure holes should be made by nipples, couplings or adapters (Fig. II-II-5) welded to the outside surface of the pipe. It is important that no part of any such fitting projects beyond the inner surface of the pipe.

II-II-11 When the primary element is in a horizontal pipe, for measuring steam the pressure holes and connecting nipples should be in the horizontal plane of the pipe center line. For measuring water and other liquids, the connections may be in the same horizontal plane or any other convenient position in the lower half of the pipe. For measuring gases, the usual position is in the vertical plane of the pipe center line, although, if necessary, other positions in the upper half of the pipe line may be used. Some of the connection positions are illustrated in Fig. II-II-6.

When measuring steam in a vertical pipe, an S nipple of at least 3/4-in. pipe size should be used at the lower connection and should be of such length that the top of the nipple is level with the straight nipple used at the upper connection. On the other hand, when measuring liquids in a vertical

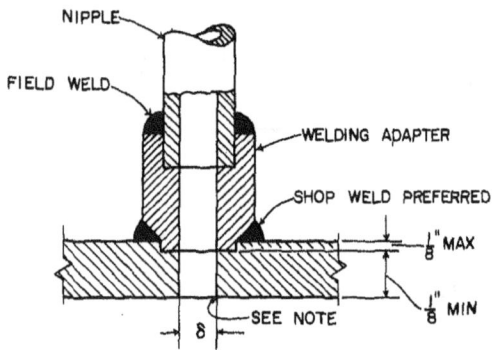

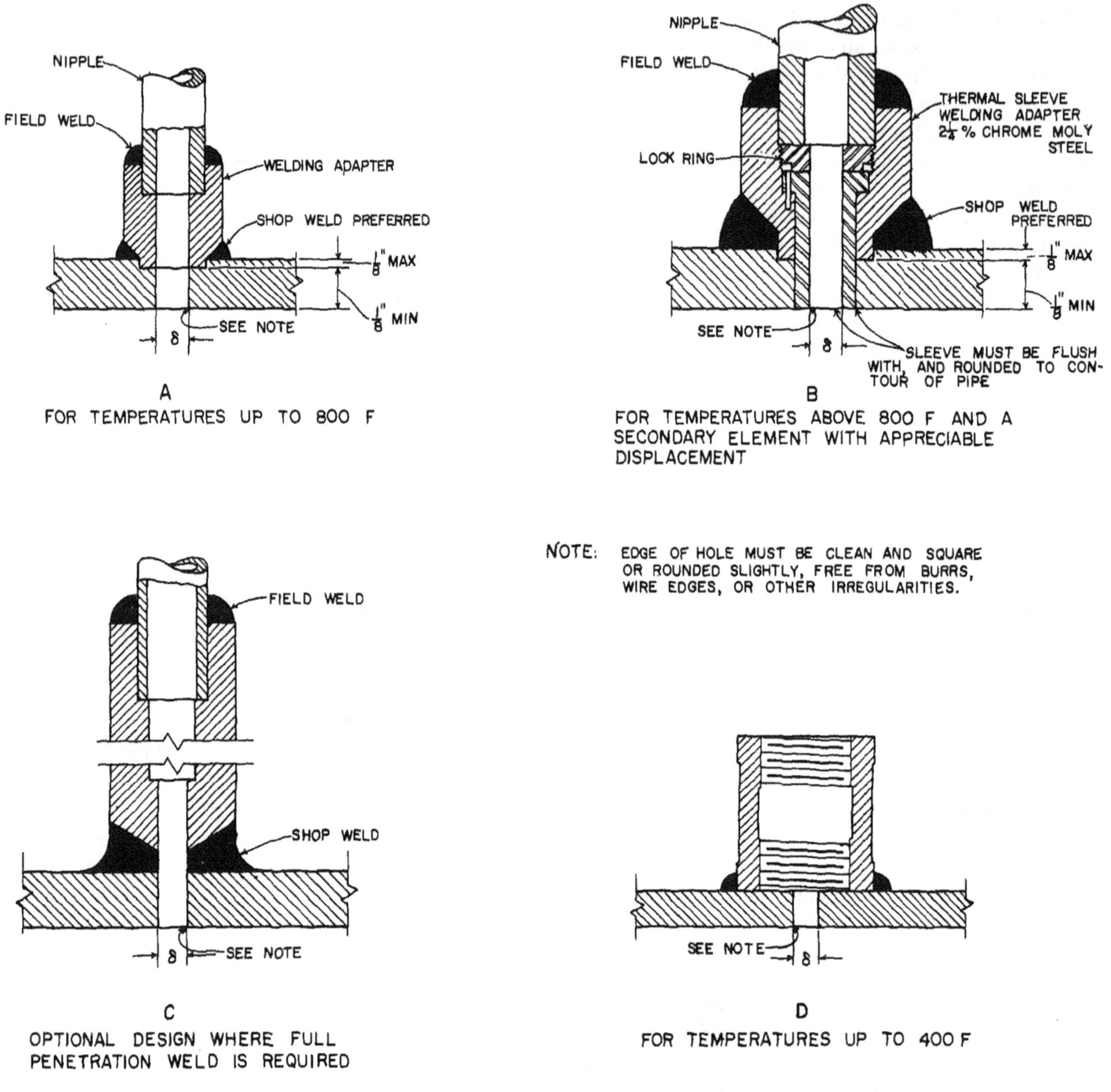

A
FOR TEMPERATURES UP TO 800 F

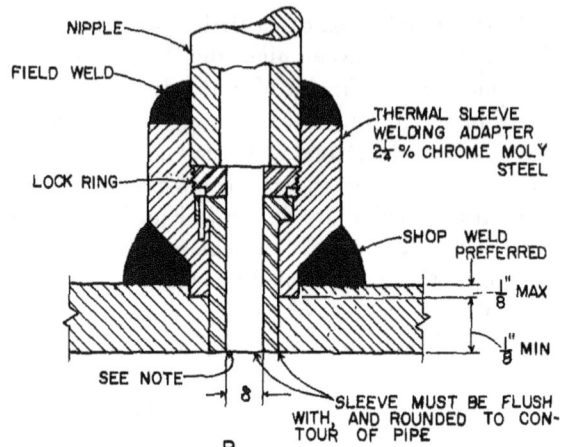

Labels in figure A:
- NIPPLE
- FIELD WELD
- WELDING ADAPTER
- SHOP WELD PREFERRED
- ⅛" MAX
- ⅛" MIN
- SEE NOTE
- δ

B
FOR TEMPERATURES ABOVE 800 F AND A
SECONDARY ELEMENT WITH APPRECIABLE
DISPLACEMENT

Labels in figure B:
- NIPPLE
- FIELD WELD
- LOCK RING
- THERMAL SLEEVE WELDING ADAPTER 2¼ % CHROME MOLY STEEL
- SHOP WELD PREFERRED
- ⅛" MAX
- ⅛" MIN
- SEE NOTE
- δ
- SLEEVE MUST BE FLUSH WITH, AND ROUNDED TO CONTOUR OF PIPE

NOTE: EDGE OF HOLE MUST BE CLEAN AND SQUARE
OR ROUNDED SLIGHTLY, FREE FROM BURRS,
WIRE EDGES, OR OTHER IRREGULARITIES.

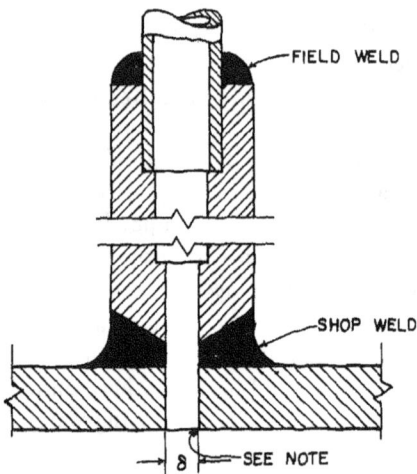

C
OPTIONAL DESIGN WHERE FULL
PENETRATION WELD IS REQUIRED

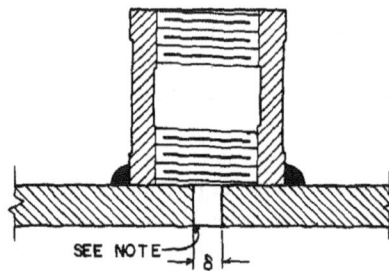

Labels in figure C:
- FIELD WELD
- SHOP WELD
- SEE NOTE
- δ

D
FOR TEMPERATURES UP TO 400 F

Labels in figure D:
- SEE NOTE
- δ

FIG. II-II-5 METHODS OF MAKING PRESSURE CONNECTIONS TO PIPES

186

ORIFICE OR
FLOW NOZZLE

RECOMMENDED
LAGGING

B

A

NAMEPLATE

FLOW

WELDING ADAPTERS

$\frac{1}{2}''$ OR $1''$ NIPPLES

$\frac{1}{2}''$ OR $1''$ VALVES SUITABLE
FOR OPERATING PRES-
SURE AND TEMPERA-
TURE

RESERVOIRS
INSTALL LEVEL LENGTHWISE
AND WITH EACH OTHER

SECTIONAL PLAN VIEW

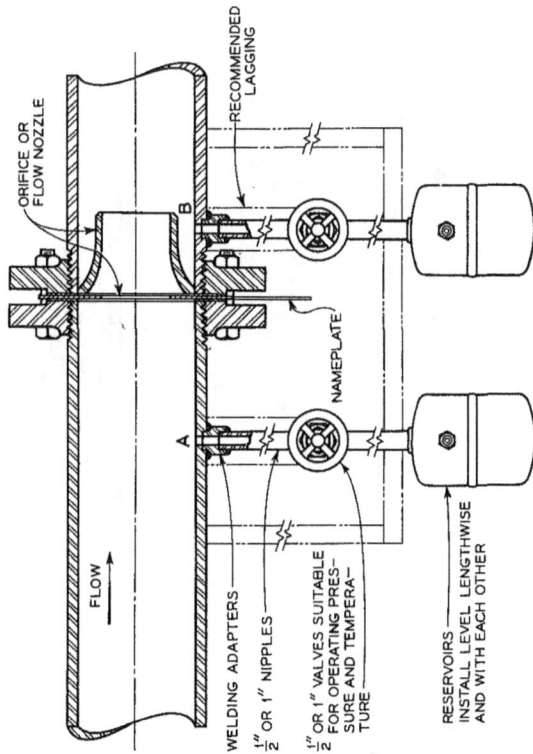

NOTE:
USE OF RESERVOIRS IS OPTIONAL WHEN
SECONDARY ELEMENT HAS NEGLIGIBLE
VOLUMETRIC DISPLACEMENT.

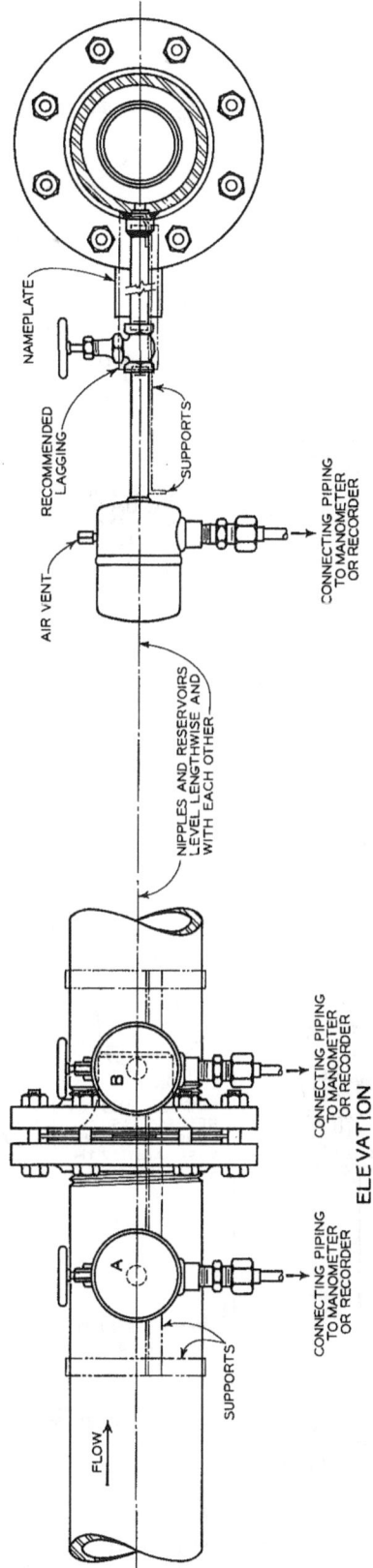

NAMEPLATE

RECOMMENDED
LAGGING

AIR VENT

SUPPORTS

CONNECTING PIPING
TO MANOMETER
OR RECORDER

NIPPLES AND RESERVOIRS
LEVEL LENGTHWISE AND
WITH EACH OTHER

B

A

FLOW

SUPPORTS

CONNECTING PIPING
TO MANOMETER
OR RECORDER

CONNECTING PIPING
TO MANOMETER
OR RECORDER

ELEVATION

FIG. II-II-6   METHOD OF CONNECTING NIPPLES, VALVES AND
RESERVOIRS TO HORIZONTAL PIPE LINES

187

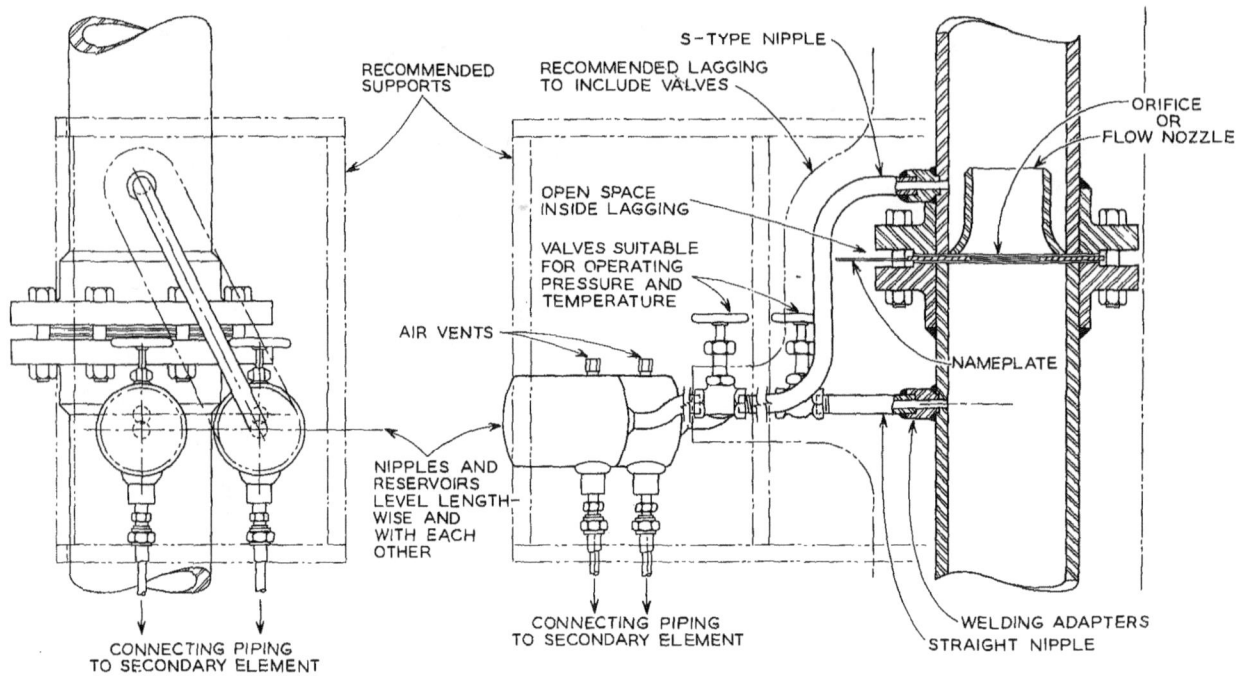

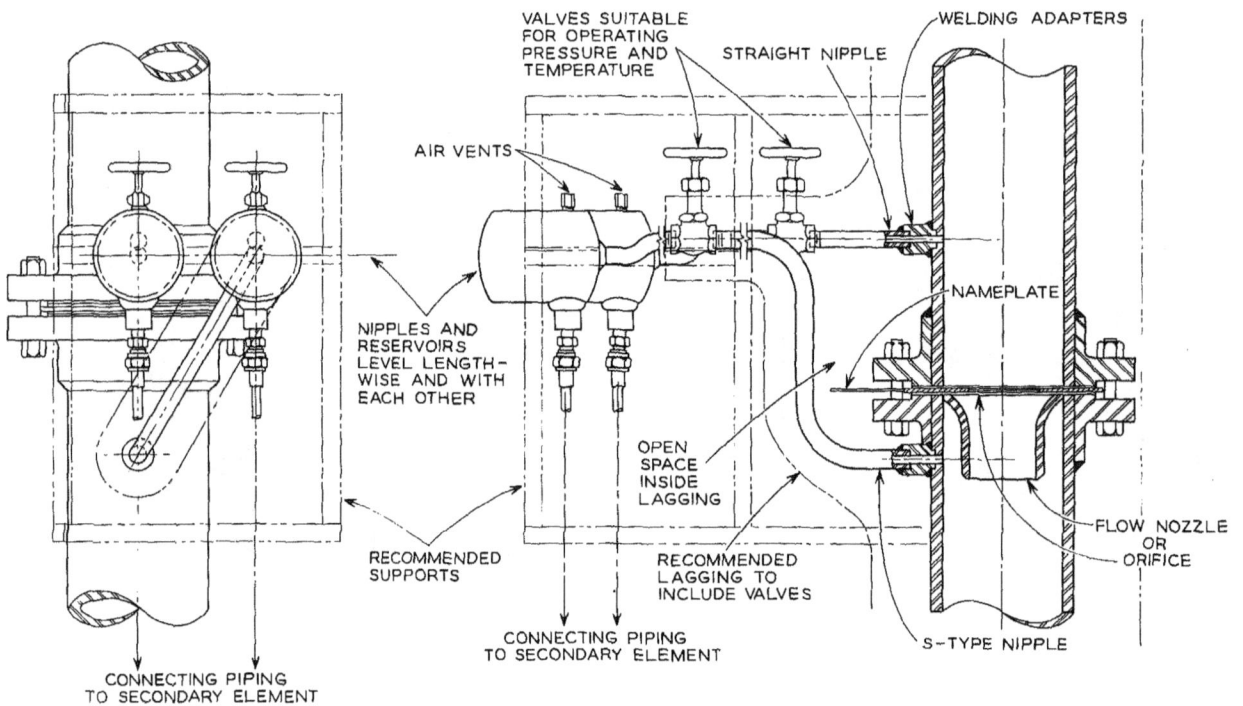

FIG. II-II-7 CONNECTING NIPPLES, VALVES AND RESERVOIRS TO
VERTICAL PIPE LINES

NOTES:

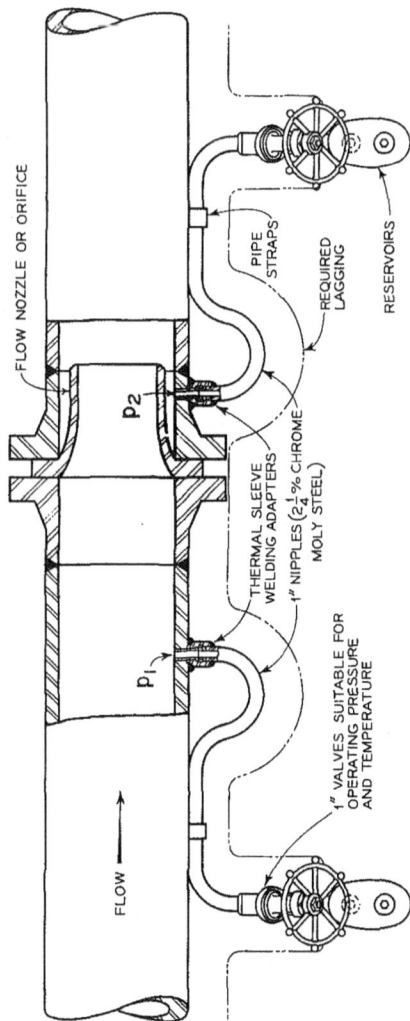

FOR INSTALLATION OF ORIFICE OR FLOW NOZZLE BETWEEN FLANGES, SEE FIGURE II-4

NIPPLES AND RESERVOIRS TO BE LEVEL LENGTHWISE AND WITH EACH OTHER.

LAGGING AROUND NIPPLES IS REQUIRED AS SHOWN.

FLOW

FLOW NOZZLE OR ORIFICE

p_2

p_1

PIPE STRAPS

REQUIRED LAGGING

RESERVOIRS

THERMAL SLEEVE WELDING ADAPTERS

1" NIPPLES ($2\frac{1}{4}$% CHROME MOLY STEEL)

1" VALVES SUITABLE FOR OPERATING PRESSURE AND TEMPERATURE

SECTIONAL PLAN VIEW

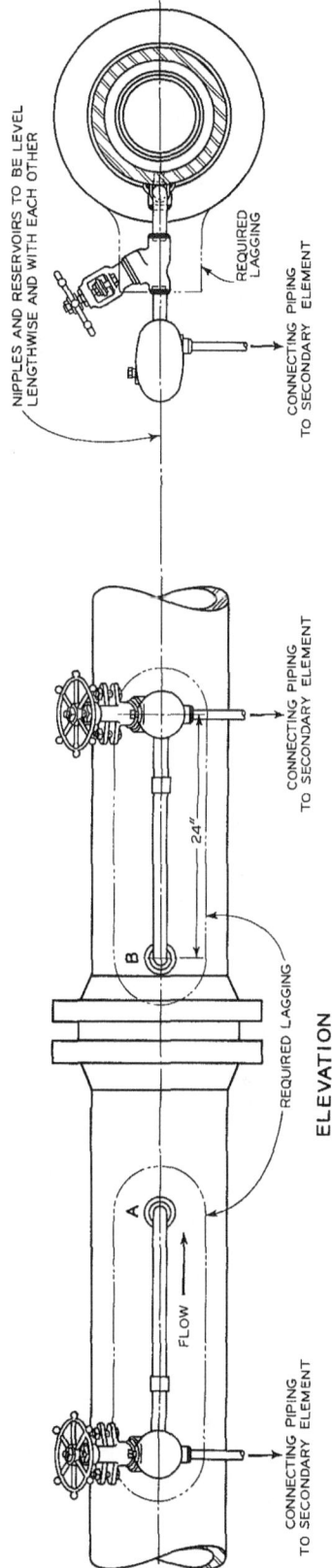

NIPPLES AND RESERVOIRS TO BE LEVEL LENGTHWISE AND WITH EACH OTHER

REQUIRED LAGGING

CONNECTING PIPING TO SECONDARY ELEMENT

CONNECTING PIPING TO SECONDARY ELEMENT

B

24"

FLOW

A

REQUIRED LAGGING

CONNECTING PIPING TO SECONDARY ELEMENT

ELEVATION

FIG. II-II-8 METHOD OF CONNECTING NIPPLES, VALVES AND RESERVOIRS TO HORIZONTAL PIPE WITH STEAM AT TEMPERATURES ABOVE 850 F, ESPECIALLY IF SECONDARY ELEMENT HAS APPRECIABLE VOLUMETRIC DISPLACEMENT

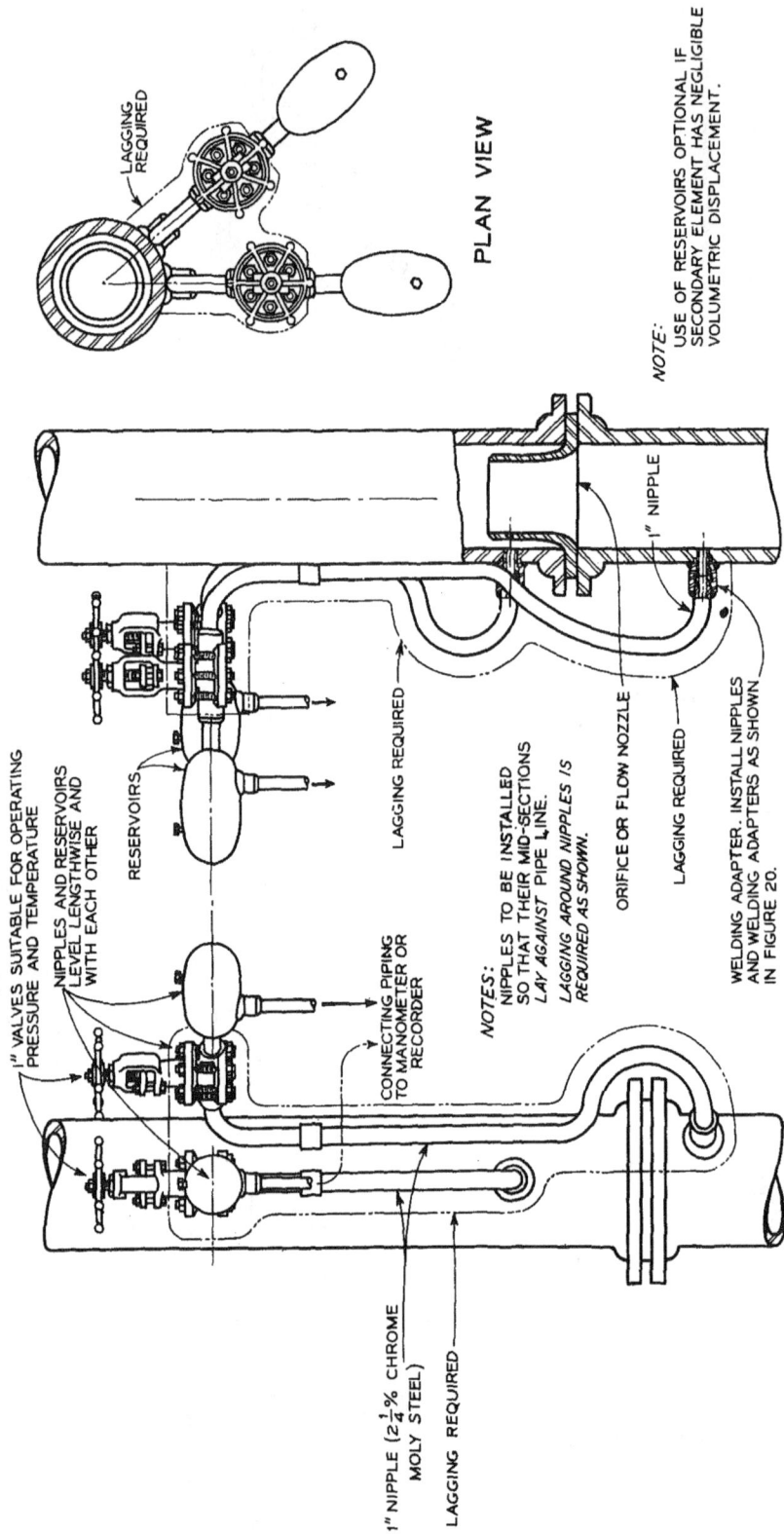

LAGGING REQUIRED

PLAN VIEW

NOTE:
USE OF RESERVOIRS OPTIONAL IF
SECONDARY ELEMENT HAS NEGLIGIBLE
VOLUMETRIC DISPLACEMENT.

1" NIPPLE

SECTIONAL VIEW – UP FLOW

1" VALVES SUITABLE FOR OPERATING
PRESSURE AND TEMPERATURE

NIPPLES AND RESERVOIRS
LEVEL LENGTHWISE AND
WITH EACH OTHER

RESERVOIRS

LAGGING REQUIRED

CONNECTING PIPING
TO MANOMETER OR
RECORDER

NOTES:
NIPPLES TO BE INSTALLED
SO THAT THEIR MID-SECTIONS
LAY AGAINST PIPE LINE.

LAGGING AROUND NIPPLES IS
REQUIRED AS SHOWN.

ORIFICE OR FLOW NOZZLE

LAGGING REQUIRED

WELDING ADAPTER. INSTALL NIPPLES
AND WELDING ADAPTERS AS SHOWN
IN FIGURE 20.

1" NIPPLE ($2\frac{1}{4}$% CHROME
MOLY STEEL)

LAGGING REQUIRED

FIG. II-II-9 CONNECTING NIPPLES, VALVES AND RESERVOIRS TO
VERTICAL STEAM PIPES WHEN STEAM TEMPERATURE
IS ABOVE 850 F AND SECONDARY ELEMENT HAS
APPRECIABLE VOLUMETRIC DISPLACEMENT

pipe, if the temperature differs from the ambient by more than 50 F, an S nipple of at least 3/4-in. pipe size should be used at the upper connection; and its length should be such that the bottom of this nipple is level with the straight nipple used at the lower connection, as shown in Fig. II-II-7. The S nipple should be lagged in with the main flow line in order to prevent condensate forming in the nipple when measuring steam and to maintain pipe temperatures in the nipple when measuring hot liquids.

For measuring steam at temperatures higher than 850 F, special bent nipples, as shown in Figs. II-II-8 and II-II-9, should be used. In both cases the entire length of these nipples as well as the shutoff valves should be lagged in with the steam line so as to convert to steam any water returning from a manometer or other gage, before it re-enters the steam pipe, and also to keep the amount of condensate to a minimum.

II-II-12 Shutoff Valves. Shutoff valves should be provided for every pressure tap connection and

should be located as close as possible to the main pipe containing the primary element. These valves must be capable of withstanding full pipe-line pressure and temperature and must be installed so as to close against the pressure in the main pipe.

II-II-13 Reservoirs. Reservoirs should be used at the ends of the inlet and outlet differential pressure connections at the primary element when measuring steam and when measuring hot water or liquids above 250 F, as shown in Figs. II-II-6 through II-II-9. These reservoirs provide water legs of equal density and elevation on both sides of a manometer or other differential pressure gages. The water volume of these reservoirs should be at least equal to the maximum displacement of the manometer or other differential pressure gage to which they are connected, and a volume two or three times this amount is preferable. The design of reservoirs and the method of connecting them should be such that they will be full of condensate at all times. Reservoirs filled partly with steam and partly with water are of little

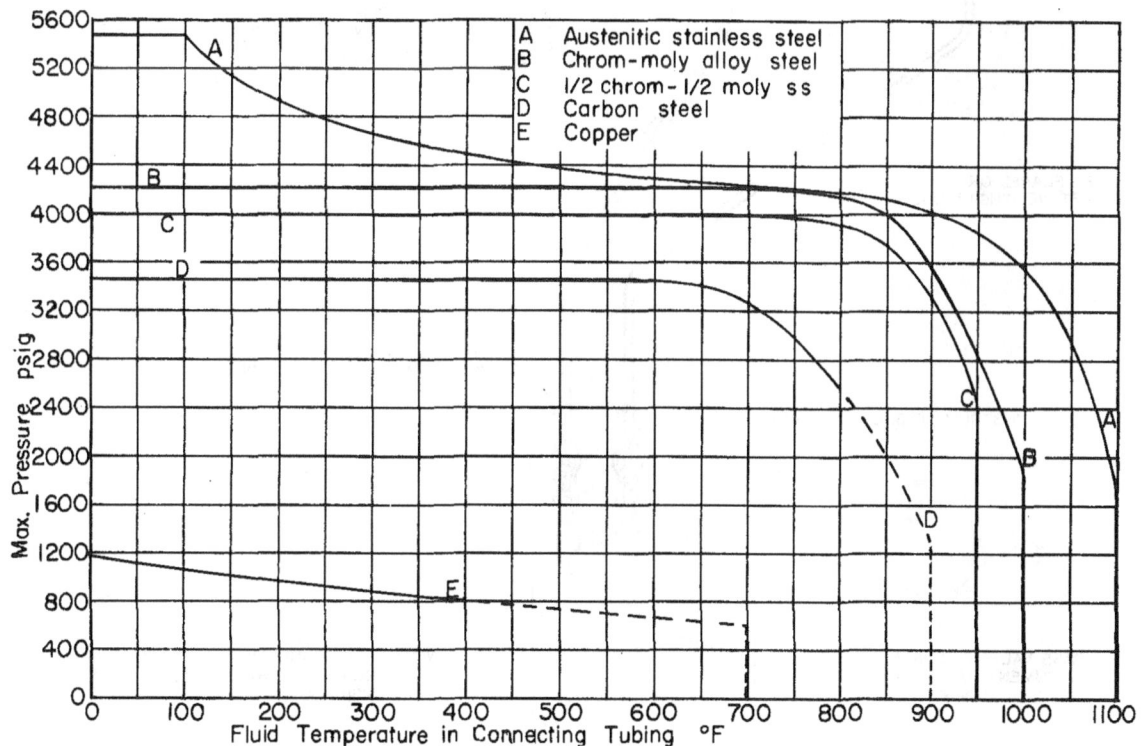

FIG. II-II-10 CHART FOR SELECTING SEAMLESS TUBING OR PIPE CONNECTING PRIMARY AND SECONDARY ELEMENTS (SELECTION MADE ACCORDING TO THE AREA WHICH CONTAINS THE POINT REPRESENTING THE OPERATING CONDITIONS. DOTTED LINES APPLY TO INTERMITTENT SERVICE. FROM ANSI STANDARD CODE FOR PRESSURE PIPING, B 31.1.0-1967.)

value. They should be installed and supported so
as to be level with each other at all times.

Reservoirs may be omitted if the differential pressure gage that is being used has zero or negligible
displacement.

II-II-14 Connecting Tubing. For connecting the
primary element to the secondary instruments, 1/2-in.
o.d. copper tubing with steel flared fittings may be
used for air, gas, steam, water, oil and most other
liquids when the operating conditions are within the
limits for copper tubing as shown in Fig. II-II-10.
For higher pressures and temperatures, 1/2-in. carbon steel, stainless steel or chrome molybdenum
steel tubing and steel flared fittings are recommend-

ed. For most gas and oil measurements, 1/2-in. steel
pipe with screwed fittings may be used.

All connecting tubing should be so arranged and
installed so as to have a slope of 1 in. per ft or more
Some representative arrangements of connecting
tubing are shown in Figs. II-II-11 through II-II-13.

II-II-15 Pressure and Temperature Instruments.
Manometers and other types of differential-pressure-
measuring gages, pressure gages, temperature-measur-
ing instruments and other instruments as needed
should be installed in accordance with the specific
instructions furnished by the manufacturer of each
instrument.

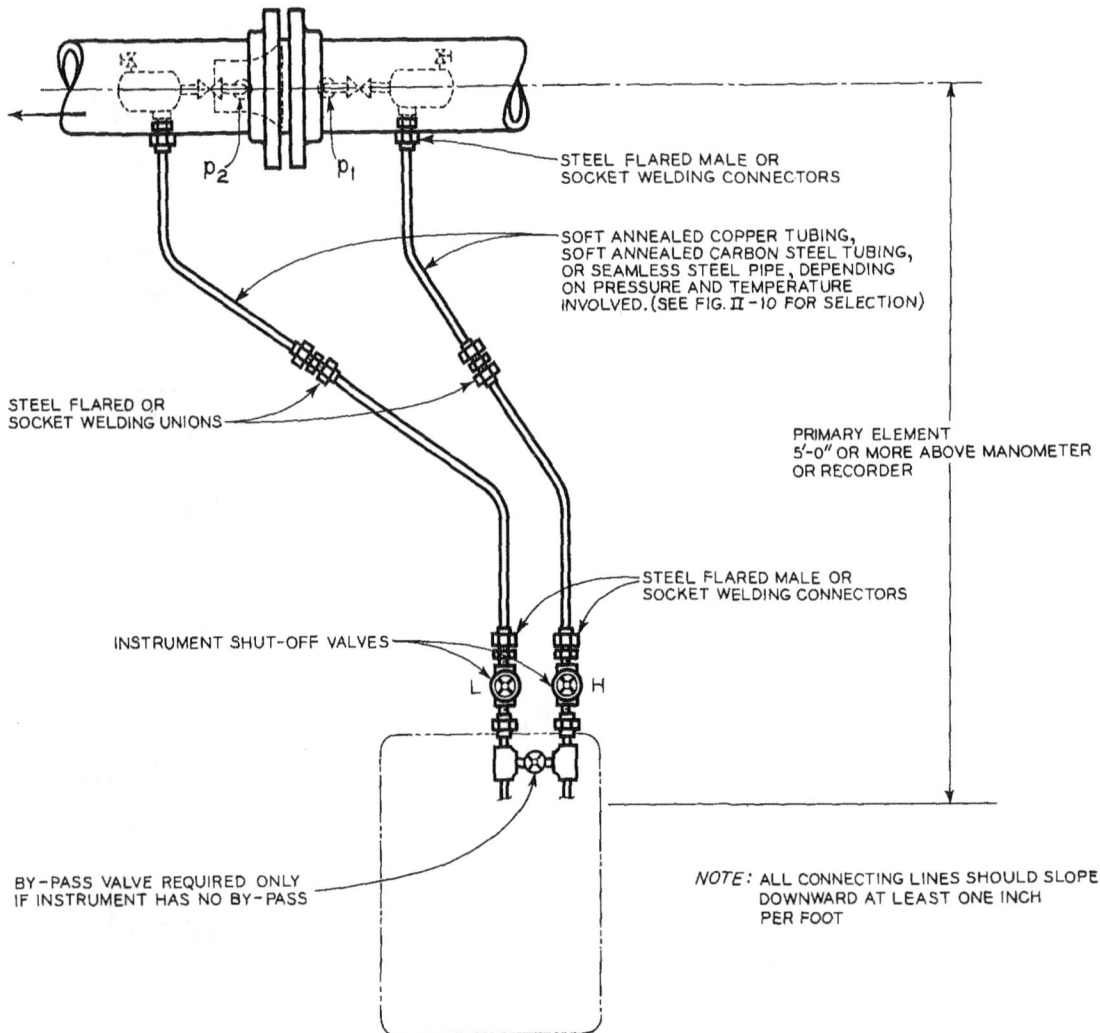

FIG. II-II-11 RECOMMENDED ARRANGEMENT OF PIPING BETWEEN
PRIMARY AND SECONDARY ELEMENTS WHEN
PRIMARY IS ABOVE SECONDARY AND METERED FLUID
IS A LIQUID, STEAM OR CONDENSABLE GAS
(SCHEMATIC)

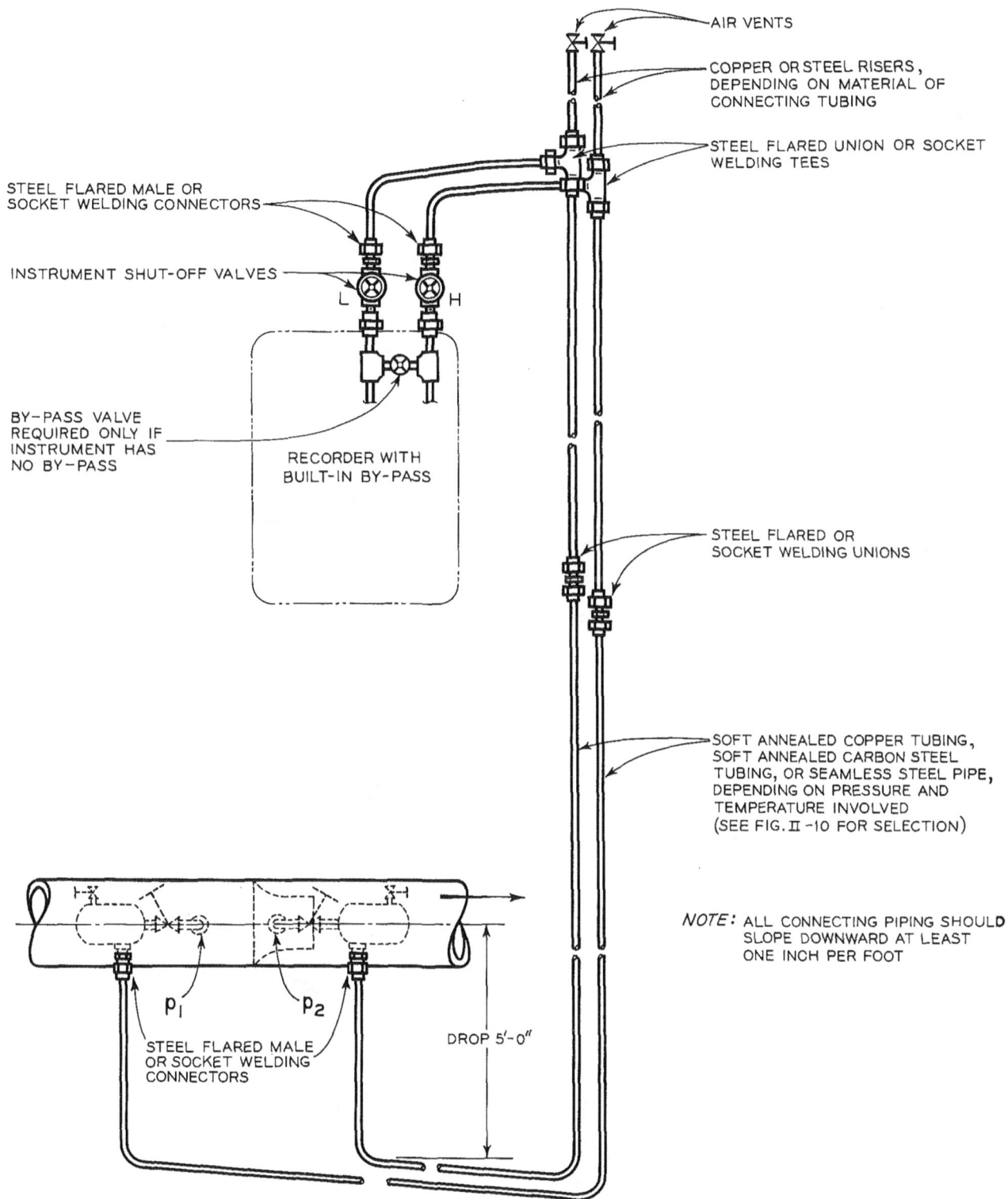

AIR VENTS

COPPER OR STEEL RISERS,
DEPENDING ON MATERIAL OF
CONNECTING TUBING

STEEL FLARED UNION OR SOCKET
WELDING TEES

STEEL FLARED MALE OR
SOCKET WELDING CONNECTORS

INSTRUMENT SHUT-OFF VALVES

L   H

BY-PASS VALVE
REQUIRED ONLY IF
INSTRUMENT HAS
NO BY-PASS

RECORDER WITH
BUILT-IN BY-PASS

STEEL FLARED OR
SOCKET WELDING UNIONS

SOFT ANNEALED COPPER TUBING,
SOFT ANNEALED CARBON STEEL
TUBING, OR SEAMLESS STEEL PIPE,
DEPENDING ON PRESSURE AND
TEMPERATURE INVOLVED
(SEE FIG. II -10 FOR SELECTION)

NOTE: ALL CONNECTING PIPING SHOULD
SLOPE DOWNWARD AT LEAST
ONE INCH PER FOOT

p_1 p_2

STEEL FLARED MALE
OR SOCKET WELDING
CONNECTORS

DROP 5'-0"

FIG. II-II-12 REPRESENTATION OF ARRANGEMENT OF PIPING
BETWEEN PRIMARY AND SECONDARY ELEMENTS
WHEN PRIMARY IS BELOW SECONDARY AND FLUID
IS A LIQUID, STEAM OR CONDENSABLE GAS

193

FIG. II-II-13

ILLUSTRATION OF ARRANGEMENTS OF PIPING BETWEEN
PRIMARY AND SECONDARY ELEMENTS WHEN METERING
AIR AND NONCONDENSING GASES

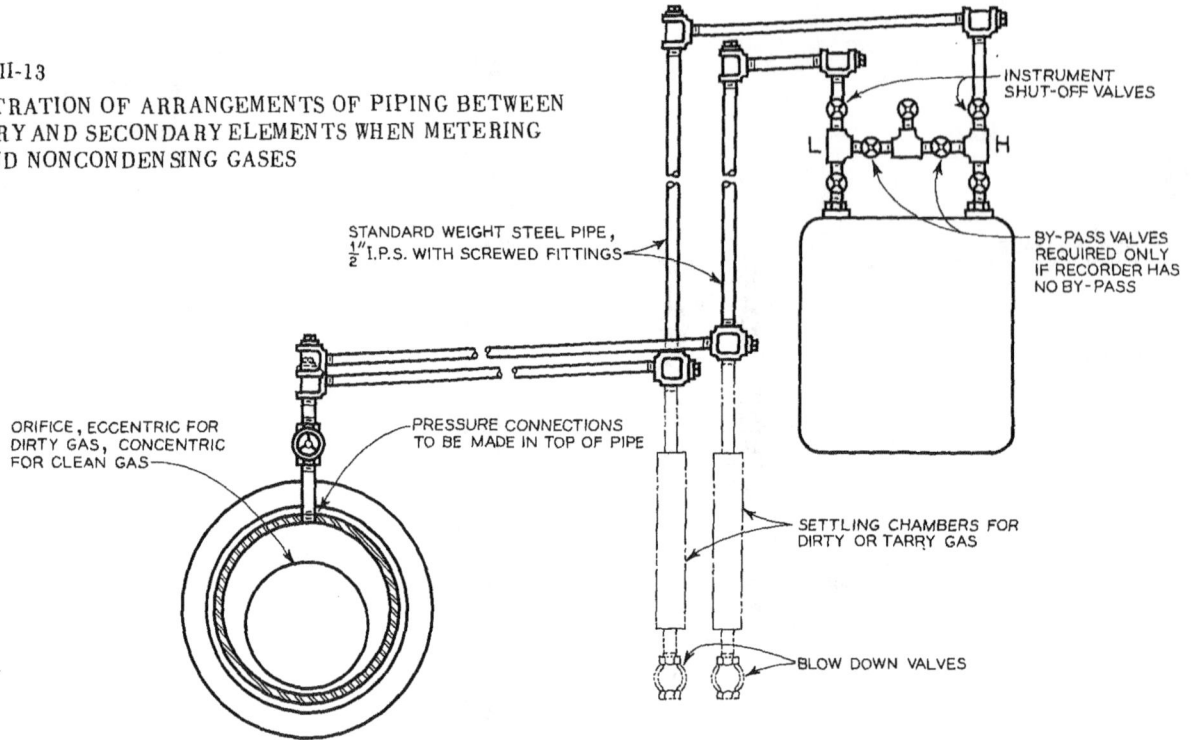

STANDARD WEIGHT STEEL PIPE,
½" I.P.S. WITH SCREWED FITTINGS

INSTRUMENT
SHUT-OFF VALVES

L   H

BY-PASS VALVES
REQUIRED ONLY
IF RECORDER HAS
NO BY-PASS

ORIFICE, ECCENTRIC FOR
DIRTY GAS, CONCENTRIC
FOR CLEAN GAS

PRESSURE CONNECTIONS
TO BE MADE IN TOP OF PIPE

SETTLING CHAMBERS FOR
DIRTY OR TARRY GAS

BLOW DOWN VALVES

SECONDARY ELEMENT ABOVE ORIFICE OR FLOW NOZZLE

NOTE: SLOPE OF ALL CONNECTING PIPES SHOULD NOT BE LESS THAN 1 INCH
PER FOOT TOWARD PRIMARY ELEMENT

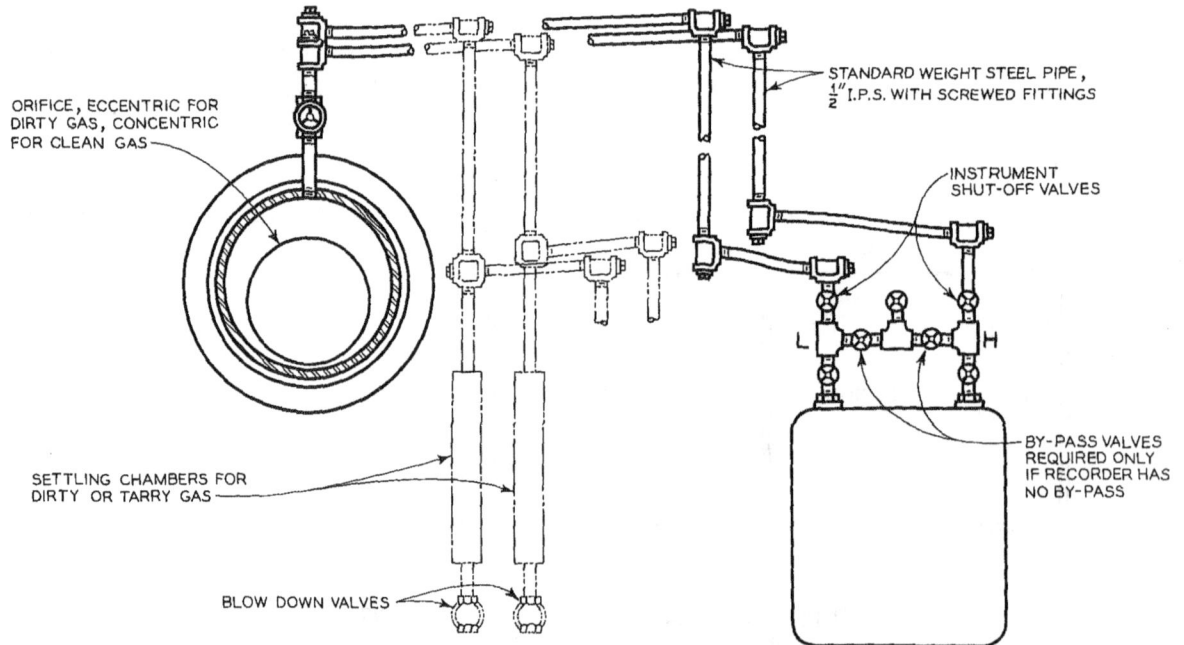

STANDARD WEIGHT STEEL PIPE,
½" I.P.S. WITH SCREWED FITTINGS

ORIFICE, ECCENTRIC FOR
DIRTY GAS, CONCENTRIC
FOR CLEAN GAS

INSTRUMENT
SHUT-OFF VALVES

L   H

BY-PASS VALVES
REQUIRED ONLY
IF RECORDER HAS
NO BY-PASS

SETTLING CHAMBERS FOR
DIRTY OR TARRY GAS

BLOW DOWN VALVES

SECONDARY ELEMENT BELOW ORIFICE OR FLOW NOZZLE

194

PROCEDURE FOR BLOWING OUT SETTLING CHAMBERS

1. OPEN BY-PASS VALVES AT SECONDARY AND CLOSE THE VALVES IN PRESSURE LINES **H** AND **L**.

2. OPEN SETTLING CHAMBER BLOW DOWN VALVES. WHEN ALL SEDIMENT IS OUT, ALLOW CLEAN WATER, STEAM OR GAS TO BLOW THROUGH FOR 10 TO 20 SECONDS.

3. CLOSE BLOW DOWN VALVES TIGHT.

4. IF MEASURING STEAM, WAIT UNTIL CHAMBERS AND CONNECTING PIPING ARE REFILLED WITH CONDENSATE.

5. SLOWLY OPEN VALVES IN LOW PRESSURE LINE **L**. CLOSE BY-PASS VALVES. SLOWLY OPEN VALVES IN HIGH PRESSURE LINE **H**.

NOTES:

SETTLING CHAMBERS SHOULD BE AT LEAST 2'-0" TO THE SIDE OF OR ABOVE THE SECONDARY ELEMENT.

CONNECTING PIPING FROM SETTLING CHAMBERS SLOPE AT LEAST 1 INCH PER FOOT TOWARD THE SECONDARY ELEMENT.

TO ALLEVIATE ACCUMULATION OF HYDROGEN GAS IN THE PIPING OR INSTRUMENT PRESSURE CASING, INSERT A MAGNESIUM ROD IN EACH SETTLING CHAMBER.

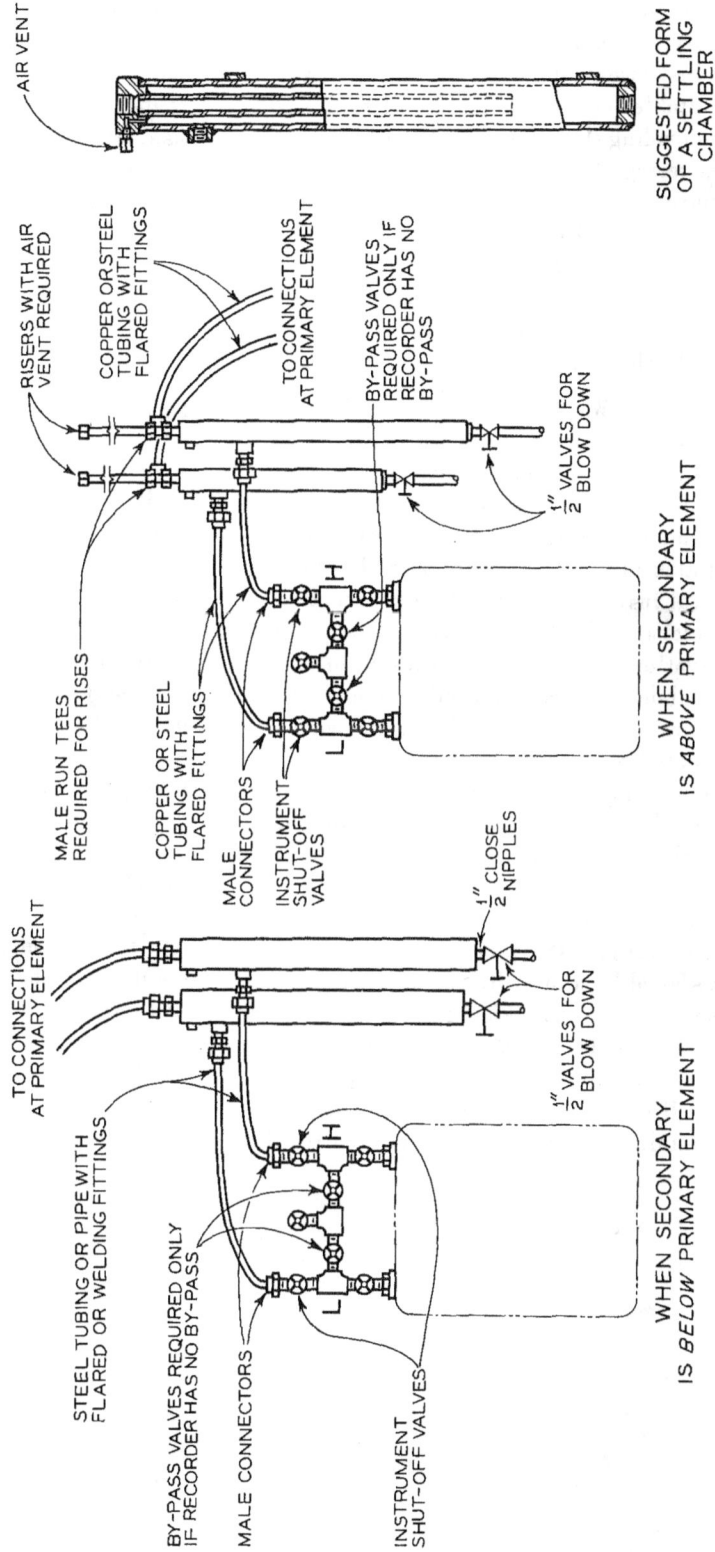

AIR VENT

SUGGESTED FORM OF A SETTLING CHAMBER

RISERS WITH AIR VENT REQUIRED

COPPER OR STEEL TUBING WITH FLARED FITTINGS

TO CONNECTIONS AT PRIMARY ELEMENT

BY-PASS VALVES REQUIRED ONLY IF RECORDER HAS NO BY-PASS

$\frac{1}{2}$" VALVES FOR BLOW DOWN

MALE RUN TEES REQUIRED FOR RISES

COPPER OR STEEL TUBING WITH FLARED FITTINGS

MALE CONNECTORS

INSTRUMENT SHUT-OFF VALVES

WHEN SECONDARY IS *ABOVE* PRIMARY ELEMENT

TO CONNECTIONS AT PRIMARY ELEMENT

$\frac{1}{2}$" CLOSE NIPPLES

$\frac{1}{2}$" VALVES FOR BLOW DOWN

STEEL TUBING OR PIPE WITH FLARED OR WELDING FITTINGS

BY-PASS VALVES REQUIRED ONLY IF RECORDER HAS NO BY-PASS

MALE CONNECTORS

INSTRUMENT SHUT-OFF VALVES

WHEN SECONDARY IS *BELOW* PRIMARY ELEMENT

FIG. II-II-14 SUGGESTED METHODS OF CONNECTING SETTLING CHAMBERS, IF NEEDED, INTO PIPING BETWEEN PRIMARY AND SECONDARY ELEMENTS

If needed, settling chambers or dirt traps may be installed as shown in Fig. II-II-14.

Shutoff valves to pressure instruments should be close to the instrument. For steam service, they should be suitable for the temperature of saturated steam corresponding to the actual line pressure. For use in the metering of other fluids, such valves should be suitable for the actual mainline pressure and temperature. For differential pressure gages, if a by-pass is not provided integrally with the instrument, such a valve should be incorporated between the shutoff valves and the gage itself, as illustrated in Figs. II-II-11 through II-II-13.

II-II-16 Drains. When measuring steam in a horizontal pipe, suitable drains or blowoffs should be provided on the under side of the pipe on the inlet and outlet sides of the primary element. If the pressures are measured through annular chambers, there should be drains in these chambers also. In other than horizontal installations, the pipe adjacent to the primary element should be drained at the point of minimum elevation. The valves or cocks used on these drains should be ones that will close tightly.

When measuring an incompressible fluid, vents should be located on the upper side of a horizontal pipe to eliminate any entrapped gas. In other than horizontal installations, the piping system should be vented at the highest point.

II-II-17 Calibrations. If a calibration of a particular differential pressure producer is desired or required, it should be made with the adjacent sections of pipe in which such primary element is to be used. The length of the actual piping to be used in the calibration should be at least that shown in the block of Fig. II-II-1, corresponding the closest to the actual installation arrangement. For best results,

any fittings which immediately precede the inlet run should be used in the calibration.

Whenever possible, the calibration range should encompass the entire range of Reynolds numbers corresponding to the rates of flow to be encountered in use. When the calibration facilities are inadequate to attain the highest Reynolds numbers to be encountered in use, the indications of the calibration may be extrapolated graphically or analytically. However, when such an extrapolation is used, the tolerance to be applied thereto should be increased to possibly double that of the direct calibration results.

Note: The extrapolation should not be extended to a condition that would correspond to a pressure ratio, p_2/p_1, below about 0.52, as a change of flow regime may occur in this pressure-ratio region. This does not apply to sonic-flow nozzles discussed latter.

II-II-18 Other Considerations. When the temperature of the fluid is above or below the ambient temperature, so that the difference in temperature might affect the fluid properties, thermal insulation of the entire meter section may be advisable.

If the meter is to be used in a special service, e.g., an acceptance test, the primary element should be sized so as to produce as high a differential pressure as operating conditions and auxiliary equipment will permit. During the time of such use, the primary element should be clean and undamaged; the inlet edge of an orifice, square and sharp; the inlet and throat sections of a flow nozzle, clean and smooth; and the inlet and throat of a Venturi tube, free of scale or incrustations. Such conditions should be established, by inspection, if possible, both before and after use.

It should be possible to read or estimate the smallest division of the scale of a manometer, pressure gage or chart of a recording gage to a value that will give the accuracy required for the service or test.

Chapter II-III

Primary Elements and Equations for Computing Rates of Flow

II-III-1 Symbols. The following symbols are used in describing the primary elements and in the equations given for computing rates of flow. Letters used to represent special factors in some equations are defined at the place of use, as also are special subscripts.

a	Area of an orifice, flow nozzle or Venturi throat	in.^{2}
C	Coefficient of discharge	ratio
c_p	Specific heat of a fluid at constant pressure	Btu/lb$_m$/°R
c_v	Specific heat of a fluid at constant volume	Btu/lb$_m$/°R
D	Diameter of pipe or meter tube	in.
d	Diameter of orifice, flow nozzle throat or Venturi throat	in.
E	Velocity of approach factor $= 1/\sqrt{1-\beta^4}$	number
F	Isentropic expansion function of a real gas (equation (I-5-124))	ratio
F_a	Area thermal expansion factor, from Fig. II-I-3	ratio
F_i	Isentropic expansion function of an ideal gas (equation (I-5-104))	ratio
G	Specific gravity; for a liquid the ratio of density of liquid to that of water at a defined temperature; for gases the ratio of the molecular weight of the gas to molecular weight of air	ratio
g	Acceleration due to gravity, local	ft/sec
g_c	Proportionality constant in the force-mass-acceleration equation $= 32.174$	number
h	Effective differential pressure	ft of fluid
h_w	Effective differential pressure	in. of water at 68 F
K	Flow coefficient $= CE$	ratio
MW	Molecular weight of a fluid	number
m	Mass rate of flow	lb$_m$/sec
p	Pressure, absolute	psia
p_t	Total or stagnation pressure	psia
q	Volume rate of flow	cu ft/sec
R	Gas constant in $pv = RT$ (here p is lb$_f$/ft^2)	ft · lb$_f$/lb$_m$·°R

R_D	Reynolds number based on D	ratio
R_d	Reynolds number based on d	ratio
r	Ratio of outlet to inlet static pressure $= p_2/p_1$	ratio
T	Absolute temperature	°R
V	Velocity	ft/sec
V_s	Velocity of sound (acoustic velocity)	ft/sec
v	Specific volume $= 1/\rho$	cu ft/lb$_m$
x	Ratio of differential pressure to inlet static pressure $= \Delta p/p_1$	ratio
Y	Expansion factor for a gas	ratio
Z	Compressibility factor for a real gas	ratio
β (beta)	Ratio of diameters $= d/D$	ratio
Γ (gamma)	Isentropic exponent of a real gas, a function of p_1, p_2 and T	number
γ (gamma)	Ratio of specific heats of a gas (ideal) $= c_p/c_v$	ratio
Δp (delta p)	Differential pressure $= p_1 - p_2$	psi
λ (lambda)	A Reynolds number reciprocal $= 1000/\sqrt{R_D}$ $= 1000/\sqrt{\beta R_d}$	ratio
μ (mu)	Absolute viscosity of a fluid	lb$_m$/ft·sec
ρ (rho)	Density	lb$_m$/cu ft
τ (tau)	Deflection of an orifice plate	in.
ϕ^* (phi)	Sonic-flow function of a real gas (equation (I-5-125))	number
ϕ_i^* (phi)	Sonic-flow function of an ideal gas (equation (I-5-105))	number

II-III-2 Thin-Plate Square-Edged Orifice: Material. The orifice plate should be stainless steel or other noncorrodible material suited to the fluid to be metered at the expected operating conditions. When the temperature of the fluid will exceed 600 F, the plate material should have a coefficient of thermal expansion no greater than that of the pipe flanges between which the plate will be installed. Whenever possible, the rate of change of temperature of the entire primary assembly should be kept as low as possible to avoid distortion of the plate from thermal stress.

The recommended thicknesses of orifice plates are given in Table II-III-1. These values are based on a maximum allowable deflection of $\tau/[0.5(D-d)]$ $\lessgtr 0.05$ (Fig. II-III-1). τ is a function of D, β, Δp and the material of the plate.

Table II-III-1 Minimum Recommended Thicknesses of Orifice Plates (inches)

Diff'l Pressure (in. H$_2$O)	Internal Diameter of Pipe (inches)				
	3 and less	6	10	20	30
	$\beta < 0.5$				
< 1000	1/8	1/8	3/16	3/8	1/2*
< 200	1/8	1/8	1/8	1/4	3/8
< 100	1/8	1/8	1/8	1/4	3/8
	$\beta > 0.5$				
< 1000	1/8	1/8	3/16	3/8	1/2
< 200	1/8	1/8	1/8	3/16	3/8
< 100	1/8	1/8	1/8	3/16	1/4

*For 1/2-in. plate in 30-in. line, maximum differential = 500 in.

II-III-3 The conventional and preferred use of orifices is to have the center of the orifice on the center line of the meter tube when installed. The use of eccentric and segmental orifices is treated later.

The width of the cylindrical surface of the orifice itself, measured normal to the plane of the inlet face of the plate, should be between $0.01 D$ and $0.02 D$ or $d/8$, whichever is smaller. If the thickness of the orifice plate exceeds the minimum of this requirement, then the outlet corner of the orifice should be beveled at an angle of about 45 deg to the face of the plate sufficiently to provide the minimum face width.

The inlet edge or corner of the orifice must be square, sharp and free from burrs, nicks, wire edge or rounding.

The inlet face of the orifice plate should be flat within 0.010 per in. of pipe diameter.

The actual diameter of the orifice hole should be carefully and accurately determined after all machine work on the plate has been completed. In doing this, particular care must be used not to damage or alter the inlet corner of the hole. Measurements should be made on at least three and preferably more diameters to determine a reliable average value of d. No diameter should differ from the average diameter by more than 0.05 per cent and preferably not more than 0.02 per cent.

II-III-4 For use in horizontal pipes, a drain hole may be provided in an orifice plate so located as to be flush with the bottom of the pipe when measuring gaseous fluids or flush with the top of the pipe when measuring liquids. If such a drain hole is provided, the diameter should be such that the hole area is

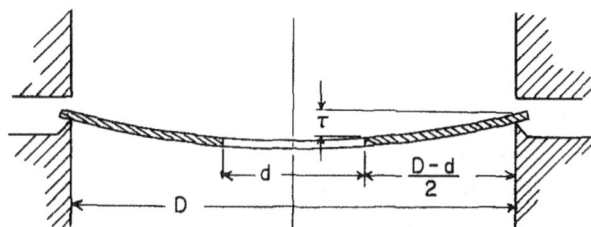

Simple Support (fitting)

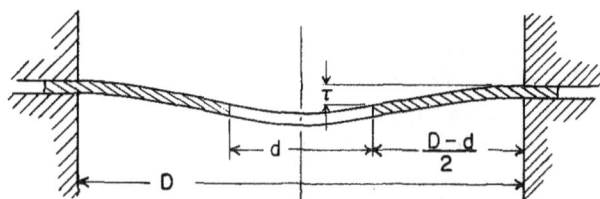

Rigid Support (flanges)

FIG. II-III-1 DEFLECTION OF AN ORIFICE PLATE BY DIFFERENTIAL PRESSURE

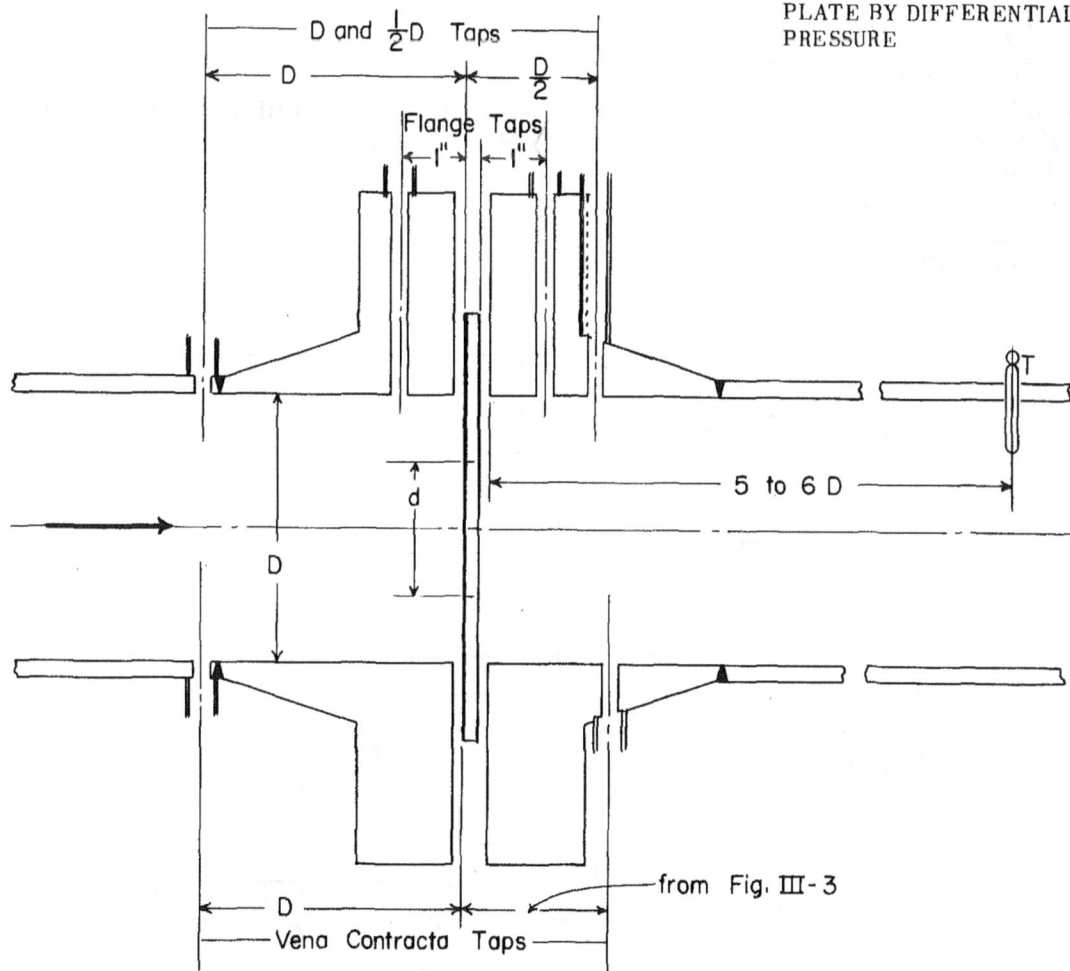

FIG. II-III-2 LOCATIONS OF PRESSURE TAPS USED WITH ORIFICES (WHEN A THERMOMETER IS REQUIRED, THE WELL FOR IT MAY BE LOCATED AS SHOWN BY T.)

less than 0.002 of the area of the orifice. In general, drain holes are considered of little value and are not recommended.

The design of the orifice plate and its outside diameter should be such as to facilitate centering the orifice accurately in the pipe line. Any method used for centering the orifice should provide that the center of the orifice is not further than 1/32 in. from the center line of the meter tube or pipe.

II-III-5 Pressure Taps. Any one of the following three pairs of pressure taps, shown by Fig. II-III-2, may be used:

1. *Flange Taps:* The centers of the inlet and outlet pressure taps are, respectively, 1 in. from the inlet and outlet faces of the orifice plate, are subject to a tolerance of ± 1/16 in. for β up to 0.40, and vary linearly to ± 1/64 in. at β = 0.75.

2. *1 D and 1/2 D Taps:* The center of the inlet pressure tap is 1 D preceding the inlet face of the orifice plate. The center of the outlet (downstream) pressure tap is 1/2 D from the *inlet* face of the orifice plate. These distances are subject to a tolerance varying linearly from ± 0.2 D at β = 0.20 to ± 0.05 D at β = 0.75.

3. *Vena Contracta Taps:* The center of the inlet tap is 1 D preceding the inlet face of the orifice plate. The distance of the outlet pressure tap from the *inlet* face of the orifice plate depends upon the diameter ratio, β, of the orifice to be used as shown by the heavy line of Fig. II-III-3. These distances are subject to a tolerance varying linearly from ± 0.2 D at β = 0.20 to ± 0.05 D at β = 0.75.

The pressure tap holes should be drilled (and preferably reamed) perpendicular to the inner surface of the meter tube or pipe in which the orifice plate is mounted. The corner of the hole with the inner surface of the pipe must be free of burrs and wire edge. It may be left square and sharp or dulled (rounded) very slightly.

II-III-6 Pressure Loss. The overall pressure loss with an orifice meter is shown by Fig. II-III-4.

II-III-7 Coefficients. Whenever possible it is desirable to calibrate an orifice in the meter-tube assembly in which it is to be used. When this is not done, the discharge coefficient to be used in computing the rate of fluid flow may be evaluated by the equation below which applies to the pressure taps used or from the value read from the appropriate table.

Values of discharge coefficients are given for flange taps in Table II-III-2, 1-D and 1/2-D taps in Table II-III-3, and vena contracta taps in Table II-III-4.

The equations and special values of the symbols by which the tables were computed are:

$$C = K/E = K \sqrt{1 - \beta^4}$$

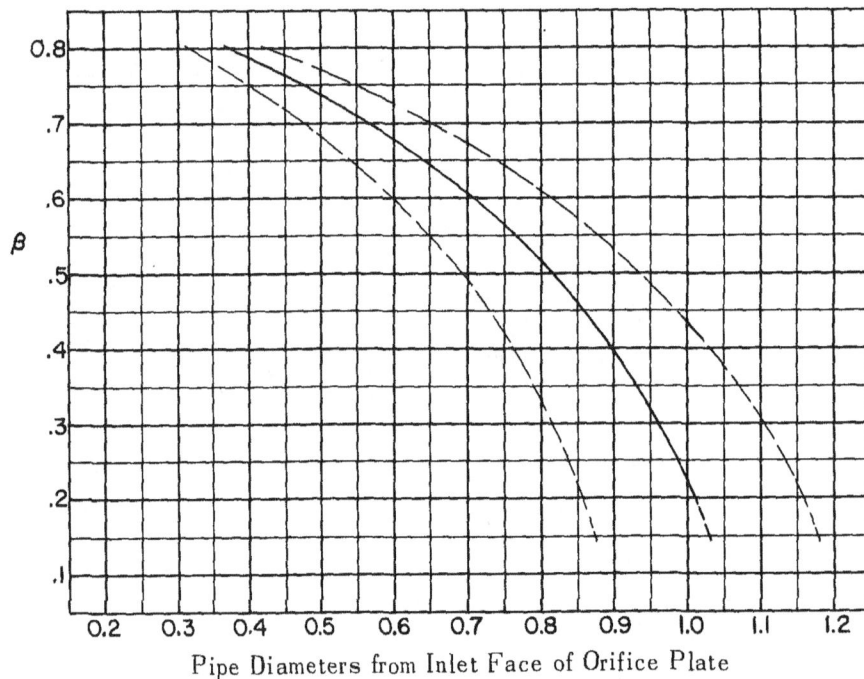

FIG. II-III-3 LOCATION OF VENA CONTRACTA OUTLET PRESSURE
TAP WITH CONCENTRIC SQUARE-EDGED ORIFICES
(BROKEN LINES SHOW MAXIMUM VARIATION LIMITS.)

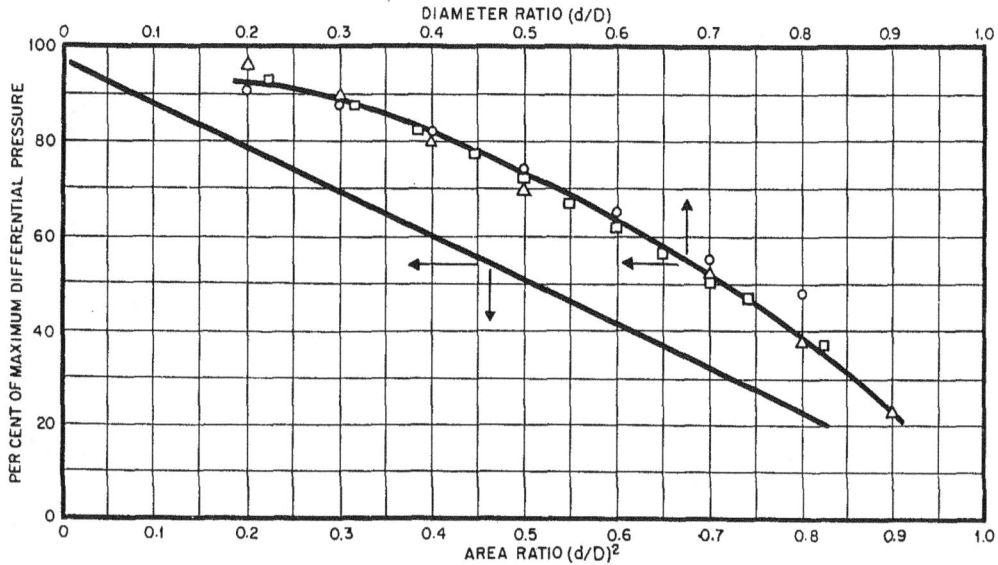

FIG. II-III-4 OVERALL PRESSURE LOSS ACROSS THIN-PLATE ORIFICES

K = Flow coefficient corresponding to any specific set of values of D, β, and R_d (or R_D)

K_o = The limiting value of K for any specific values of D and β when R_d (or R_D) becomes infinitely large

$R_D = \beta R_d$

For flange taps,

K_e = The particular value of K for any specific values of D and β when $R_d = (10^6\, d)/15$

$K = K_o \left(1 + \dfrac{A}{R_d}\right)$

$K_o = K_e \left(\dfrac{10^6 d}{10^6 d + 15A}\right)$

$K_e = 0.5993 + \dfrac{0.007}{D} + \left(0.364 + \dfrac{0.076}{\sqrt{D}}\right)\beta^4$

$\qquad + 0.4\left(1.6 - \dfrac{1}{D}\right)^5 \left[\left(0.07 + \dfrac{0.5}{D}\right) - \beta\right]^{5/2}$

$\qquad - \left(0.009 + \dfrac{0.034}{D}\right)(0.5 - \beta)^{3/2}$

$\qquad + \left(\dfrac{65}{D^2} + 3\right)(\beta - 0.7)^{3/2}$ ⟨circled⟩ $\qquad$ (II-III-1)

Page 201 — Equation (II-III-1):
Last factor should be $(\beta - 0.7)^{5/2}$

and

$A = d\left(830 - 5000\beta + 9000\dot\beta^2 - 4200\beta^3 \right.$

$\qquad\qquad \left. + \dfrac{530}{\sqrt{D}}\right)$ $\qquad$ (II-III-2)

Note: In equation (II-III-1), each of the last three terms for some value of β reduces to the form of $x\sqrt{-1}$, i.e., to an "imaginary" number. In such cases the term is to be dropped.

For 1-D and 1/2-D taps and vena contracta taps,

$$K = K_o + b\lambda$$

and

$$\lambda = 1000/\sqrt{R_D} = 1000/\sqrt{\beta R_d}$$

For the 1-D and 1/2-D taps,

$K_o = (0.6014 - 0.01352D^{-1/4})$

$\qquad + (0.3760 + 0.07257D^{-1/4})$

$\qquad \left(\dfrac{0.00025}{D^2\beta^2 + 0.0025D} + \beta^4 + 1.5\beta^{16}\right)$ (II-III-3)

and

$b = \left(0.0002 + \dfrac{0.0011}{D}\right) + \left(0.0038 + \dfrac{0.0004}{D}\right)$

$\qquad\qquad [\beta^2 + (16.5 + 5D)\beta^{16}]$ (II-III-4)

For vena contracts taps,

$K_o = 0.5992$

$\qquad + 0.4252 \left(\dfrac{0.0006}{D^2\beta^2 + 0.01D}\right.$

$\qquad\qquad\qquad \left. + \beta^4 + 1.25\beta^{16}\right)$ (II-III-5)

and

$b = 0.00025 + 0.002325(\beta + 1.75\beta^4$

$\qquad\qquad + 10\beta^{12} + 2D\beta^{16})$ (II-III-6)

Table II-III-2 (a) Flange Taps: Discharge Coefficients, C, for Square-Edged Orifices

(2-in. Pipe, D = 2.067 in.)

β \ R_d	10,000	12,000	14,000	16,000	18,000	20,000	25,000	30,000	40,000	50,000	75,000	100,000	500,000	1,000,000
.1500	.6109	.6089	.6075	.6065	.6056	.6050	.6038	.6030	.6020	.6014	.6006	.6002	.5993	.5992
.2000	.6098	.6077	.6061	.6050	.6041	.6034	.6021	.6012	.6001	.5995	.5986	.5982	.5972	.5970
.2500	.6104	.6081	.6064	.6052	.6043	.6035	.6021	.6012	.6001	.5994	.5985	.5980	.5969	.5968
.3000	.6125	.6100	.6083	.6070	.6059	.6051	.6037	.6027	.6015	.6007	.5998	.5993	.5981	.5979
.3500	.6156	.6130	.6110	.6096	.6085	.6076	.6060	.6049	.6036	.6028	.6017	.6012	.5999	.5997
.4000	.6197	.6166	.6145	.6128	.6115	.6105	.6087	.6075	.6059	.6050	.6038	.6032	.6017	.6015
.4500	.6249	.6213	.6187	.6168	.6153	.6140	.6119	.6104	.6086	.6075	.6061	.6053	.6036	.6034
.5000	.6314	.6270	.6238	.6215	.6196	.6182	.6155	.6138	.6116	.6102	.6085	.6076	.6055	.6052
.5500	.6391	.6337	.6298	.6269	.6246	.6228	.6195	.6174	.6147	.6130	.6109	.6098	.6072	.6068
.5750	.6434	.6374	.6331	.6298	.6273	.6253	.6217	.6193	.6163	.6145	.6121	.6109	.6080	.6076
.6000	.6479	.6412	.6365	.6329	.6301	.6279	.6239	.6212	.6179	.6158	.6132	.6118	.6086	.6082
.6250	.6525	.6451	.6399	.6359	.6328	.6304	.6259	.6230	.6193	.6171	.6141	.6126	.6091	.6087
.6500	.6571	.6489	.6431	.6388	.6354	.6327	.6278	.6245	.6205	.6180	.6148	.6131	.6092	.6087
.6750	.6614	.6525	.6461	.6413	.6376	.6346	.6292	.6256	.6212	.6185	.6149	.6131	.6088	.6083
.7000	.6652	.6554	.6484	.6432	.6391	.6359	.6300	.6261	.6212	.6183	.6144	.6124	.6077	.6071
.7250	.6697	.6591	.6515	.6458	.6413	.6378	.6314	.6271	.6218	.6186	.6143	.6122	.6071	.6065
.7500	.6788	.6672	.6589	.6526	.6478	.6439	.6369	.6323	.6265	.6230	.6183	.6160	.6104	.6097

Table II-III-2(b) Flange Taps: Discharge Coefficients, C, for Square-Edged Orifices

(4-in. Pipe, D = 4.026 in.)

β \ R_d	10,000	12,000	14,000	16,000	18,000	20,000	25,000	30,000	40,000	50,000	75,000	100,000	500,000	1,000,000
.1500	.6126	.6094	.6071	.6054	.6041	.6030	.6011	.5998	.5982	.5973	.5960	.5954	.5939	.5937
.2000	.6147	.6114	.6090	.6072	.6058	.6047	.6026	.6013	.5996	.5986	.5973	.5966	.5950	.5948
.2500	.6167	.6133	.6108	.6090	.6076	.6064	.6044	.6030	.6013	.6003	.5989	.5983	.5966	.5964
.3000	.6187	.6152	.6127	.6108	.6093	.6082	.6061	.6047	.6029	.6019	.6005	.5998	.5981	.5979
.3500		.6178	.6152	.6132	.6116	.6103	.6081	.6066	.6047	.6036	.6021	.6013	.5995	.5993
.4000		.6219	.6188	.6166	.6148	.6133	.6108	.6091	.6069	.6056	.6039	.6031	.6010	.6008
.4500		.6278	.6241	.6213	.6192	.6175	.6144	.6123	.6097	.6082	.6061	.6051	.6026	.6023
.5000			.6311	.6276	.6249	.6228	.6189	.6163	.6131	.6111	.6086	.6073	.6042	.6038
.5500			.6398	.6354	.6320	.6292	.6243	.6210	.6169	.6144	.6112	.6095	.6056	.6051
.5750			.6448	.6398	.6359	.6329	.6273	.6236	.6190	.6162	.6125	.6107	.6062	.6057
.6000			.6501	.6445	.6401	.6367	.6304	.6263	.6211	.6179	.6138	.6117	.6067	.6061
.6250				.6493	.6445	.6406	.6336	.6289	.6231	.6196	.6149	.6126	.6070	.6063
.6500				.6541	.6487	.6444	.6366	.6313	.6248	.6209	.6157	.6131	.6069	.6061
.6750				.6587	.6527	.6479	.6392	.6334	.6262	.6219	.6161	.6132	.6062	.6054
.7000				.6628	.6562	.6508	.6413	.6349	.6269	.6221	.6157	.6125	.6049	.6039
.7250					.6594	.6535	.6430	.6360	.6272	.6220	.6150	.6115	.6031	.6020
.7500					.6634	.6571	.6456	.6379	.6283	.6226	.6149	.6111	.6019	.6007

Table II-III-2(c) Flange Taps: Discharge Coefficients, C, for Square-Edged Orifices

(8-in. Pipe, D = 7.981 in.)

β \ R_d	14,000	16,000	18,000	20,000	25,000	30,000	40,000	50,000	75,000	100,000	500,000	1,000,000
.1500	.6166	.6137	.6114	.6096	.6064	.6042	.6015	.5999	.5977	.5967	.5941	.5937
.2000	.6184	.6155	.6132	.6114	.6081	.6060	.6032	.6016	.5994	.5983	.5957	.5954
.2500	.6190	.6162	.6140	.6122	.6091	.6070	.6044	.6028	.6007	.5997	.5971	.5968
.3000	.6197	.6170	.6148	.6131	.6101	.6080	.6055	.6039	.6019	.6009	.5984	.5981
.3500	.6217	.6189	.6167	.6149	.6117	.6096	.6070	.6054	.6032	.6022	.5996	.5993
.4000	.6261	.6228	.6203	.6183	.6147	.6123	.6093	.6074	.6050	.6038	.6009	.6006
.4500	.6334	.6294	.6263	.6238	.6193	.6164	.6126	.6104	.6074	.6059	.6023	.6019
.5000	.6443	.6390	.6350	.6318	.6259	.6220	.6172	.6142	.6104	.6084	.6037	.6032
.5500	.6586	.6518	.6464	.6421	.6344	.6292	.6228	.6190	.6138	.6112	.6051	.6043
.5750		.6592	.6531	.6482	.6393	.6334	.6260	.6216	.6157	.6127	.6056	.6047
.6000		.6674	.6603	.6547	.6446	.6378	.6294	.6243	.6175	.6142	.6061	.6050
.6250			.6680	.6616	.6501	.6424	.6328	.6270	.6193	.6155	.6062	.6051
.6500			.6759	.6687	.6556	.6469	.6361	.6295	.6208	.6165	.6061	.6048
.6750				.6757	.6610	.6513	.6391	.6317	.6220	.6171	.6053	.6039
.7000				.6823	.6660	.6551	.6415	.6333	.6224	.6170	.6039	.6023
.7250					.6705	.6584	.6433	.6343	.6222	.6162	.6017	.5999
.7500					.6749	.6617	.6451	.6351	.6219	.6153	.5993	.5973

Table II-III-2(d) Flange Taps: Discharge Coefficients, C, for Square-Edged Orifices

(16-in. Pipe, D = 15.25 in.)

β \ R_d	18,000	20,000	25,000	30,000	40,000	50,000	75,000	100,000	500,000	1,000,000
.1500	.6244	.6213	.6158	.6122	.6076	.6049	.6012	.5994	.5950	.5944
.2000	.6249	.6220	.6167	.6131	.6087	.6060	.6025	.6007	.5965	.5960
.2500	.6236	.6210	.6161	.6129	.6089	.6065	.6032	.6016	.5977	.5973
.3000			.6157	.6128	.6091	.6068	.6039	.6024	.5989	.5984
.3500			.6167	.6138	.6101	.6079	.6050	.6035	.5999	.5995
.4000			.6202	.6168	.6127	.6102	.6068	.6052	.6011	.6006
.4500			.6269	.6226	.6172	.6140	.6098	.6076	.6025	.6018
.5000				.6314	.6241	.6197	.6139	.6110	.6039	.6031
.5500				.6433	.6332	.6272	.6191	.6150	.6054	.6042
.5750				.6504	.6386	.6315	.6221	.6173	.6060	.6046
.6000				.6581	.6444	.6362	.6252	.6197	.6065	.6049
.6250				.6664	.6505	.6410	.6283	.6220	.6068	.6049
.6500				.6749	.6568	.6459	.6314	.6241	.6067	.6045
.6750				.6836	.6630	.6506	.6341	.6259	.6061	.6036
.7000				.6919	.6687	.6548	.6363	.6270	.6047	.6020
.7250				.6999	.6740	.6585	.6378	.6275	.6026	.5995
.7500				.7077	.6791	.6619	.6390	.6276	.6001	.5966

Table II-III-3 (a) Taps at 1 D and 1/2 D: Discharge Coefficients, C, for Square-Edged Orifices

(2-in. Pipe, D = 2.067 in.)

β \ R_d	10,000	12,000	14,000	16,000	18,000	20,000	25,000	30,000	40,000	50,000	75,000	100,000	500,000	1,000,000
.1500	.6125	.6106	.6092	.6080	.6071	.6063	.6047	.6035	.6019	.6008	.5990	.5980	.5943	.5934
.2000	.6109	.6092	.6078	.6067	.6058	.6051	.6036	.6025	.6009	.5999	.5982	.5973	.5938	.5930
.2500	.6107	.6090	.6076	.6066	.6057	.6049	.6035	.6024	.6009	.5998	.5982	.5973	.5938	.5930
.3000	.6114	.6096	.6083	.6072	.6063	.6056	.6041	.6030	.6015	.6004	.5988	.5978	.5943	.5935
.3500	.6129	.6111	.6097	.6086	.6077	.6069	.6053	.6042	.6026	.6015	.5999	.5989	.5953	.5944
.4000	.6151	.6132	.6118	.6106	.6097	.6088	.6072	.6061	.6044	.6033	.6015	.6005	.5967	.5958
.4500	.6181	.6161	.6146	.6133	.6123	.6115	.6098	.6085	.6068	.6056	.6038	.6027	.5987	.5978
.5000	.6216	.6196	.6180	.6167	.6156	.6147	.6129	.6116	.6098	.6085	.6066	.6054	.6013	.6003
.5500	.6257	.6235	.6218	.6205	.6193	.6184	.6165	.6151	.6132	.6119	.6098	.6086	.6042	.6032
.5750	.6279	.6256	.6239	.6225	.6213	.6203	.6184	.6170	.6150	.6136	.6115	.6103	.6058	.6047
.6000	.6301	.6278	.6260	.6245	.6233	.6223	.6203	.6189	.6168	.6154	.6133	.6120	.6073	.6062
.6250	.6323	.6299	.6281	.6266	.6253	.6243	.6222	.6207	.6186	.6172	.6149	.6136	.6088	.6077
.6500	.6346	.6321	.6302	.6286	.6273	.6262	.6241	.6226	.6204	.6189	.6165	.6151	.6102	.6090
.6750	.6369	.6343	.6323	.6307	.6293	.6282	.6260	.6243	.6220	.6204	.6180	.6165	.6113	.6101
.7000	.6395	.6367	.6345	.6328	.6314	.6302	.6278	.6260	.6236	.6219	.6193	.6178	.6122	.6109
.7250	.6424	.6394	.6371	.6352	.6336	.6323	.6298	.6278	.6252	.6234	.6205	.6188	.6128	.6114
.7500	.6463	.6429	.6403	.6382	.6364	.6349	.6320	.6299	.6269	.6249	.6217	.6198	.6130	.6114

Table II-III-3 (b) Taps at 1 D and 1/2 D: Discharge Coefficients, C, for Square-Edged Orifices

(4-in. Pipe, D = 4.026 in.)

β \ R_d	10,000	12,000	14,000	16,000	18,000	20,000	25,000	30,000	40,000	50,000	75,000	100,000	500,000	1,000,000
.1500	.6067	.6054	.6044	.6037	.6030	.6024	.6014	.6006	.5994	.5987	.5975	.5968	.5943	.5937
.2000	.6063	.6051	.6041	.6033	.6027	.6022	.6011	.6003	.5993	.5985	.5974	.5967	.5942	.5936
.2500	.6068	.6055	.6046	.6038	.6031	.6026	.6015	.6007	.5996	.5989	.5977	.5970	.5945	.5939
.3000	.6080	.6067	.6056	.6048	.6041	.6036	.6025	.6016	.6005	.5997	.5984	.5977	.5951	.5945
.3500		.6084	.6073	.6064	.6057	.6051	.6039	.6030	.6018	.6009	.5996	.5989	.5961	.5954
.4000		.6107	.6095	.6086	.6078	.6072	.6059	.6050	.6036	.6027	.6013	.6005	.5975	.5968
.4500		.6136	.6124	.6114	.6105	.6098	.6084	.6074	.6060	.6050	.6035	.6026	.5994	.5986
.5000			.6157	.6146	.6138	.6130	.6115	.6104	.6089	.6078	.6062	.6052	.6018	.6009
.5500			.6195	.6183	.6174	.6166	.6150	.6138	.6121	.6110	.6093	.6082	.6045	.6036
.5750			.6215	.6203	.6193	.6184	.6168	.6156	.6138	.6127	.6109	.6098	.6059	.6050
.6000			.6235	.6223	.6212	.6203	.6186	.6173	.6156	.6143	.6124	.6113	.6073	.6063
.6250				.6243	.6232	.6223	.6205	.6191	.6173	.6160	.6140	.6128	.6086	.6076
.6500				.6263	.6251	.6242	.6223	.6209	.6189	.6176	.6155	.6142	.6098	.6087
.6750				.6284	.6272	.6261	.6241	.6226	.6205	.6191	.6168	.6155	.6107	.6096
.7000				.6308	.6294	.6283	.6261	.6244	.6221	.6205	.6181	.6166	.6114	.6102
.7250					.6321	.6308	.6283	.6265	.6239	.6221	.6193	.6177	.6118	.6105
.7500					.6356	.6341	.6312	.6290	.6260	.6239	.6207	.6188	.6120	.6103

Table II-III-3 (c) Taps at 1 D and 1/2 D: Discharge Coefficients, C, for Square-Edged Orifices

(8-in. Pipe, $D = 7.981$ in.)

β \ R_d	14,000	16,000	18,000	20,000	25,000	30,000	40,000	50,000	75,000	100,000	500,000	1,000,000
.1500	.6028	.6022	.6017	.6012	.6004	.5998	.5990	.5984	.5975	.5970	.5950	.5946
.2000	.6029	.6023	.6018	.6014	.6005	.5999	.5991	.5985	.5976	.5971	.5951	.5947
.2500	.6036	.6030	.6025	.6020	.6012	.6005	.5996	.5990	.5981	.5975	.5955	.5950
.3000	.6049	.6042	.6036	.6031	.6022	.6015	.6006	.5999	.5989	.5983	.5961	.5956
.3500	.6066	.6059	.6053	.6047	.6037	.6030	.6019	.6012	.6001	.5994	.5971	.5965
.4000	.6089	.6081	.6074	.6068	.6057	.6049	.6038	.6030	.6018	.6010	.5984	.5978
.4500	.6117	.6108	.6101	.6094	.6082	.6073	.6061	.6052	.6039	.6031	.6002	.5995
.5000	.6150	.6140	.6132	.6125	.6112	.6102	.6088	.6079	.6064	.6055	.6024	.6017
.5500	.6186	.6176	.6167	.6160	.6145	.6134	.6119	.6109	.6093	.6083	.6049	.6041
.5750		.6195	.6186	.6178	.6162	.6151	.6135	.6125	.6108	.6098	.6062	.6054
.6000		.6214	.6205	.6196	.6180	.6168	.6152	.6140	.6123	.6112	.6075	.6066
.6250			.6224	.6215	.6198	.6186	.6168	.6156	.6137	.6126	.6086	.6077
.6500			.6245	.6236	.6217	.6204	.6185	.6172	.6152	.6140	.6097	.6087
.6750				.6258	.6238	.6223	.6202	.6188	.6166	.6152	.6105	.6094
.7000				.6285	.6262	.6245	.6221	.6205	.6180	.6165	.6111	.6099
.7250					.6293	.6273	.6245	.6226	.6197	.6179	.6115	.6100
.7500					.6337	.6313	.6278	.6254	.6218	.6196	.6118	.6099

Table II-III-3 (d) Taps at 1 D and 1/2 D: Discharge Coefficients, C, for Square-Edged Orifices

(16-in. Pipe, $D = 15.25$ in.)

β \ R_d	18,000	20,000	25,000	30,000	40,000	50,000	75,000	100,000	500,000	1,000,000
.1500	.6015	.6012	.6005	.6000	.5993	.5988	.5980	.5976	.5959	.5956
.2000	.6018	.6015	.6008	.6002	.5995	.5990	.5982	.5978	.5961	.5957
.2500	.6026	.6022	.6015	.6009	.6001	.5996	.5987	.5982	.5965	.5960
.3000			.6026	.6020	.6011	.6005	.5996	.5990	.5971	.5966
.3500			.6041	.6034	.6024	.6018	.6008	.6002	.5980	.5975
.4000			.6061	.6053	.6042	.6035	.6024	.6017	.5993	.5987
.4500			.6085	.6077	.6065	.6057	.6044	.6037	.6010	.6003
.5000				.6104	.6091	.6082	.6068	.6060	.6031	.6024
.5500				.6135	.6121	.6111	.6096	.6087	.6054	.6046
.5750				.6152	.6137	.6126	.6110	.6100	.6066	.6058
.6000				.6169	.6153	.6142	.6125	.6114	.6077	.6069
.6250				.6188	.6170	.6158	.6139	.6128	.6088	.6079
.6500				.6208	.6189	.6175	.6155	.6142	.6098	.6088
.6750				.6233	.6210	.6195	.6171	.6157	.6107	.6095
.7000				.6265	.6238	.6220	.6191	.6174	.6114	.6099
.7250				.6310	.6276	.6253	.6217	.6196	.6120	.6102
.7500				.6376	.6332	.6302	.6255	.6227	.6128	.6104

205

Table II-III-4 (a) Vena Contracta Taps: Discharge Coefficients, C, for Square-Edged Orifices

(2-in. Pipe, D = 2.067 in.)

β \ R_d	10,000	12,000	14,000	16,000	18,000	20,000	25,000	30,000	40,000	50,000	75,000	100,000	500,000	1,000,000
.1500	.6100	.6086	.6076	.6067	.6060	.6054	.6043	.6034	.6022	.6014	.6001	.5994	.5996	.5960
.2000	.6099	.6085	.6074	.6065	.6058	.6051	.6039	.6030	.6018	.6009	.5996	.5988	.5960	.5953
.2500	.6105	.6090	.6079	.6070	.6062	.6055	.6043	.6033	.6020	.6011	.5998	.5989	.5960	.5953
.3000	.6117	.6101	.6089	.6079	.6071	.6065	.6051	.6041	.6028	.6018	.6004	.5995	.5964	.5956
.3500	.6134	.6118	.6105	.6095	.6086	.6079	.6065	.6054	.6040	.6030	.6014	.6005	.5972	.5964
.4000	.6157	.6140	.6126	.6115	.6106	.6099	.6084	.6072	.6057	.6046	.6030	.6020	.5985	.5977
.4500	.6187	.6168	.6154	.6142	.6132	.6124	.6108	.6096	.6080	.6069	.6051	.6041	.6003	.5994
.5000	.6223	.6203	.6187	.6175	.6164	.6156	.6138	.6126	.6108	.6096	.6077	.6066	.6026	.6017
.5500	.6264	.6243	.6226	.6213	.6201	.6192	.6174	.6160	.6141	.6128	.6108	.6096	.6053	.6042
.5750	.6287	.6264	.6247	.6233	.6221	.6212	.6192	.6178	.6158	.6145	.6124	.6111	.6066	.6056
.6000	.6310	.6287	.6269	.6254	.6242	.6232	.6212	.6197	.6176	.6162	.6140	.6127	.6080	.6069
.6250	.6334	.6309	.6290	.6275	.6262	.6251	.6230	.6215	.6193	.6178	.6155	.6142	.6093	.6081
.6500	.6357	.6331	.6311	.6295	.6282	.6271	.6249	.6232	.6209	.6194	.6170	.6155	.6103	.6091
.6750	.6380	.6353	.6332	.6315	.6301	.6289	.6265	.6248	.6224	.6207	.6182	.6166	.6112	.6099
.7000	.6403	.6374	.6351	.6333	.6318	.6305	.6280	.6262	.6236	.6218	.6191	.6174	.6116	.6102
.7250	.6425	.6394	.6369	.6349	.6333	.6319	.6292	.6273	.6245	.6226	.6196	.6179	.6116	.6101
.7500	.6446	.6412	.6385	.6364	.6346	.6331	.6302	.6280	.6250	.6229	.6197	.6178	.6109	.6093

Table II-III-4 (b) Vena Contracta Taps: Discharge Coefficients, C, for Square-Edged Orifices

(4-in. Pipe, D = 4.026 in.)

β \ R_d	10,000	12,000	14,000	16,000	18,000	20,000	25,000	30,000	40,000	50,000	75,000	100,000	500,000	1,000,000
.1500	.6084	.6071	.6060	.6052	.6045	.6039	.6027	.6018	.6006	.5998	.5986	.5978	.5951	.5944
.2000	.6089	.6075	.6064	.6055	.6048	.6042	.6030	.6021	.6008	.6000	.5987	.5979	.5951	.5944
.2500	.6099	.6084	.6072	.6063	.6055	.6049	.6036	.6027	.6014	.6005	.5991	.5983	.5953	.5946
.3000	.6112	.6097	.6085	.6075	.6067	.6060	.6047	.6037	.6023	.6014	.5999	.5990	.5959	.5952
.3500	.6131	.6114	.6101	.6091	.6083	.6075	.6061	.6051	.6036	.6026	.6011	.6002	.5969	.5961
.4000	.6155	.6137	.6124	.6113	.6104	.6096	.6081	.6070	.6054	.6044	.6027	.6018	.5982	.5974
.4500		.6166	.6152	.6140	.6130	.6122	.6106	.6094	.6078	.6066	.6049	.6038	.6001	.5992
.5000		.6201	.6186	.6173	.6163	.6154	.6137	.6124	.6106	.6094	.6076	.6064	.6024	.6015
.5500		.6242	.6225	.6211	.6200	.6191	.6172	.6159	.6140	.6127	.6106	.6094	.6051	.6041
.5750			.6246	.6232	.6220	.6210	.6191	.6177	.6157	.6144	.6123	.6110	.6065	.6054
.6000			.6268	.6253	.6241	.6231	.6211	.6196	.6175	.6161	.6139	.6126	.6079	.6068
.6250			.6290	.6274	.6262	.6251	.6230	.6214	.6192	.6178	.6155	.6141	.6092	.6080
.6500				.6295	.6282	.6270	.6248	.6232	.6209	.6193	.6169	.6154	.6103	.6090
.6750				.6315	.6301	.6289	.6266	.6248	.6224	.6207	.6181	.6166	.6111	.6098
.7000				.6335	.6319	.6307	.6281	.6263	.6237	.6219	.6191	.6175	.6116	.6102
.7250					.6336	.6322	.6295	.6275	.6247	.6227	.6197	.6180	.6116	.6100
.7500					.6352	.6337	.6307	.6285	.6254	.6232	.6200	.6180	.6110	.6093

Table II-III-4 (c) Vena Contracta Taps: Discharge Coefficients, C, for Square-Edged Orifices

(8-in. Pipe, D = 7.981 in.)

R_d / β	14,000	16,000	18,000	20,000	25,000	30,000	40,000	50,000	75,000	100,000	500,000	1,000,000
.1500	.6055	.6047	.6040	.6034	.6022	.6014	.6002	.5994	.5981	.5973	.5946	.5940
.2000	.6061	.6052	.6045	.6039	.6027	.6018	.6006	.5997	.5984	.5976	.5948	.5941
.2500	.6071	.6061	.6054	.6047	.6035	.6025	.6012	.6003	.5989	.5981	.5952	.5945
.3000	.6083	.6074	.6066	.6059	.6045	.6036	.6022	.6012	.5998	.5989	.5950	.5951
.3500	.6101	.6090	.6082	.6074	.6060	.6050	.6035	.6025	.6010	.6001	.5968	.5960
.4000	.6123	.6112	.6103	.6095	.6080	.6069	.6054	.6043	.6027	.6017	.5982	.5973
.4500	.6151	.6139	.6130	.6121	.6106	.6094	.6077	.6066	.6048	.6038	.6001	.5992
.5000	.6185	.6173	.6162	.6153	.6136	.6124	.6106	.6094	.6075	.6064	.6024	.6014
.5500	.6225	.6211	.6200	.6191	.6172	.6158	.6139	.6126	.6106	.6094	.6051	.6041
.5750	.6246	.6232	.6220	.6210	.6191	.6177	.6157	.6144	.6122	.6110	.6065	.6054
.6000		.6253	.6241	.6231	.6211	.6196	.6175	.6161	.6139	.6126	.6079	.6067
.6250		.6275	.6262	.6251	.6230	.6215	.6193	.6178	.6155	.6141	.6091	.6080
.6500			.6283	.6272	.6249	.6233	.6210	.6194	.6169	.6155	.6103	.6090
.6750			.6304	.6292	.6268	.6250	.6226	.6209	.6183	.6167	.6111	.6098
.7000				.6311	.6285	.6266	.6240	.6221	.6193	.6176	.6116	.6102
.7250				.6330	.6302	.6281	.6252	.6232	.6201	.6183	.6117	.6101
.7500					.6318	.6295	.6262	.6240	.6206	.6185	.6112	.6095

Table II-III-4 (d) Vena Contracta Taps: Discharge Coefficients, C, Square-Edged Orifices

(16-in. Pipe, D = 15.25 in.)

R_d / β	18,000	20,000	25,000	30,000	40,000	50,000	75,000	100,000	500,000	1,000,000
.1500	.6039	.6033	.6021	.6013	.6001	.5992	.5980	.5972	.5945	.5939
.2000	.6044	.6038	.6026	.6017	.6005	.5996	.5983	.5975	.5947	.5940
.2500	.6053	.6047	.6034	.6025	.6012	.6003	.5989	.5981	.5951	.5944
.3000	.6065	.6058	.6045	.6035	.6022	.6012	.5997	.5989	.5958	.5950
.3500		.6074	.6060	.6050	.6035	.6025	.6010	.6000	.5968	.5960
.4000		.6095	.6080	.6069	.6053	.6043	.6026	.6017	.5982	.5973
.4500			.6105	.6094	.6077	.6066	.6048	.6038	.6000	.5992
.5000			.6136	.6124	.6106	.6094	.6075	.6064	.6024	.6014
.5500				.6159	.6139	.6126	.6106	.6094	.6051	.6041
.5750				.6177	.6157	.6144	.6123	.6110	.6065	.6054
.6000				.6196	.6176	.6161	.6139	.6126	.6079	.6067
.6250				.6216	.6194	.6179	.6155	.6141	.6092	.6080
.6500				.6235	.6212	.6196	.6171	.6156	.6103	.6091
.6750				.6254	.6229	.6212	.6185	.6169	.6112	.6099
.7000				.6273	.6245	.6227	.6198	.6180	.6118	.6103
.7250				.6292	.6262	.6241	.6208	.6189	.6120	.6103
.7500				.6313	.6278	.6255	.6218	.6195	.6117	.6098

It is believed that the tolerances applicable to the coefficients above and to the right of the heavy stepped lines in Tables II-III-2, II-III-3 and II-III-4 do not exceed ± 1.0 per cent. The values below and to the left of this line are, for the most part, extrapolations outside the range of the test data and are subject to a larger tolerance (see Table II-V-1). Similar tolerance values will apply to the coefficients computed by the equations for other sizes of pipe than those given in the tables, particularly pipes larger than 16 in., and corresponding values of β and R_d or R_D [1-3].

Linear interpolation may be used within the tables; however, the use of the equations for interpolating is preferable.

Note 1: If the equations are used for pipes smaller than 2 in., the tolerances are to be doubled.

Note 2: As in previous editions, the coefficients in the tables are given to four significant figures so that, in using tabular values to compute a flow, two or more parties can obtain results agreeing within 1 or 2 in the fourth significant figure, although the actual uncertainty may be about 1 per cent, as indicated by the statement above and the tolerances given in Table II-V-1.

II-III-8 Expansion Factors. When metering compressible fluids, air, fuel gas, steam, etc., if the static pressure is measured at the inlet pressure tap, i.e., p_1, the expansion factor to be used in computing the rate of flow may be read from Fig. II-III-5 or evaluated by the equation

$$Y = 1 - (0.41 + 0.35\beta^4)\, x/\gamma \quad \text{(II-III-7)}$$

in which

x = Differential pressure ratio, $\Delta p/p_1$

γ = Ratio of specific heats of the gas, assuming it to be an ideal gas

β = Ratio of diameters, d/D

If the static pressure is measured at the outlet pressure tap, i.e., p_2, a common practice when flange taps are used, then the corresponding expansion factor, Y_2, is to be computed by the equation

$$Y_2 = \sqrt{1 + x_2} - (0.41 + 0.35\beta^4)\frac{x_2}{\gamma}\frac{1}{\sqrt{1 + x_2}} \quad \text{(II-III-8)}$$

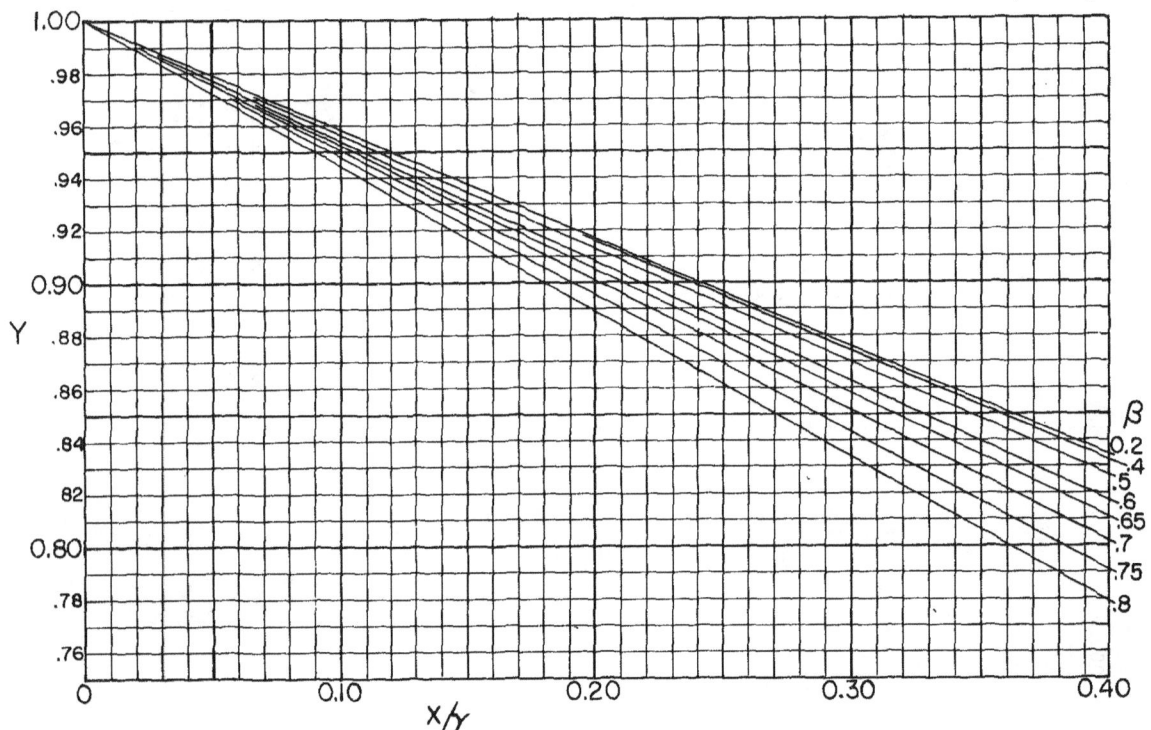

FIG. II-III-5 EXPANSION FACTORS FOR THIN-PLATE SQUARE-EDGED ORIFICES WITH FLANGE TAPS, D AND $1/2\, D$ TAPS AND VENA CONTRACTA TAPS (STATIC PRESSURE MEASURED FROM UPSTREAM PRESSURE TAP.
$Y = 1 - (0.41 + 0.35\beta^4)\, x/\gamma$.)

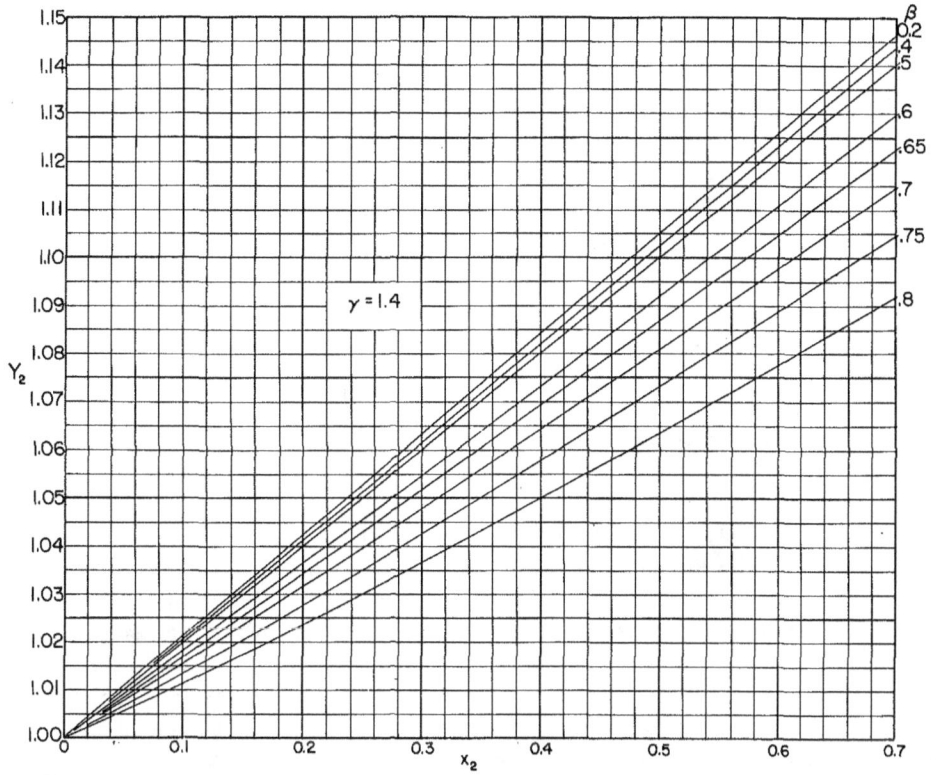

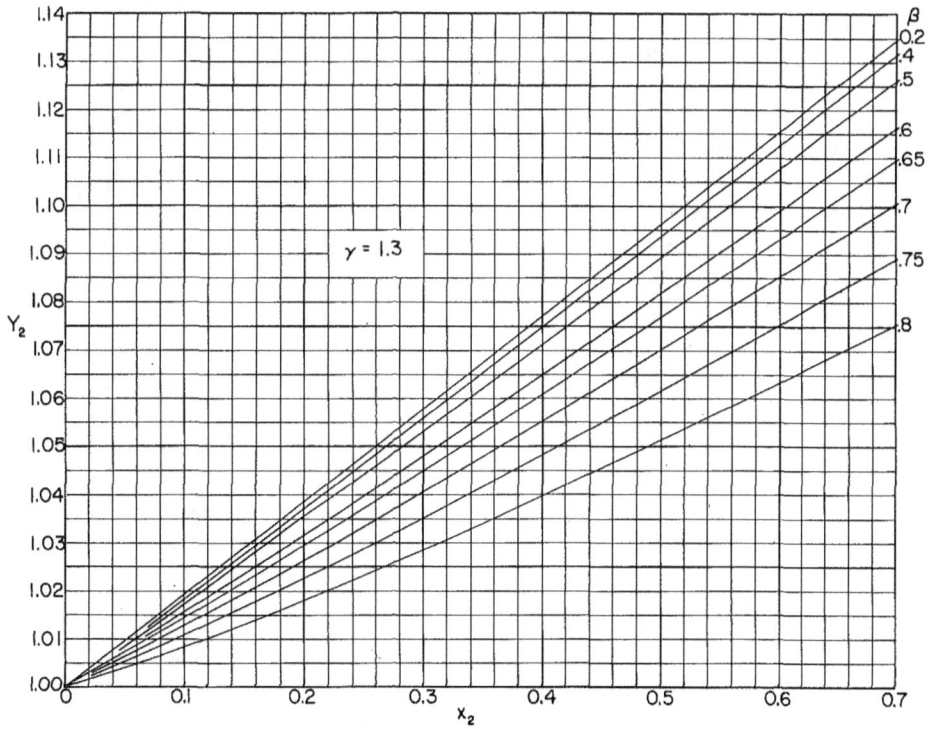

FIG. II-III-6 EXPANSION FACTORS FOR THIN-PLATE SQUARE-
EDGED ORIFICES WHEN THE STATIC PRESSURE IS
MEASURED FROM THE DOWNSTREAM PRESSURE TAP
$(Y_2 = \sqrt{1 + x_2} - (0.41 + 0.35\beta^4)\, x_2/\gamma \sqrt{1 + x_2}.)$

or read from Fig. II-III-6. Here $x_2 = \Delta p/p_2$, the ratio of the differential pressure to the downstream static pressure.

As explained in Chapter I-5, Par. I-5-32, these expansion factors *must* be used in conjunction with the flow coefficient, K, or the equivalent, $C/\sqrt{1-\beta^4}$.

Note: $x_2 = x/(1 - x)$, and also

$$Y_2 = Y\sqrt{1 + x_2} \qquad \text{(II-III-9)}$$

as developed in Chapter I-5, Par. I-5-33.

Eccentric and Segmental Orifices

II-III-9 These orifice forms (Fig. II-III-7) are suitable when the fluid being metered carries a considerable amount of sediment or material in suspension.

II-III-10 Material and Installation. The material for these orifice plates should be the same as for concentric orifices. Also, the same criterion for the plate thickness as given earlier may be used. The diameter, d, or area, a, should be determined as described for concentric orifices.

As with concentric orifices, the inlet edge of the opening must be square, sharp and free of burrs, wire edge or the slightest amount of rounding.

When an eccentric orifice is installed between flanges, the circumference of the hole should lack being tangent with the pipe surface by 1 per cent of the pipe diameter. Also, the radius of the circular part of a segmental orifice should be 1 per cent less than that of the pipe and so installed that this

ECCENTRIC

SEGMENTAL

FIG. II-III-7 ECCENTRIC AND SEGMENTAL ORIFICES

amount of difference is uniform around the circular portion.

II-III-11 Pressure Taps. Both flange and vena contracta pressure taps are used with these orifices, with the latter predominating. For flange taps, the locations are the same as with concentric orifices, namely, 1 in. from the adjacent face of the orifice plate. For vena contracta taps, the inlet pressure tap is one pipe diameter preceding the inlet face of the plate. The positions of the outlet pressure tap, as measured from the inlet face of the orifice plate, are given by the curves of Fig. II-III-8.

With eccentric orifices the pressure taps should, if possible, be in the side of the pipe diametrically opposite the point at which the orifice is substantially tangent to the pipe surface. However, there may be cases where the installation of the pipe does not provide room for this location of the taps, in which case the taps may be moved circumferentially up to but not more than 90 deg and the same coefficients used. With segmental orifices, the pressure taps are always in the element of the pipe that is normal to the straight edge of the orifice.

II-III-12 Coefficients and Expansion Factors. The discharge coefficients for eccentric orifices are given by Fig. II-III-9 and for segmental orifices, by Fig. II-III-10.

The curves in Fig. II-III-9 were developed for use with the pipe Reynolds number, R_D, which is the manner in which they were originally reported [4]. When the throat Reynolds number, R_d, is evaluated, the corresponding value of R_D by which the coefficient may be selected is $R_D = \beta R_d$.

To illustrate the method of interpolating between the curves for $R_D = 10^4$ and 10^6: Assume an eccentric orifice in a 6-in. pipe with flange taps, a diameter ration $\beta = 0.64$, and an orifice Reynolds number, $R_d = 62,500$. Then,

$R_D = 0.64 \times 62,500 = 40,000$, and $\sqrt{R_D} = 200$, $1/200 = .005$. At $\beta = 0.64$ and $R_D = 10^4$ $C = 0.642$; also $1/\sqrt{R_D} = .01$. At $R_D = 10^6$ $C = 0.627$; also $1/\sqrt{10^6} = .001$. $0.642 - 0.627 = .015$, and $.015\,[(.01 - .005)/(.01 - .001)] = .015 \times 4/9 = .0067$. Therefore, $C = 0.642 - .0067 = 0.6353$, or 0.635.

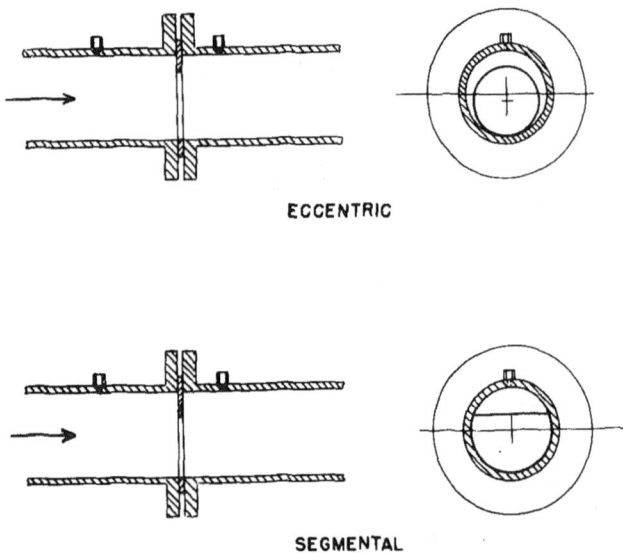

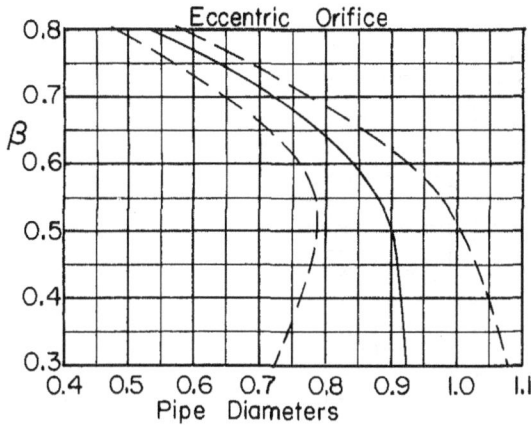

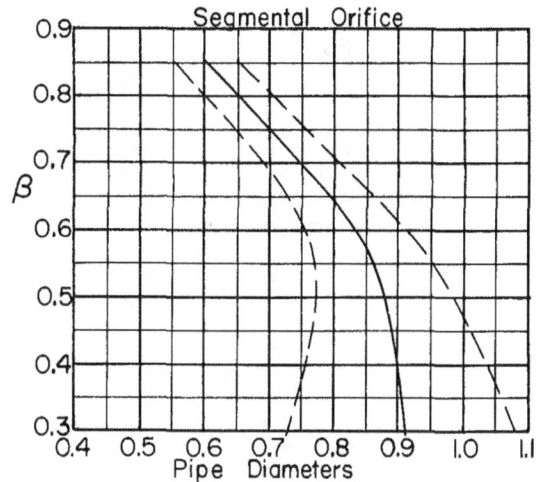

FIG. II-III-8 LOCATIONS OF OUTLET PRESSURE TAPS FOR VENA CONTRACTA
TAPS WITH ECCENTRIC AND SEGMENTAL ORIFICES

With segmental orifices the pipe Reynolds number, R_D, is usually the value computed. Figure II-III-10 gives coefficients at only one value of R_D, as coefficients at other values of R_D have not been established.

II-III-13 Expansion factors to be used with eccentric or segmental orifices when metering compressible fluids are given by Figs. II-III-11 and II-III-12.

II-III-14 When using the discharge coefficient curves for segmental orifices (Fig. II-III-10) and the expansion factors (Fig. II-III-11), it must be remembered that $\beta = \sqrt{m}$. That is, for these orifices the area ratio, m, is the primary dimensional ratio, and the only significance of β is that of the diameter ratio of an equivalent circular orifice.

Note: The users of differential pressure meters are cautioned to be very careful to select the proper discharge coefficient or flow coefficient and the appropriate expansion factor for the particular differential producer being used. Carelessness in selecting these factors may introduce errors which would be difficult to trace.

Small Precision Bore Orifice Meters

II-III-15 Meter tubes of diameters smaller than 1.5 in. can be produced and duplicated provided special manufacturing care and procedures are used [5].

II-III-16 Meter Tubes. The meter tubes should be carefully selected from thick-wall stainless-steel stock of a type suitable for the service in which it will be used. Each tube should be welded to the orifice-holding flange, bored, ground and honed to a uniform inside diameter within a tolerance of ± 0.001 in. and to a surface finish of 5 to 10 microinch.

II-III-17 Orifice Plates. For most services orifice plates are made from stainless-steel sheet of 1/8-in. nominal thickness. Since surface roughness, flatness, orifice-edge squareness and thickness have relatively very large effects on reproducibility and flow-measurement reliability, the plate must be prepared very carefully. The plate should be flat within 0.001 in., and the surface roughness

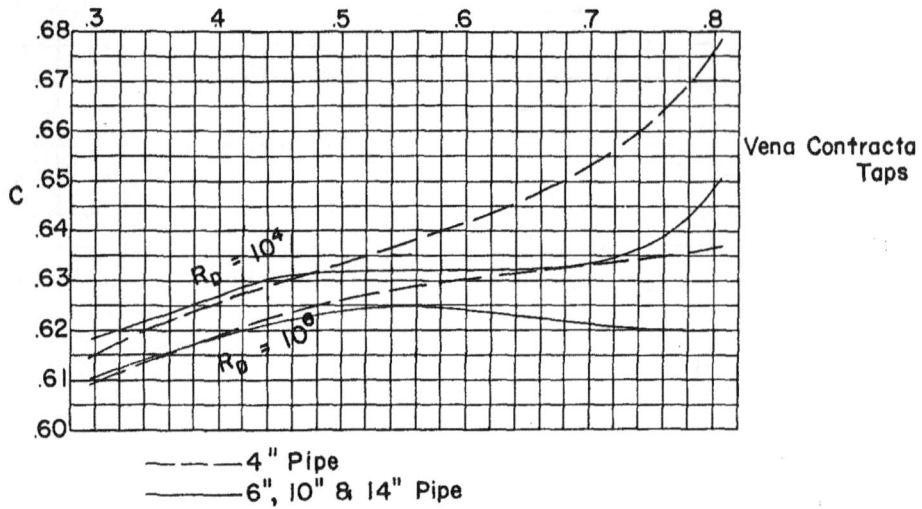

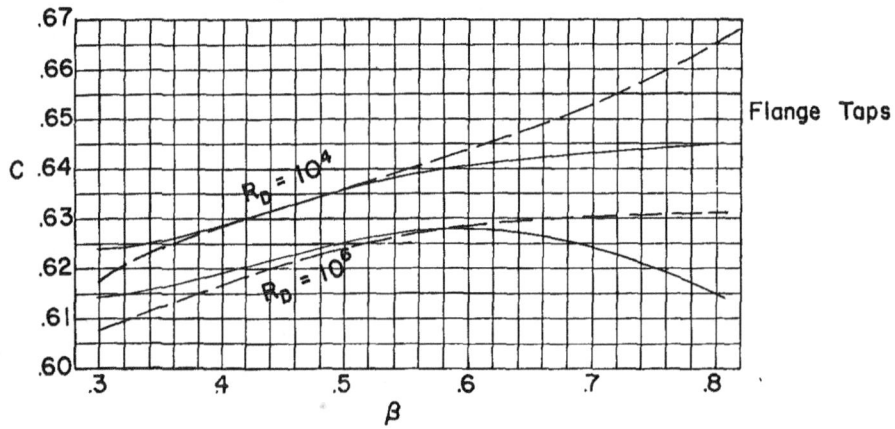

FIG. II-III-9 DISCHARGE COEFFICIENTS OF ECCENTRIC SQUARE-
EDGED ORIFICES

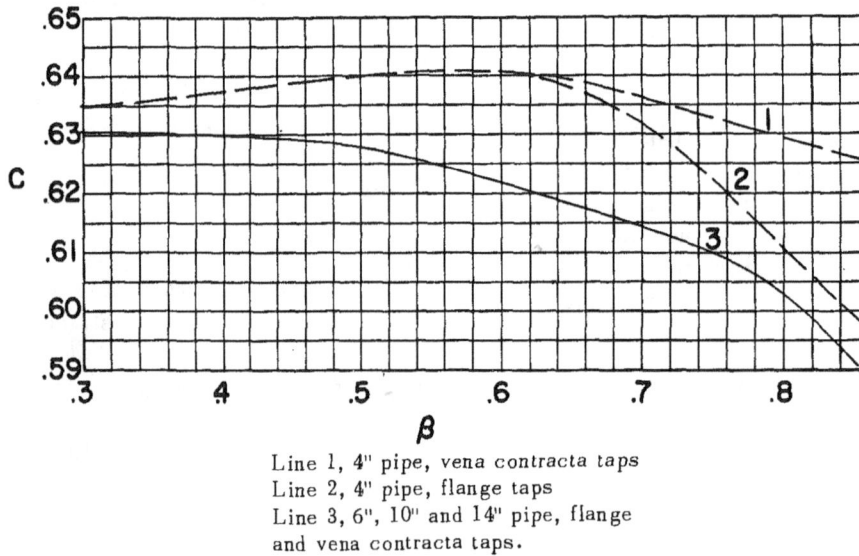

Line 1, 4" pipe, vena contracta taps
Line 2, 4" pipe, flange taps
Line 3, 6", 10" and 14" pipe, flange
and vena contracta taps.

FIG. II-III-10 DISCHARGE COEFFICIENTS FOR SEGMENTAL
ORIFICES FOR VALUES OF $R_D = 10^4$

212

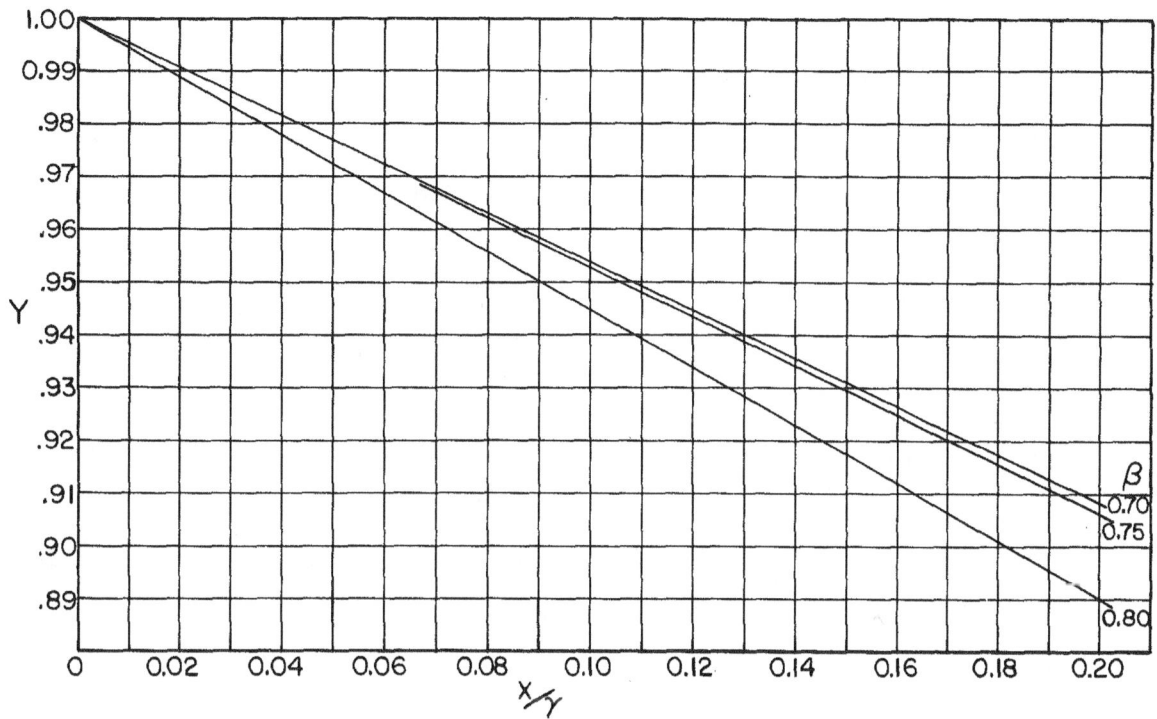

FIG. II-III-11 EXPANSION FACTORS FOR THIN-PLATE ECCENTRIC
ORIFICES (STATIC PRESSURE MEASURED FROM
UPSTREAM TAP)

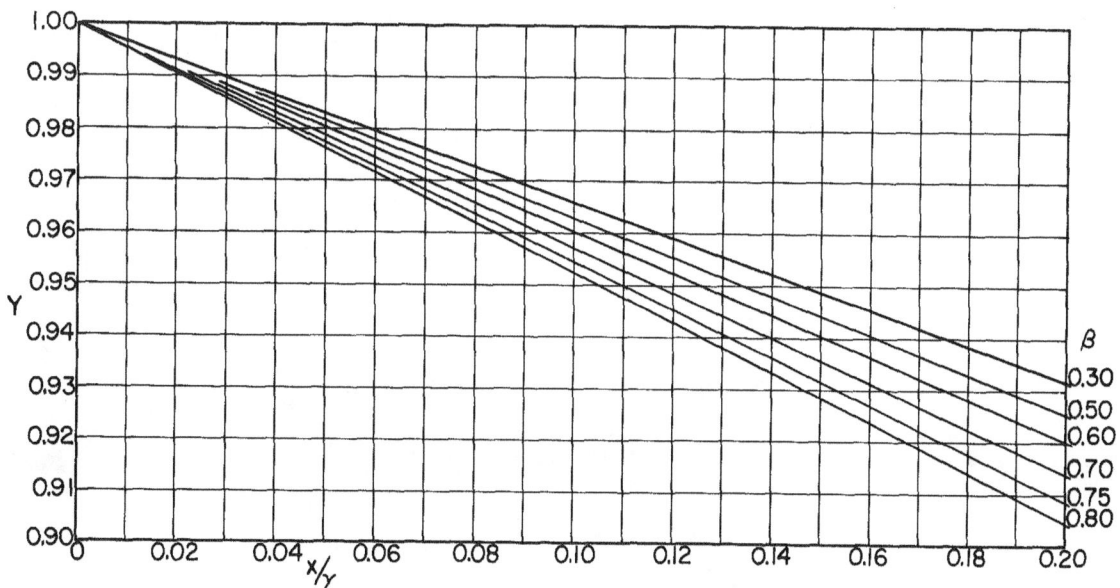

FIG. II-III-12 EXPANSION FACTORS FOR METERING COMPRESSIBLE
FLUIDS WITH THIN-PLATE SEGMENTAL ORIFICES

213

should not exceed 20 microinch. The thickness of
the cylindrical face of the orifice should not exceed
$0.02\ D$ or $1/8\ d$, whichever is smaller. The upstream edge, or corner, must be square and sharp.
Because of the thinness of the cylindrical face of
the orifice, the safest way to measure the orifice
diameter, d, without damaging the orifice edge, is
with an optical comparator.

II-III-18 Pressure Taps. Pressures are measured
from annular grooves on each side of the plate, as
shown in Fig. II-III-13. This form of pressure opening is used since a 1-in. location from the downstream orifice face would be equivalent to one to
two pipe diameters in the 1-in. and smaller tubes
and, thus, in the varying pressure-recovery region.
On the other hand, placing the downstream tap at
$1/2\ D$ would be impractical in these small sizes.
At the upstream annular pressure groove there is a
slight pressure buildup; and, although a 1-in. location could be used, the use of the groove makes for
uniformity in manufacture.

II-III-19 Coefficients. The flow coefficient,
K, for these small orifice meters may be computed
by the equation

where

K = Flow coefficient

D = Inside diameter of meter tube (in.)

R_D = Pipe Reynolds number

R_d = Orifice Reynolds number

β = Ratio, (orifice diameter)/(tube diameter)

$\lambda = 1000/\sqrt{R_D} = 1000/\sqrt{\beta R_d}$

This equation has been found to give coefficients
within ± 0.75 per cent of the values obtained from a
calibration when pressures are measured from corner
grooves as described above and when 0.5 in. $\bar{<} D$
$\bar{<} 1.5$ in., $0.1 < \beta < 0.8$, and $R_D > 1000$.

The equation that is to be used for computing the
flow coefficient, when the meter tube has flange
pressure taps located in the conventional (1-in.)
position, is

$$K = 0.5980 + 0.468\ (\beta^4 + 10\ \beta^{12})$$
$$+ (0.00087 + 0.0081\ \beta^4)\ \lambda \quad \text{(II-III-11)}$$

and is applicable if 1 in. $< D < 1\ 1/2$ in., $0.15 < \beta$
< 0.7, and $R_D > 1000$.

$$K = \left[0.5991 + \frac{0.0044}{D} + \left(0.3155 + \frac{0.0175}{D} \right)\ (\beta^4 + 2\beta^{16}) \right]$$
$$+ \left[\frac{0.00052}{D} - 0.000192 + \left(0.01648 - \frac{0.00116}{D} \right)\ (\beta^4 + 4\beta^{16}) \right]\lambda \quad \text{(II-III-10)}$$

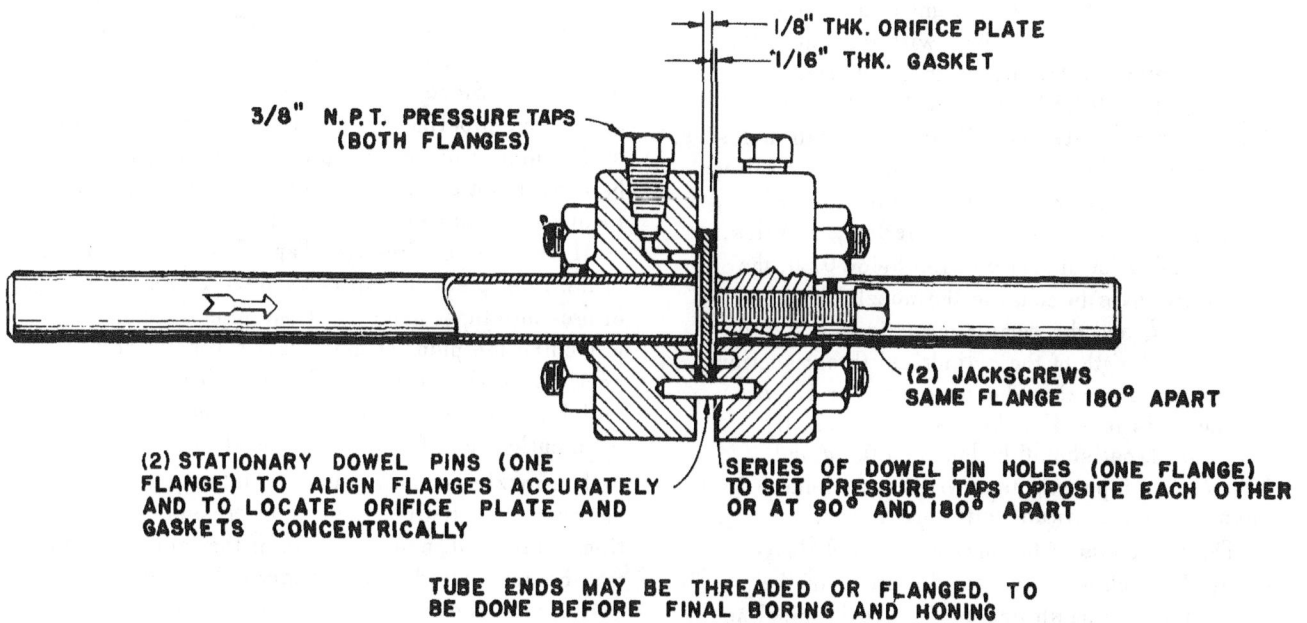

FIG. II-III-13 HONED SMALL-BORE ORIFICE FLOW SECTION

215

Flow Nozzles

II-III-20 Material. Flow nozzles should be made of a corrosion-resistant material. For high-temperature service and many other services, stainless steel should be used. For water at temperatures below 400 F and pressures below about 250 psig, as well as for some other fluids including oils and gases, bronze may be used. Aluminum may be used for air, some other gases, and also liquids, if free of aluminum corroding elements, where temperatures and pressures do not exceed about 200 F and 200 psig.

II-III-21 Nozzle Form. The recommended form of flow nozzle is the "long-radius" or elliptical inlet nozzle, in which the curvature of the inlet to the nozzle throat is the quadrant of an ellipse. The proportions of the ellipse with respect to the pipe diameter and the nozzle throat diameter are shown by Fig. II-III-14. For the high β nozzles, diameter ratios between 0.45 and 0.80 both inclusive, the entrance curvature is the quadrant of an ellipse having a semi-major axis of $1/2\ D$ and a semi-minor axis of $1/2\ (D - d)$. For the low β nozzles, recommended for diameter ratios below 0.50, the semi-major axis is equal to the nozzle throat diameter, d, and the semi-minor axis is $5/8\ d$ to $2/3\ d$. The length of the cylindrical throat section of the high β nozzles should be $0.6\ d$ or $1/3\ D$, whichever is less. For the low β nozzles, the length of throat should be between $0.6\ d$ and $0.75\ d$ when pipe-wall taps will be used and $0.75\ d$ when the nozzle is made with throat taps.

The thickness of the nozzle wall and flange should be such as to prevent distortion of the nozzle throat from strains caused by the pipeline temperature and pressure, flange bolting or other methods of installing the nozzle in the pipeline. The outside diameter of the nozzle flange or the design of the flange facing should be such that the nozzle throat can be centered accurately in the pipe. (See Chapter II-II, Par. II-II-6.)

II-III-22 The throat of a flow nozzle should be as nearly cylindrical as possible. Any taper should not exceed the following negative amounts:

1. -0.001 in. for $d \lesseqqgtr 3.00$ in.
2. -0.0015 in. for $3.01 \lesseqqgtr d \lesseqqgtr 6.00$ in.
3. -0.002 in. for $d \gtreqqless 6.01$ in.

That is, any taper should be such that the throat diameter *decreases* toward the outlet end. There must be no bell mouth or diameter increase near the outlet end, especially within the last 1/4 in.

Any out-of-roundness of the nozzle throat should not exceed:

1. ± 0.002 in. for $d \lesseqqgtr 3.00$ in.
2. ± 0.003 in. for $3.01 \lesseqqgtr d \lesseqqgtr 6.00$ in.
3. ± 0.004 in. for $d \gtreqqless 6.01$ in.

The actual diameter of the nozzle throat should be determined by careful measurements after all machining and finishing has been completed. Measurements should be made on three or more diameters and desirably in two or more cross sections. When these measurements are made, care is necessary in order not to scratch or otherwise alter the surface of the throat.

II-III-23 Pressure Taps. Two pairs of pressure-tap locations are used with the ASME long-radius flow nozzles, namely, pipe-wall taps and nozzle-throat taps. Since flow nozzles may be used in a continuous pipeline, at the end of a pipe section or at the inlet or outlet of a plenum chamber, it will be convenient to describe the inlet and outlet tap locations separately rather than in pairs.

II-III-24 Inlet Pressure Tap. The same location of the inlet pressure tap is used with both pairs of pressure taps, namely, at one pipe diameter, D, preceding the plane of the nozzle elliptical inlet section, as shown in Fig. II-III-15. The same location applies when a nozzle is mounted at the open outlet end of a pipe section (Fig. II-III-16).

If a nozzle is installed at the inlet to a plenum chamber (Fig. II-III-17), no inlet pressure connection is required, and the value of the inlet pressure may be considered as atmospheric. If a nozzle is installed at the outlet of a plenum chamber, the inlet pressure (to the nozzle) will be the pressure in the plenum chamber.

II-III-25 Outlet Pressure Tap. When a high β nozzle is to be used in a continuous pipeline, the outlet pressure tap is located in the pipe wall $1/2\ D$ following the plane of the beginning of the elliptical inlet section of the nozzle. For a low β nozzle, if a pipe-wall tap is to be used, it should be located a distance of $1\text{-}1/2\ d$ (throat diameter) following the entrance plane of the nozzle. Likewise, for nozzles made with throat taps, the location of the tap or taps is $1\text{-}1/2\ d$ following the entrance plane.

If a nozzle is installed at the outlet end of a pipe section or the outlet of a plenum chamber, the outlet pressure may be measured with a barometer

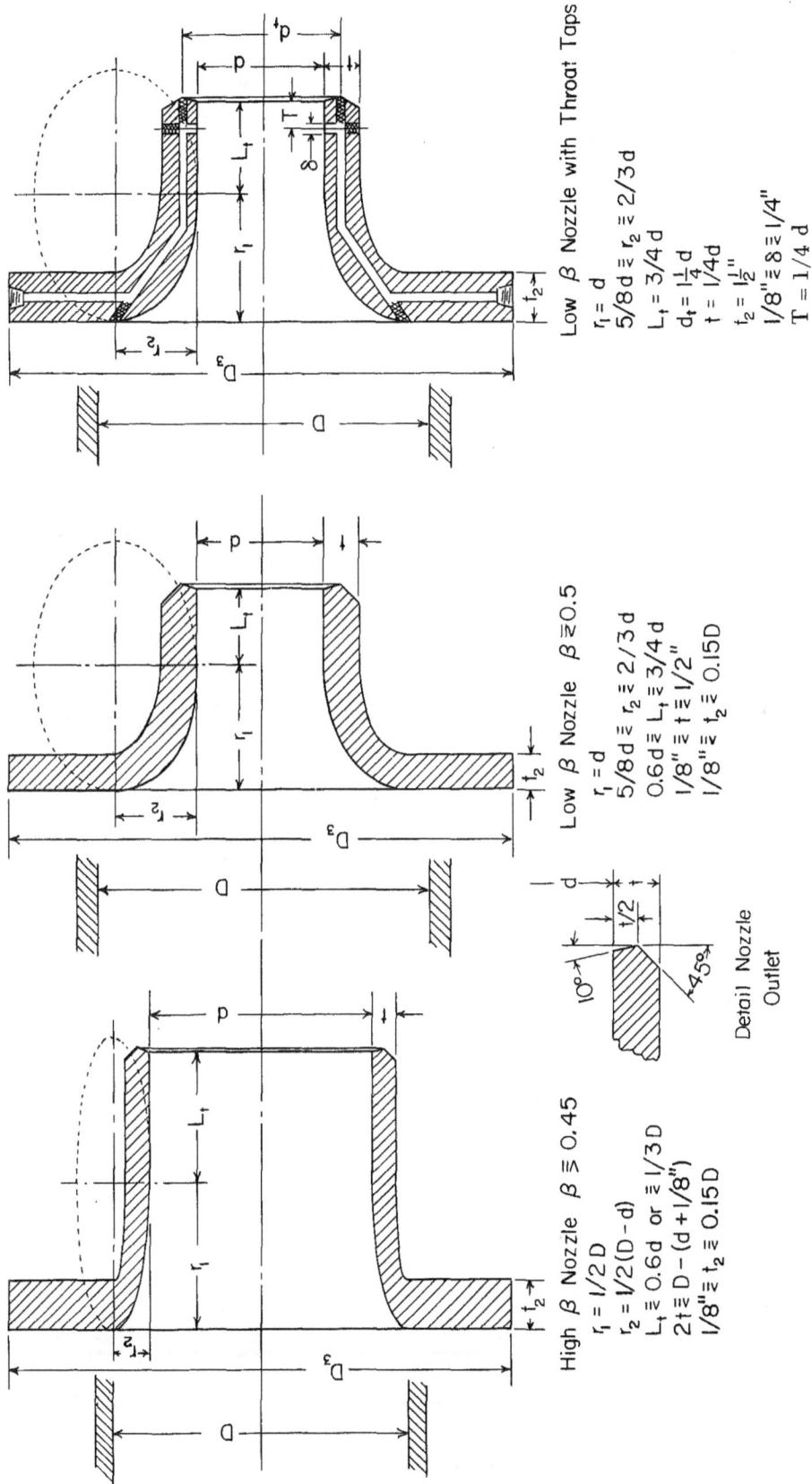

Low β Nozzle with Throat Taps

$r_1 = d$

$5/8\,d \leqq r_2 \leqq 2/3\,d$

$L_t = 3/4\,d$

$d_t = 1\frac{1}{4}\,d$

$t = 1/4\,d$

$t_2 = 1\frac{1}{2}''$

$1/8'' \leqq 8 \leqq 1/4''$

$T = 1/4\,d$

Low β Nozzle $\beta \leqq 0.5$

$r_1 = d$

$5/8\,d \leqq r_2 \leqq 2/3\,d$

$0.6\,d \leqq L_t \leqq 3/4\,d$

$1/8'' \leqq t \leqq 1/2''$

$1/8'' \leqq t_2 \leqq 0.15D$

Detail Nozzle
Outlet

High β Nozzle $\beta \geqq 0.45$

$r_1 = 1/2\,D$

$r_2 = 1/2(D-d)$

$L_t \leqq 0.6\,d$ or $\geqq 1/3\,D$

$2t \leqq D-(d+1/8'')$

$1/8'' \leqq t_2 \leqq 0.15D$

FIG. II-III-14 RECOMMENDED PROPORTIONS OF ASME LONG-RADIUS
FLOW NOZZLES

217

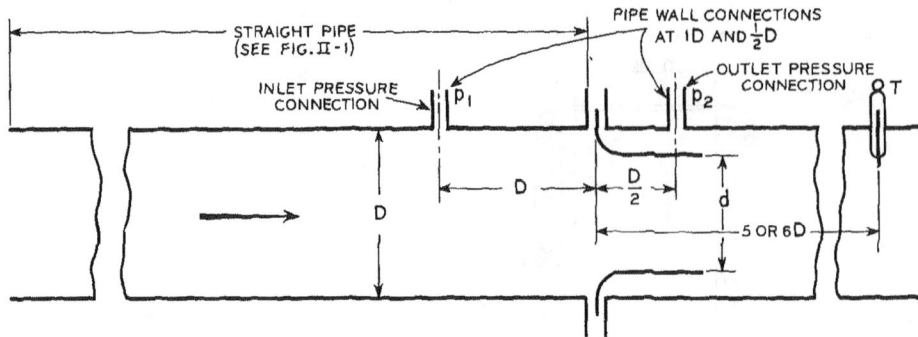

HIGH β NOZZLES
WITH PIPE WALL TAPS

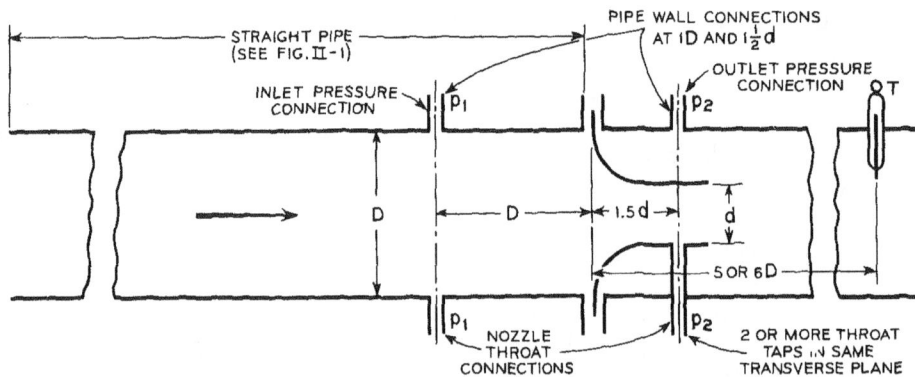

LOW β NOZZLES
WITH PIPE WALL TAPS
OR THROAT TAPS (OPTIONAL)

FIG. II-III-15 LOCATIONS OF PRESSURE TAPS USED WITH ASME LONG-RADIUS FLOW NOZZLES
WHEN IN A CONTINUOUS PIPELINE (WHEN A THERMOMETER IS REQUIRED,
THE WELL FOR IT MAY BE LOCATED AS SHOWN BY T.)

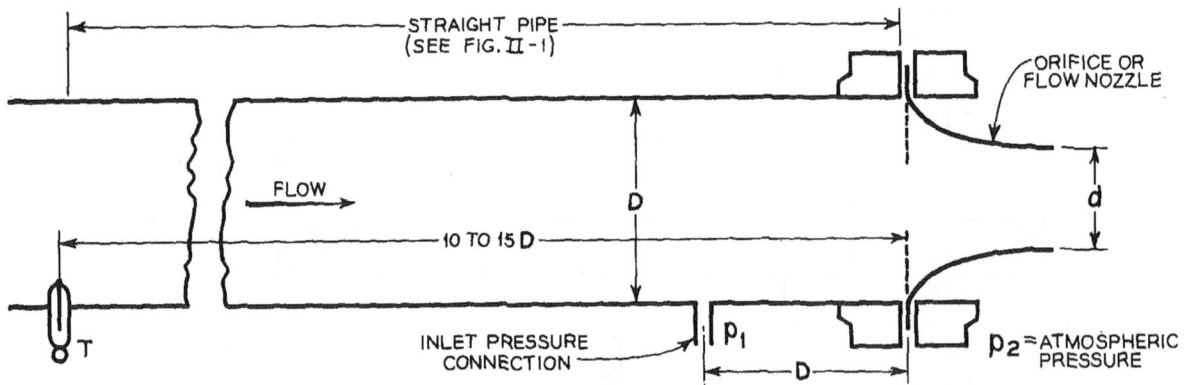

FIG. II-III-16 LOCATION OF INLET PRESSURE TAP FOR A NOZZLE
MOUNTED AT OPEN OUTLET END OF PIPE

located as near the nozzle outlet as possible, but not in the path of the emerging fluid. Alternately, if the nozzle is equipped with throat taps, the pressure in the throat may be measured with a manometer. If a nozzle is installed at the inlet to a plenum chamber, the pressure in the plenum chamber may be used as the nozzle outlet pressure. Of course, if the nozzle has throat taps, a special connection to these may be brought out from the chamber.

II-III-26 Construction of Pressure Taps. Pressure taps should be drilled (and preferably reamed) radially with respect to the pipe in which they are made. This drilling should be done after any coupling or other fitting for attaching of pressure tubing has been welded to the pipe. The hole where it breaks through the inner surface of the pipe must be free of burrs or wire edge, and the corner or edge of the hole left square and sharp or dulled (rounded) very slightly.

Special care and procedures are required for making pressure taps in the throat of a flow nozzle (Fig. II-III-14). The holes should be drilled and reamed before the final boring and polishing of the throat section. A plug sized for a press fit is pressed into the hole. The plug should be made with provision for removing it after the final machining and polishing of the nozzle surface. After removal of the plug, the edge of the hole should be free of burrs and square and sharp with the throat surface. Any slight burr may be removed by rolling a tapered piece of maple wood around the pressure tap hole.

It is recommended that throat tap nozzles be made with two or four tap holes; this is because these nozzles are used frequently in test work, and the additional taps are useful for multiple instrumentation. In such cases, it will be helpful to have a like number of inlet pressure taps.

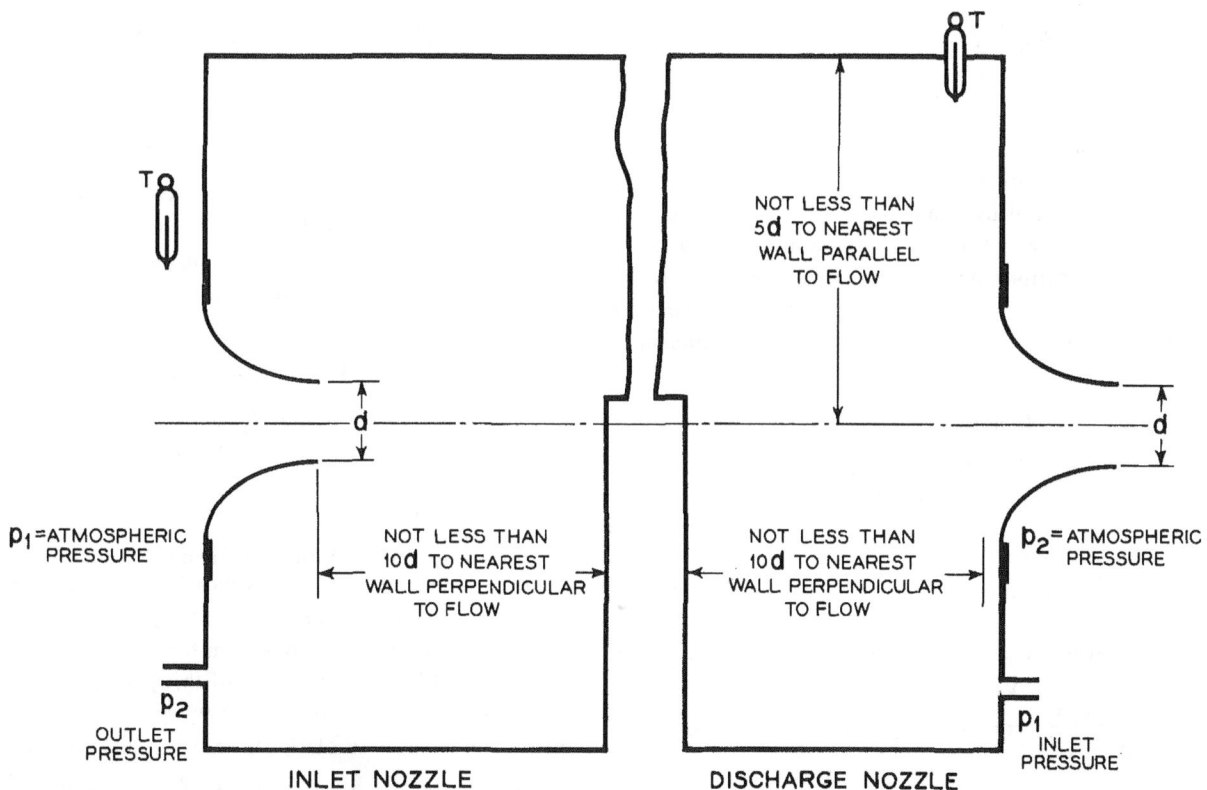

FIG. II-III-17 PRESSURE TAP LOCATIONS WHEN NOZZLES ARE USED
AT INLET OR OUTLET OF A PLENUM CHAMBER

II-III-27 Pressure Loss. The overall pressure loss with a flow nozzle is shown in Fig. II-III-18.

II-III-28 Coefficients. As with orifice meters, it is desirable whenever possible to calibrate a flow nozzle with the pipe section in which it is to be used. When a calibration is not made, the coefficient of discharge may be computed by equation (II-III-12) or obtained from Table II-III-5.

$$C = 0.99622 + 0.00059\,D$$

$$- (6.36 + 0.13D - 0.24\beta^2)\frac{1}{\sqrt{R_d}} \quad \text{(II-III-12)}$$

For low β flow nozzles with throat pressure taps, the discharge coefficient may be read from Fig. II-III-19 if the nozzle has not been calibrated.

II-III-29 Expansion Factors. When metering a compressible fluid with a flow nozzle, the expansion factor, Y_a, to be used in computing the flow, given by equation (I-5-26), is

$$Y_a = \left[r^{2/\gamma}\left(\frac{\gamma}{\gamma-1}\right)\left(\frac{1 - r^{(\gamma-1)/\gamma}}{1-r}\right)\left(\frac{1-\beta^4}{1-\beta^4 r^{2/\gamma}}\right) \right]^{1/2} \quad \text{(II-III-13)}$$

Values of Y_a for two values of γ are given by Figs. II-III-20 and II-III-21 and also by Tables II-III-6 and II-III-7.

FLUID METERS

Their Theory and Application

ERRATA

Page 220 — Nozzle Coefficients — This correction should be attached to page 220.

The tables of coefficients of discharge for long-radius nozzles with pipe wall taps, Tables II-III-5 (a-d) are computed by equation (II-III-12). This equation was developed from data determined during commercial calibrations and a research program [11]. This resulted in a preponderance of the data being for relatively small pipe sizes and low Reynolds numbers. The test limits were:

$$D \text{ between 2 and 15.75 inches}$$
$$R_d \text{ between } 10^4 \text{ and } 10^6$$
$$\beta \text{ between 0.15 and 0.75}$$

Within these limits, equation (II-III-12) may be used for interpolation but *must never* be used for extrapoltion outside of these ranges.

In the present state of the art it is suggested that an equation based on the flat plate boundary layer theory would be appropriate. A general form of such an equation is

$$C = A - B\left(\frac{R_{dt}}{R_d}\right)^a \quad \text{(II-III-41)}$$

where $a = \tfrac{1}{2}$ *for* R_d *less than* R_{dt}

$$= 1/5 \text{ for } R_d \text{ greater than } R_{dt}$$

R_{dt} = throat Reynolds where the boundary layer changes from laminar to turbulent.

R_{dt} is more of a zone than a point since it may range from about 300,000 to about 3,000,000. It must be determined experimentally the same as A and B.

Where calibration data are not available the following equation has been suggested [12].

$$C = 0.9975 - 0.00653\,(10^6/R_d)^a$$
$$a = \tfrac{1}{2} \text{ for } R_d < 10^6$$
$$= 1/5 \text{ for } R_d > 10^6 \quad \text{(II-III-42)}$$

Slight variations in form and dimension of either pipe or nozzle may affect the observed pressures, and thus cause the values of the exponent, a, and slope term, B, to diverge considerably from the values in equation (II-III-42) [13]. In view of this, it is possible that a tolerance greater than $\pm 2.0\%$ should be applied when any one of D, β, or R_d is outside the range of values listed above. Representative values of C by equation (II-III-42) are:

R_d	10,000	20,000	50,000	100,000	200,000
C	0.9322	0.9513	0.9683	0.9768	0.9829
R_d	500,000	10^6	5×10^6	10^7	10^8
C	0.9883	0.9910	0.9928	0.9934	0.9949

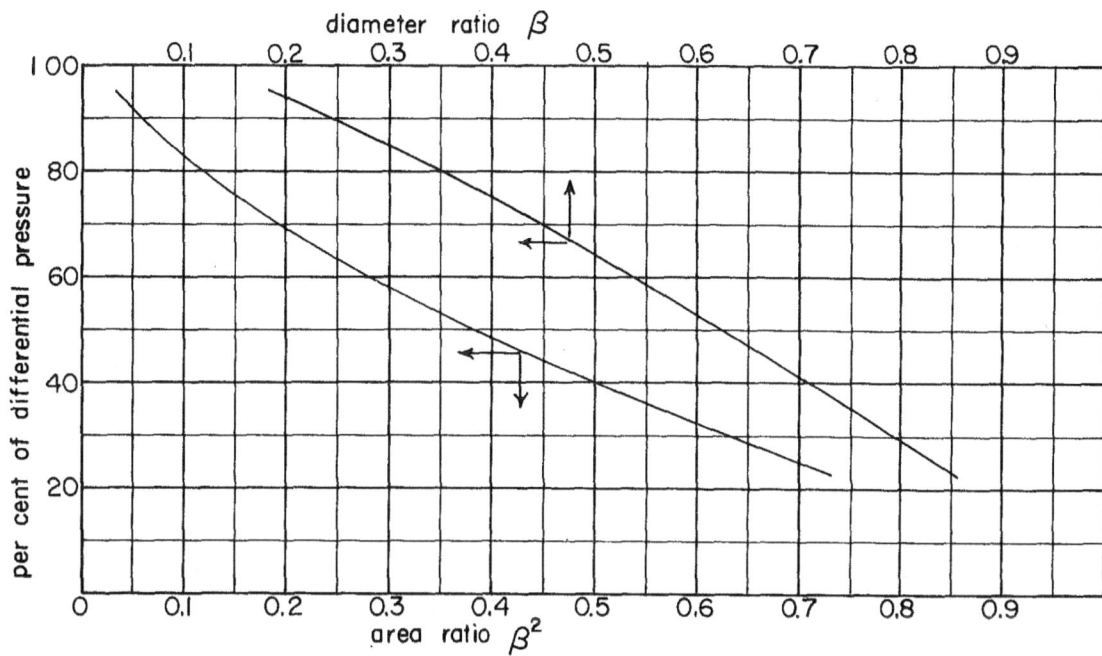

FIG. II-III-18 OVERALL PRESSURE LOSS ACROSS FLOW NOZZLES

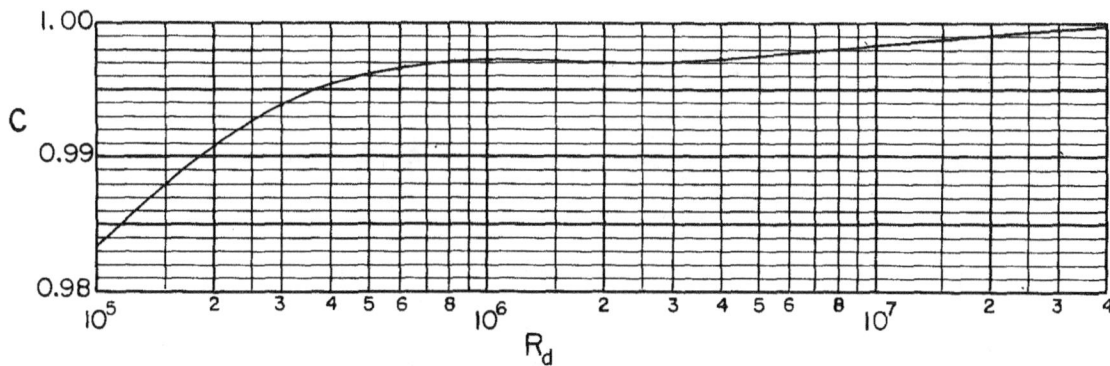

FIG. II-III-19 TYPICAL CALIBRATION CURVE OF A LOW– β FLOW
NOZZLE WITH THROAT PRESSURE TAPS

Table II-III-5 (a) Long-Radius Flow Nozzles: Discharge Coefficients, C, with Pipe Taps at 1 D and 1/2 D

(2-in. Pipe, D = 2.067 in.)

β \ R_d	10,000	12,000	14,000	16,000	18,000	20,000	25,000	30,000	40,000	50,000	75,000	100,000	500,000	1,000,000
.1500	.9312	.9369	.9414	.9450	.9480	.9506	.9555	.9592	.9643	.9678	.9732	.9765	.9881	.9908
.2000	.9312	.9370	.9415	.9451	.9481	.9506	.9556	.9592	.9643	.9678	.9733	.9765	.9881	.9908
.2500	.9313	.9370	.9415	.9451	.9481	.9506	.9556	.9592	.9644	.9678	.9733	.9765	.9881	.9908
.3000	.9313	.9371	.9416	.9452	.9482	.9507	.9556	.9593	.9644	.9679	.9733	.9765	.9881	.9908
.3500	.9314	.9372	.9416	.9452	.9482	.9508	.9557	.9593	.9644	.9679	.9733	.9766	.9881	.9908
.4000	.9315	.9373	.9417	.9453	.9483	.9508	.9557	.9594	.9645	.9680	.9734	.9766	.9881	.9908
.4500	.9316	.9373	.9418	.9454	.9484	.9509	.9558	.9594	.9645	.9680	.9734	.9766	.9881	.9909
.5000	.9317	.9375	.9419	.9455	.9485	.9510	.9559	.9595	.9646	.9681	.9734	.9767	.9881	.9909
.5500	.9319	.9376	.9420	.9456	.9486	.9511	.9560	.9596	.9646	.9681	.9735	.9767	.9882	.9909
.5750	.9319	.9376	.9421	.9456	.9486	.9511	.9560	.9596	.9647	.9681	.9735	.9767	.9882	.9909
.6000	.9320	.9377	.9421	.9457	.9487	.9512	.9561	.9597	.9647	.9682	.9735	.9767	.9882	.9909
.6250	.9321	.9378	.9422	.9458	.9487	.9512	.9561	.9597	.9648	.9682	.9736	.9768	.9882	.9909
.6500	.9322	.9378	.9423	.9458	.9488	.9513	.9561	.9597	.9648	.9682	.9736	.9768	.9882	.9909
.6750	.9322	.9379	.9423	.9459	.9488	.9513	.9562	.9598	.9648	.9683	.9736	.9768	.9882	.9909
.7000	.9323	.9380	.9424	.9460	.9489	.9514	.9563	.9598	.9649	.9683	.9737	.9768	.9882	.9909
.7250	.9324	.9381	.9425	.9460	.9490	.9515	.9563	.9599	.9649	.9684	.9737	.9769	.9882	.9909
.7500	.9325	.9382	.9426	.9461	.9490	.9515	.9564	.9599	.9650	.9684	.9737	.9769	.9883	.9909

Table II-III-5 (b) Long-Radius Flow Nozzles: Discharge Coefficients, C, with Pipe Taps at 1 D and 1/2 D

(4-in. Pipe, D = 4.026 in.)

β \ R_d	10,000	12,000	14,000	16,000	18,000	20,000	25,000	30,000	40,000	50,000	75,000	100,000	500,000	1,000,000
.1500	.9298	.9358	.9404	.9442	.9473	.9499	.9551	.9589	.9642	.9678	.9735	.9768	.9889	.9917
.2000	.9298	.9358	.9405	.9442	.9473	.9500	.9551	.9589	.9642	.9678	.9735	.9768	.9889	.9917
.2500	.9299	.9359	.9405	.9443	.9474	.9500	.9551	.9589	.9642	.9679	.9735	.9769	.9889	.9917
.3000	.9299	.9359	.9406	.9443	.9474	.9500	.9552	.9590	.9643	.9679	.9735	.9769	.9889	.9917
.3500	.9300	.9360	.9406	.9444	.9475	.9501	.9552	.9590	.9643	.9679	.9736	.9769	.9889	.9917
.4000	.9301	.9361	.9407	.9445	.9475	.9502	.9553	.9591	.9644	.9680	.9736	.9769	.9889	.9917
.4500	.9302	.9362	.9408	.9445	.9476	.9502	.9553	.9591	.9644	.9680	.9736	.9770	.9889	.9918
.5000	.9303	.9363	.9409	.9446	.9477	.9503	.9554	.9592	.9645	.9681	.9737	.9770	.9889	.9918
.5500	.9305	.9364	.9410	.9447	.9478	.9504	.9555	.9593	.9645	.9681	.9737	.9770	.9890	.9918
.5750		.9365	.9411	.9448	.9479	.9505	.9555	.9593	.9646	.9682	.9737	.9771	.9890	.9918
.6000		.9365	.9411	.9448	.9479	.9505	.9556	.9593	.9646	.9682	.9738	.9771	.9890	.9918
.6250			.9412	.9449	.9480	.9506	.9556	.9594	.9646	.9682	.9738	.9771	.9890	.9918
.6500			.9413	.9450	.9480	.9506	.9557	.9594	.9647	.9683	.9738	.9771	.9890	.9918
.6750				.9450	.9481	.9507	.9557	.9595	.9647	.9683	.9739	.9772	.9890	.9918
.7000				.9451	.9482	.9507	.9557	.9595	.9648	.9683	.9739	.9772	.9890	.9918
.7250					.9482	.9508	.9558	.9596	.9648	.9684	.9739	.9772	.9890	.9918
.7500					.9483	.9509	.9559	.9596	.9648	.9684	.9739	.9772	.9890	.9918

Table II-III-5 (c) Long-Radius Flow Nozzles: Discharge Coefficients, *C*, with Pipe Taps at 1 *D* and 1/2 *D*

(8-in. Pipe, *D* = 7.981 in.)

β \ R_d	12,000	14,000	16,000	18,000	20,000	25,000	30,000	40,000	50,000	75,000	100,000	500,000	1,000,000
.1500	.9334	.9384	.9424	.9458	.9486	.9541	.9582	.9639	.9678	.9739	.9775	.9905	.9935
.2000	.9334	.9384	.9425	.9458	.9486	.9542	.9582	.9640	.9679	.9739	.9775	.9905	.9935
.2500	.9335	.9385	.9425	.9459	.9487	.9542	.9583	.9640	.9679	.9739	.9776	.9905	.9935
.3000	.9335	.9385	.9426	.9459	.9487	.9542	.9583	.9640	.9679	.9740	.9776	.9905	.9935
.3500	.9336	.9386	.9426	.9460	.9488	.9543	.9584	.9641	.9680	.9740	.9776	.9905	.9935
.4000	.9337	.9387	.9427	.9460	.9489	.9544	.9584	.9641	.9680	.9740	.9776	.9905	.9936
.4500	.9338	.9388	.9428	.9461	.9489	.9544	.9585	.9642	.9680	.9741	.9777	.9905	.9936
.5000	.9339	.9389	.9429	.9462	.9490	.9545	.9585	.9642	.9681	.9741	.9777	.9905	.9936
.5500		.9390	.9430	.9463	.9491	.9546	.9586	.9643	.9681	.9742	.9777	.9906	.9936
.5750		.9390	.9430	.9464	.9492	.9546	.9587	.9644	.9682	.9742	.9778	.9906	.9936
.6000			.9431	.9464	.9492	.9547	.9587	.9644	.9682	.9742	.9778	.9906	.9936
.6250			.9432	.9465	.9493	.9547	.9587	.9644	.9682	.9742	.9778	.9906	.9936
.6500				.9465	.9493	.9548	.9588	.9644	.9683	.9743	.9778	.9906	.9936
.6750				.9466	.9494	.9548	.9588	.9645	.9683	.9743	.9779	.9906	.9936
.7000					.9494	.9549	.9589	.9645	.9684	.9743	.9779	.9906	.9936
.7250					.9495	.9549	.9589	.9646	.9684	.9744	.9779	.9906	.9936
.7500						.9550	.9590	.9646	.9684	.9744	.9779	.9906	.9937

Table II-III-5 (d) Long-Radius Flow Nozzles: Discharge Coefficients, *C*, with Pipe Taps at 1 *D* and 1/2 *D*

(16-in. Pipe, *D* = 15.25 in.)

β \ R_d	14,000	16,000	18,000	20,000	25,000	30,000	40,000	50,000	75,000	100,000	500,000	1,000,000
.1500	.9347	.9392	.9430	.9462	.9524	.9570	.9635	.9679	.9747	.9788	.9934	.9969
.2000	.9347	.9393	.9430	.9462	.9525	.9571	.9635	.9679	.9748	.9788	.9934	.9969
.2500	.9348	.9393	.9431	.9463	.9525	.9571	.9635	.9679	.9748	.9788	.9934	.9969
.3000	.9348	.9394	.9431	.9463	.9525	.9571	.9636	.9680	.9748	.9789	.9934	.9969
.3500	.9349	.9394	.9432	.9464	.9526	.9572	.9636	.9680	.9748	.9789	.9934	.9969
.4000	.9350	.9395	.9433	.9464	.9526	.9572	.9637	.9680	.9749	.9789	.9934	.9969
.4500	.9351	.9396	.9433	.9465	.9527	.9573	.9637	.9681	.9749	.9790	.9935	.9969
.5000	.9352	.9397	.9434	.9466	.9528	.9574	.9638	.9681	.9749	.9790	.9935	.9969
.5500		.9398	.9435	.9467	.9529	.9574	.9638	.9682	.9750	.9790	.9935	.9969
.5750		.9398	.9436	.9467	.9529	.9575	.9639	.9682	.9750	.9791	.9935	.9969
.6000			.9436	.9468	.9530	.9575	.9639	.9683	.9750	.9791	.9935	.9969
.6250			.9437	.9468	.9530	.9576	.9639	.9683	.9751	.9791	.9935	.9969
.6500				.9469	.9531	.9576	.9640	.9683	.9751	.9791	.9935	.9969
.6750				.9470	.9531	.9576	.9640	.9684	.9751	.9792	.9936	.9970
.7000					.9532	.9577	.9641	.9684	.9752	.9792	.9936	.9970
.7250					.9532	.9577	.9641	.9684	.9752	.9792	.9936	.9970
.7500						.9578	.9641	.9685	.9752	.9792	.9936	.9970

Table II-III-6 Expansion Factors for Flow Nozzles and Venturi Tubes

$$Y_a = \left(r^{2/\gamma}\ \frac{\gamma}{\gamma-1}\ \frac{1-r^{(\gamma-1)/\gamma}}{1-r}\ \frac{1-\beta^4}{1-\beta^4 r^{2/\gamma}} \right)^{1/2}$$

$$\gamma = 1.4$$

β	β^4	0.95	0.90	0.85	0.80	0.75	0.70	0.65	0.60	0.55
0.2	0.0016	0.9728	0.9448	0.9160	0.8863	0.8556	0.8238	0.7908	0.7565	0.7207
.3	.0081	.9726	.9444	.9154	.8855	.8546	.8227	.7896	.7552	.7193
.4	.0256	.9719	.9432	.9137	.8833	.8520	..8198	.7864	.7517	.7156
0.50	.0625	.9706	.9405	.9099	.8785	.8464	.8133	.7793	.7441	.7076
.55	.0915	.9694	.9383	.9067	.8745	.8416	.8080	.7734	.7378	.7010
.60	.1296	.9678	.9352	.9023	.8690	.8351	.8006	.7653	.7292	.6920
0.65	.1785	.9655	.9309	.8962	.8613	.8261	.7905	.7543	.7175	.6798
.70	.2401	.9622	.9247	.8876	.8506	.8136	.7765	.7392	.7016	.6633
.725	.2763	.9600	.9207	.8819	.8436	.8056	.7676	.7297	.6915	.6530
0.75	.3164	.9573	.9158	.8751	.8353	.7960	.7571	.7184	.6797	.6409
.775	.3608	.9540	.9097	.8669	.8252	.7845	.7445	.7050	.6657	.6266
.80	.4096	.9498	.9022	.8566	.8128	.7705	.7292	.6889	.6491	.6097
0.82	.4521	.9457	.8947	.8466	.8009	.7570	.7147	.6736	.6334	.5939
.84	.4979	.9405	.8856	.8344	.7864	.7409	.6975	.6557	.6152	.5755
.86	.5470	.9338	.8740	.8194	.7688	.7215	.6769	.6344	.5936	.5541

Table II-III-7 Expansion Factors for Flow Nozzles and Venturi Tubes

$$Y_a = \left(r^{2/\gamma}\ \frac{\gamma}{\gamma-1}\ \frac{1-r^{(\gamma-1)/\gamma}}{1-r}\ \frac{1-\beta^4}{1-\beta^4 r^{2/\gamma}} \right)^{1/2}$$

$$\gamma - 1.3$$

β	β^4	.095	0.90	0.85	0.80	0.75	0.70	0.65	0.60	0.55
0.2	0.0016	0.9707	0.9407	0.9099	0.8781	0.8454	0.8117	0.7768	0.7406	0.7030
.3	.0081	.9705	.9402	.9092	.8773	.8445	.8106	.7756	.7393	.7016
.4	.0256	.9698	.9390	.9074	.8750	.8417	.8075	.7722	.7357	.6978
0.50	.0625	.9683	.9362	.9034	.8700	.8358	.8008	.7648	.7278	.6896
.55	.0915	.9671	.9338	.9001	.8658	.8309	.7952	.7588	.7214	.6829
.60	.1296	.9654	.9305	.8954	.8599	.8240	.7876	.7505	.7126	.6738
0.65	.1785	.9629	.9259	.8889	.8519	.8146	.7771	.7392	.7007	.6614
.70	.2401	.9594	.9193	.8798	.8406	.8016	.7627	.7237	.6844	.6447
.725	.2763	.9570	.9150	.8739	.8333	.7933	.7535	.7139	.6742	.6343
0.75	.3164	.9542	.9098	.8667	.8246	.7833	.7426	.7023	.6622	.6221
.775	.3608	.9507	.9034	.8580	.8141	.7714	.7297	.6886	.6481	.6077
.80	.4096	.9462	.8955	.8473	.8013	.7570	.7141	.6723	.6313	.5908
0.82	.4521	.9418	.8876	.8368	.7888	.7431	.6992	.6568	.6155	.5750
.84	.4979	.9362	.8779	.8241	.7739	.7266	.6817	.6387	.5971	.5567
.86	.5470	.9292	.8658	.8084	.7557	.7067	.6608	.6172	.5756	.5353

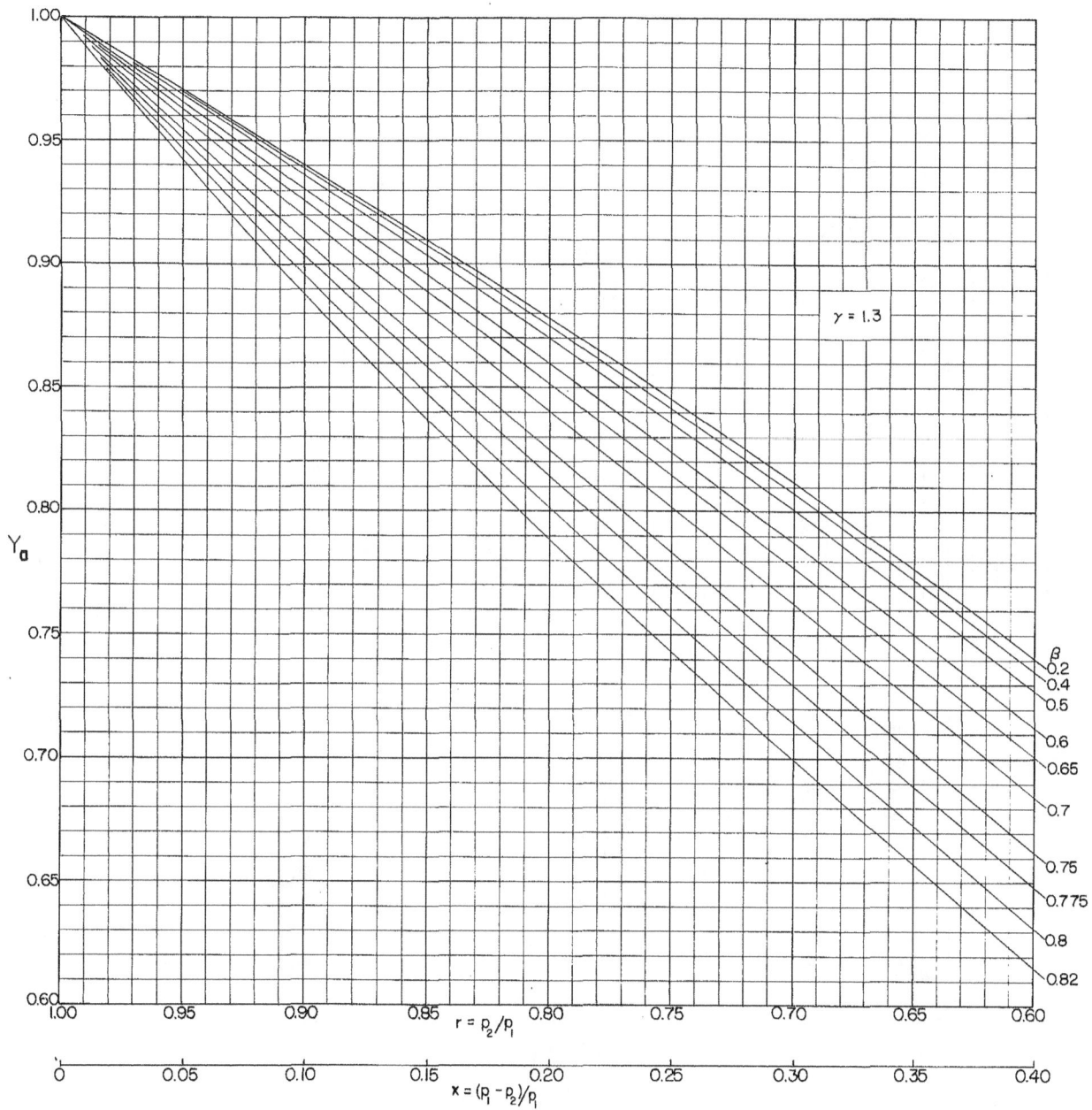

FIG. II-III-20 EXPANSION FACTORS FOR FLOW NOZZLES AND
VENTURI TUBES, $\gamma = 1.3$

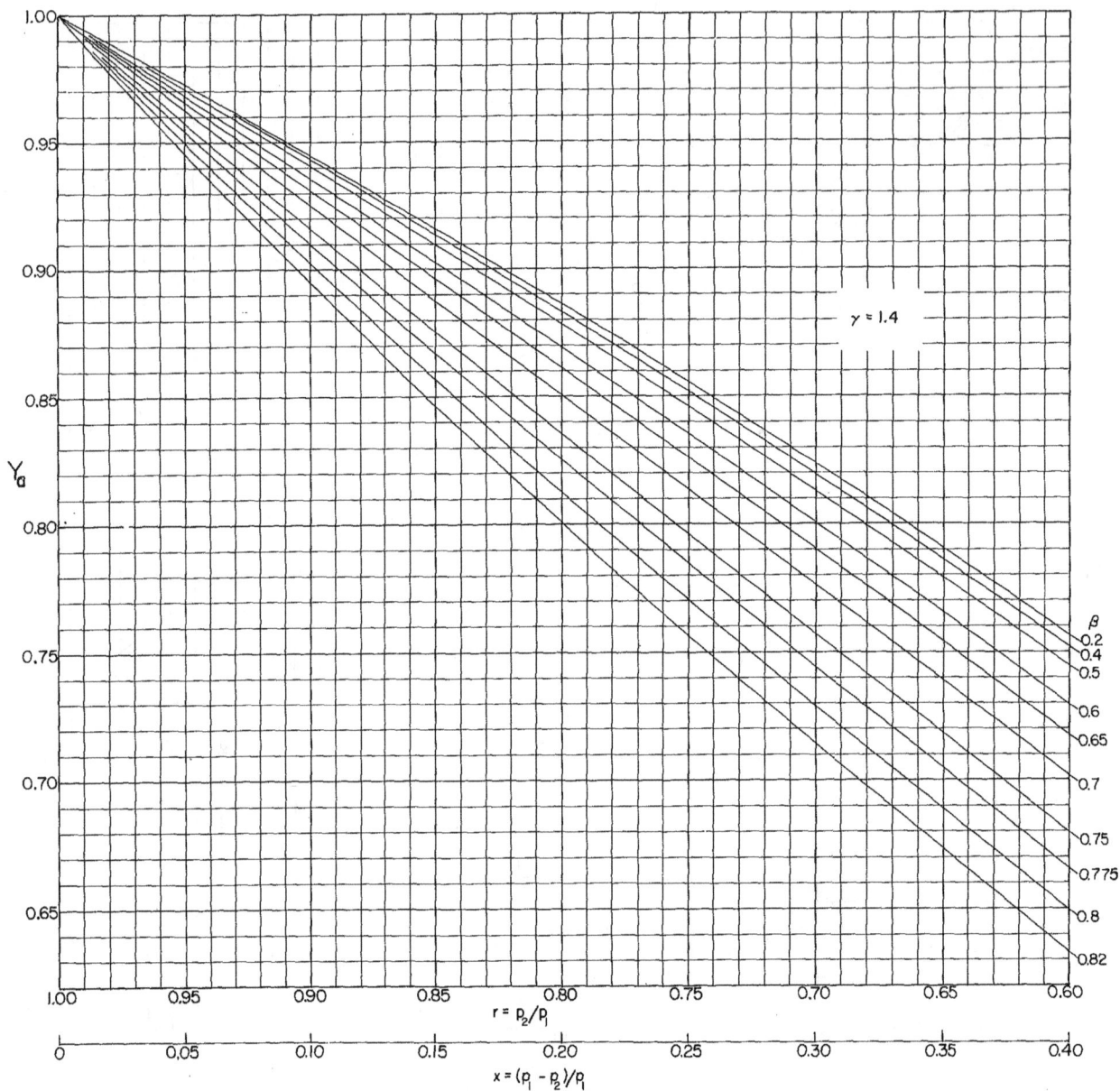

FIG. II-III-21 EXPANSION FACTORS FOR FLOW NOZZLES AND
VENTURI TUBES, $\gamma = 1.4$

II-III-30 1932 ISA Nozzle. Figure II-III-22 shows the form and proportions of the 1932 ISA flow nozzle (i.e., the flow nozzle adopted by the International Standards Association in 1932). It is used extensively abroad but hardly at all in this country. Pressures are measured from corner taps as shown by the figure.

Representative values of the flow coefficient, K, as a function of the pipe Reynolds number, R_D, are shown in Fig. II-III-23.

With compressible fluids, the expansion factor given by equation (II-III-13) applies.

For more detailed data on these flow nozzles, including tables of coefficients and correction factors, reference should be made to ISO Recommendation R541 [6].

II-III-31 Calibration of a Subsonic-Flow Nozzle. A Pitot tube or impact tip may be used to determine the discharge coefficient of a flow nozzle when other methods cannot be used. The procedure employs the use of a movable impact tip to measure the velocity pressure of the jet from the nozzle wall toward the center along one to four or more radial paths. The open end of the traversing tip should be very small and thin walled. Unless the nozzle is very large (e.g., over 12-in. throat diameter), a tip diameter of about 0.02 in. is a convenient size. The tip should be mounted so that, as it is moved along a radial path, it just clears the outlet end of the nozzle. Also, the mounting for the tip should incorporate some means for indicating the position of the center of the tip with reference to the axis of the nozzle or the inner surface of the nozzle throat. The amount of travel required in any test will depend some on the jet velocity, but it will not be necessary that the travel of the tip should be equal to the nozzle radius. (See Fig. II-III-24.)

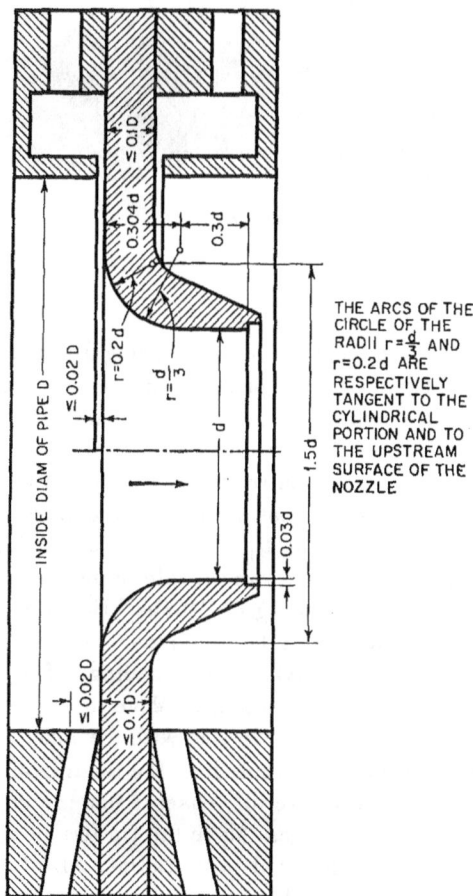

FIG. II-III-22 1932 FLOW NOZZLE OF INTERNATIONAL STANDARDS ASSOCIATION

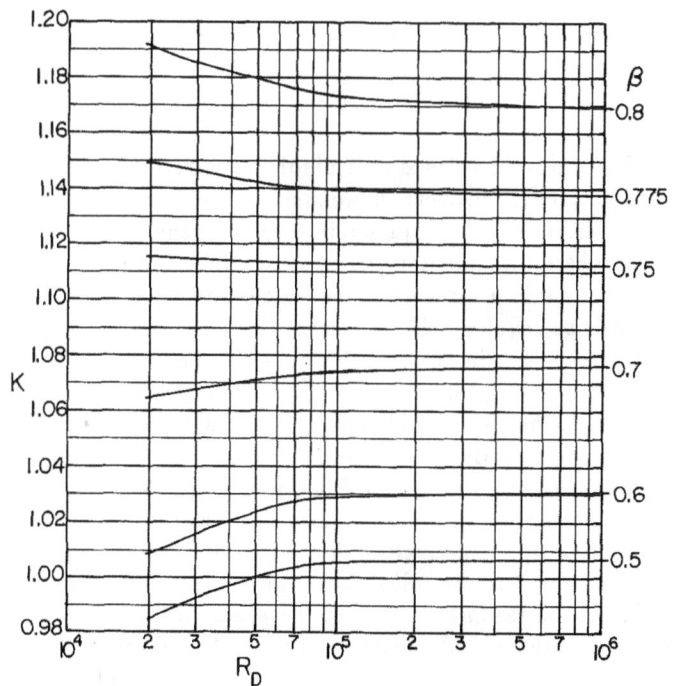

FIG. II-III-23 FLOW COEFFICIENTS, K, FOR THE 1932 ISA FLOW NOZZLE AS FUNCTION OF PIPE REYNOLDS NUMBER, R_D

FIG. II-III-24 METHOD OF MOUNTING IMPACT TIPS FOR CALIBRATING
A FLOW NOZZLE

If the nozzle is used as a discharge nozzle from a large plenum chamber, a pressure connection to the chamber will be required. In the more general case where the nozzle is mounted on the end of a pipe or within a pipe, a stationary central impact tip will be required. This tip must be mounted so that it does not interfere with the movement of the small traversing tip. The connection from the large tip is branched — one side going to a static pressure gage, usually a manometer. The other branch is connected to a differential pressure gage, the other side of which is connected to the small traversing tip. The range of this differential gage must be equal to the full velocity pressure (or impact pressure) of the nozzle jet, and the precision with which it can be read should be of the order of 0.5 per cent or less of the full range. Also, because of the smallness of the traversing-tip opening, the capacity of the connections between the differential gage and traversing tip should be kept as small as possible.

II-III-32 Procedure. Before starting a test, the traversing tip must be adjusted in its mounting so that when its center is exactly aligned with the (inner) surface of the nozzle throat the scale indicating the tip position reads "zero" (or an amount equal to the nozzle-throat radius). At the start of a test have the traversing tip withdrawn so that its inner edge is in line with the nozzle surface, i.e., the position scale reading is "minus" the value of the tip radius (assuming that a "plus" reading is toward the nozzle axis). Take readings of both the static and differential gages. Move the tip into the jet until its center is in line with the inner surface of the nozzle throat, i.e., the reading of the positioning scale is "zero;" observe and record the readings of the gages. Move the tip further into the jet by an amount equal to its radius. All of the tip is now just within the nozzle circumference. Observe and record the gage readings. Moving the tip by steps equal to its diameter, take two or three more sets of readings. Increase the amount of the tip movement to about four or five times its diameter, and continue taking readings of the pressure gages until a zero reading of the differential gage is obtained.

If the traversing tip can be moved to other positions around the nozzle to permit traverses at other meridians, these should be made in the manner as described, using the same tip positions as in the first traverse.

II-III-33 Computations. If traverses have been made at more than one position, compute the averages of the differential pressure for each separate radial setting of the traversing tip. The central impact pressure should have remained constant or very nearly constant throughout the series of readings. Now let

B = Radius of the cylindrical throat section of the nozzle

b = Radial distance between the geometrical axis of the nozzle throat and the geometrical center of the small traversing tip for any setting

c = Radius of the small traversing tip

Δ = Impact pressure measured by the larger central impact tip or the static pressure in the inlet chamber

δ = Impact pressure at the mouth of the smaller tip

$\quad$ = Δ − (the observed, or average, differential pressure)

Next, for each of the several positions of the traversing tip compute δ, $\sqrt{\delta/\Delta}$ and $(b/B)^2$. Plot the values of $\sqrt{\delta/\Delta}$ as ordinates against the corresponding values of $(b/B)^2$ as abscissae using suitable scales which will permit covering the entire range from 0 to 1.0 on the ordinate and a sufficient portion of the full abscissa scale to include all of the values of $(b/B)^2$. Figure II-III-25 shows a representative plot of such data.

II-III-34 If the diameter of the traversing tip could be diminished to a mere point, then when the axis of the tip is the plane of the nozzle wall, i.e., when $b = B$, the jet velocity would be zero and therefore δ would be 0.0 also. Hence, the ideal curve should pass through the point, (1.0, 0.0), which has been designated M in Fig. II-III-25. Compute $B - c$ and $[(B-c)/B]^2$, and draw a short line across the curve at this value of $[(B - c)/B]^2$. The point of intersection is designated N in Fig. II-III-25. Connect N with M with a straight line. Finally, let L represent the point vertically above M, i.e., at (1.0, 1.0), and let U represent the point where the curve becomes tangent to the $\sqrt{\delta/\Delta} = 1.0$ line. With a planimeter or by counting

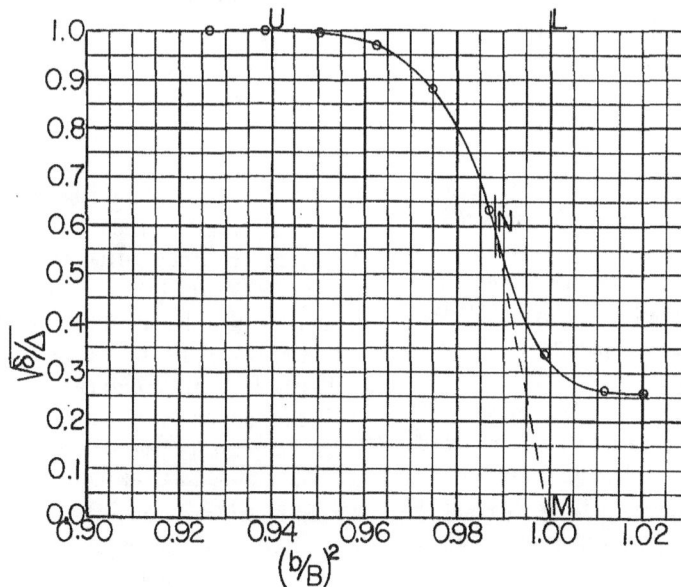

FIG. II-III-25 TYPE OF CURVE OBTAINED IN DETERMINING DISCHARGE COEFFICIENT OF FLOW NOZZLE WITH IMPACT TIP

the squares, determine the area $\overline{MLUNM}$. Then if the area bounded by the four lines, $(b/B)^2 = 0$, $(b/B)^2 = 1.0$, $\sqrt{\delta/\Delta} = 0$ and $\sqrt{\delta/\Delta} = 1.0$, is called 1, the coefficient of the nozzle will be

$$C = 1 - (\text{area } \overline{MLUNM}) \qquad \text{(II-III-14)}$$

In Fig. II-III-25 the area $\overline{MLUNM} = 48$ squares very nearly. The area representing the full theoretical flow comprises 4000 squares. Hence, for the test there represented the coefficient is

$$C = \frac{4000 - 48}{4000} = 0.988$$

Note: For a detailed discussion of this procedure see Appendix B [7, 8].

Venturi Tubes

II-III-35 Fabrication. The classical or Herschel Venturi tube, as used in this country, is usually made of cast iron or cast steel in the smaller sizes. In very large sizes, the tubes may be made of rough-welded sheet metal [9].

The grouping of Venturi tubes used by the International Standards Organization Committee on Flow Measurement, is:

1. Tubes with a rough-cast unmachined surface of the entrance converging cone and recommended for use in 4- to 32-in. pipes.

2. Tubes with a machined converging entrance cone and seldom used in pipes larger than 10 in.

3. Tubes with a rough-welded sheet metal converging entrance cone and suitable for use in pipes up to 48 in.

Since the Venturi tubes used in this country are almost exclusively in group (1), this group will be the only one discussed in detail.

II-III-36 Venturi-Tube Proportions. As shown in Fig. II-III-26, the inlet section consists of a short cylindrical section joined by an easy curvature to a truncated cone having an included angle of 21 ±1 deg. The inlet cone is joined by another smooth curve to a short cylindrical section called the throat. The exit from this throat section leads by another easy curve into the exit or diffuser cone, the recommended included angle of which is 7 to 8 deg. If the inlet and throat sections are not a single unit or casting, the joint between them when assembled should be smooth, with neither step nor protruding gasket.

II-III-37 Pressure Taps. In the inlet and throat sections there should be four or more pressure holes leading into annular chambers, to which chambers the pipes to the secondary instruments (pressure gages) are connected. The cross-sectional area of the annular chambers, or tubes, should be not less than half the sum of the areas of the respective pressure holes. The recommended size of the pressure holes is between 5/32 and 25/64 in., inclusive, but never greater than 0.1 D or 0.13 d, respectively. Furthermore, within these limits the holes should be as small as suitable for use with the fluid being metered. The edge of the pressure holes with the inner surfaces of the inlet and throat must be free from burrs or nicks and may be square and sharp or rounded very slightly. (If rounded, the radius of rounding should be less than 0.1 the taphole diameter.)

II-III-38 Throat. The throat section may be lined with bronze or other corrosion-resistant material. It should be machined after being installed in the throat-section casting. When the tube is to be used at elevated temperatures, the thermal-expansion characteristics of the liner material should be as nearly as possible the same as that of the throat casting material.

It is recommended that the machining of the throat should produce a surface finish (roughness) of (5×10^{-6}) d arithmetical mean deviation from the mean line of the profile (equivalent approximately to a 50-microinch finish). The machining should include the short curvature leading from the converging entrance section into the throat.

The throat diameter, d, should be measured very carefully in the plane of the throat pressure taps. The diameters should be near each pair of pressure taps and between the taps, with a minimum of four measurements. To determine whether the throat is cylindrical, diameters in other planes than that of the pressure taps should be measured. The same limitations regarding out-of-roundness as given for flow nozzles (Par. II-III-22) may be applied to the throat of a Venturi tube. The mean value of all diameters is to be used as the value of "d" in calculations of flow.

II-III-39 Other Features. In some cases it may be necessary to install a drain cock or vent in the pipe immediately preceding the Venturi tube to provide for removal of deposits or gases. These should be closed normally, especially when any important measurement is being made.

The diverging outlet cone angle may be as much as 15 deg, but the overall pressure loss will be greater, as shown in Fig. II-III-27. However, the 7-deg cone may be shortened, at the downstream end,

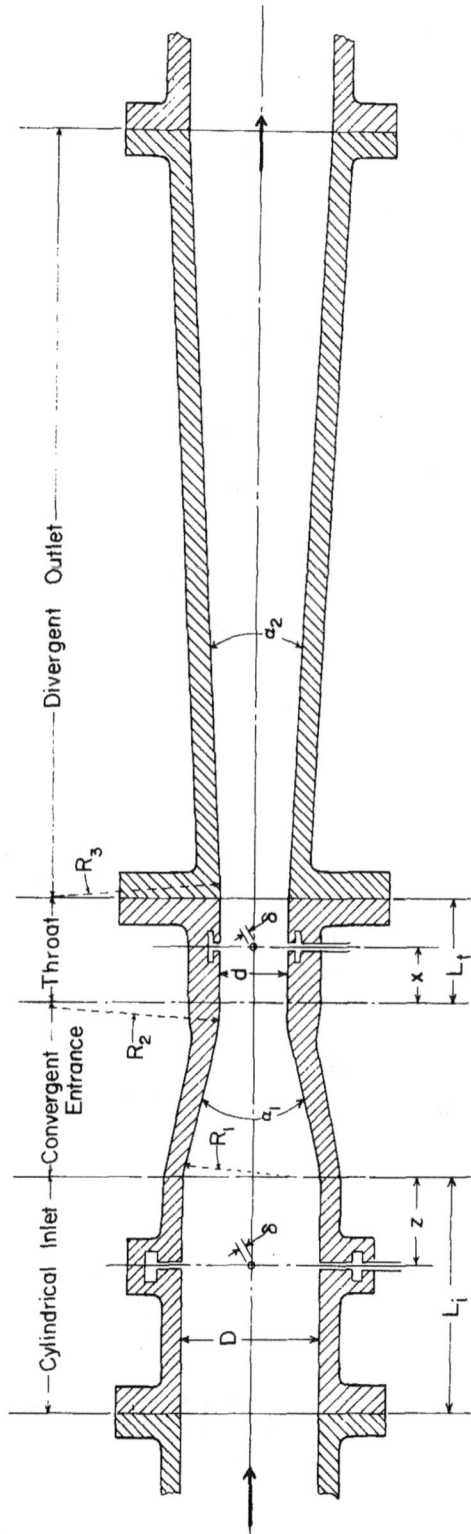

$L_i \stackrel{\sim}{=} D$ or $L_i \stackrel{\sim}{<} (D/4 + 10'')$

$z \stackrel{\sim}{<} D/2 \pm D/4$ for $4'' \stackrel{\sim}{<} D \stackrel{\sim}{<} 6''$

$D/4 \stackrel{\sim}{<} z \stackrel{\sim}{<} D/2$ for $6'' \stackrel{\sim}{<} D \stackrel{\sim}{<} 32''$

$L_t \stackrel{\sim}{=} d/3$

$y \stackrel{\sim}{<} d/6$

$5/32'' \stackrel{\sim}{<} \delta \stackrel{\sim}{<} 25/64''$ and

$\delta < 0.1 D$ or $0.13 d$

$R_1 = 1.375 D \pm 20\%$

$R_2 = 3.625 d \pm 0.125 d$

$5d \stackrel{\sim}{<} R_3 \stackrel{\sim}{<} 15 d$

$a_1 = 21° \pm 1°$

$7° \stackrel{\sim}{<} a_2 \stackrel{\sim}{<} 8°$ or $7° \stackrel{\sim}{<} a \stackrel{\sim}{<} 15°$

FIG. II-III-26 DIMENSIONAL PROPORTIONS OF CLASSICAL (HERSCHEL) VENTURI TUBES WITH A ROUGH-CAST, CONVERGENT INLET CONE

231

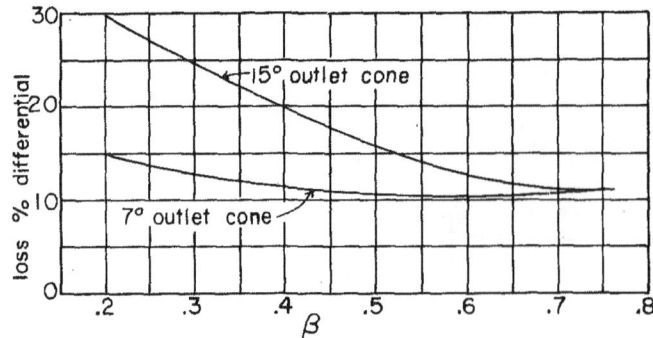

FIG. II-III-27 OVERALL PRESSURE LOSS
THROUGH VENTURI TUBES

i.e., truncated by about 35 per cent of the normal length with only a very slight effect on the pressure loss.

As stated before, the dimensional relations shown in Fig. II-III-26 are those recommended when the converging entrance cone has a rough-cast surface. The proportions for tubes with a machined entrance cone or a rough-welded sheet-metal entrance cone are slightly different. However, it seems doubtful that the small dimensional differences would have any significant effect on flow measurements. To be sure, the character of the surface of the entrance cone does have an effect, as also the radius of curvature between the inlet cone and throat section.

II-III-40 Overall Pressure Loss. The overall pressure loss for Venturi tubes is shown by Fig. II-III-27.

II-III-41 Expansion Factors. The expansion factors to be used with Venturi tubes when measuring compressible fluids are the same as used with flow nozzles, as given by the curves of Figs. II-III-20 and II-III-21 and by Tables II-III-6 and II-III-7.

II-III-42 Discharge Coefficients. For a classical Venturi tube with a rough-cast entrance cone,

$C = 0.984$ subject to a tolerance of $\pm 0.70\%$
when
$$4 \text{ in. (100 mm)} \gtreqless D \gtreqless 32 \text{ in. (800 mm)}$$
$$0.3 \gtreqless \beta \gtreqless 0.75$$
$$2 \times 10^5 \gtreqless R_D \gtreqless 2 \times 10^6$$

For tubes with a machined entrance cone,

$C = 0.995$ subject to a tolerance of $\pm 1.00\%$
when
$$2 \text{ in. (50 mm)} \gtreqless D \gtreqless 10 \text{ in. (250 mm)}$$
$$0.4 \gtreqless \beta \gtreqless 0.75$$
$$2 \times 10^5 \gtreqless R_D \gtreqless 1 \times 10^6$$

For tubes with a rough-welded sheet-metal entrance cone,

$C = 0.985$ subject to a tolerance of $\pm 1.50\%$
when
$$8 \text{ in. (200 mm)} \gtreqless D \gtreqless 48 \text{ in. (1200 mm)}$$
$$0.4 \gtreqless \beta \gtreqless 0.70$$
$$2 \times 10^5 \gtreqless R_D \gtreqless 2 \times 10^6$$

II-III-43 Nozzle-Venturi. The nozzle-Venturi (Fig. II-III-28) is used very little in this country. As made and used abroad, the cylindrical inlet and reducing conical sections of the classical Venturi are replaced with a single, short, curved inlet. The curvature of this inlet section is the same as that of the ISA 1932 nozzle, as shown in Fig. II-III-22. Beyond (downstream of) the throat of this inlet-nozzle section, there is an additional cylindrical section $0.4\,d$ long, before the beginning of the

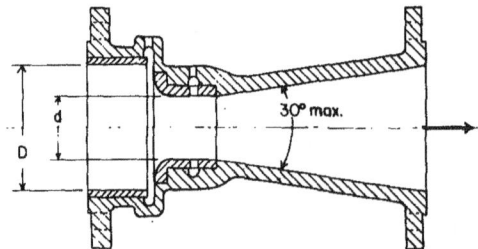

FIG. II-III-28 NOZZLE-VENTURI

diverging outlet section. (Thus, the total length of the cylindrical "throat" is 0.7 d.) The included angle of the divergent outlet section may be 30 deg or less. The inlet pressure tap is a corner tap or annular slit at the intersection of the inlet face of the nozzle section with the pipe wall. The outlet or throat pressure taps are located 0.3 d from the entrance plane of the nozzle-throat portion and 0.4 D preceding the outlet end of the total throat section. For further data and coefficients for these nozzle-Venturis, refer to ISO 781 [9].

II-III-44 Computation of Rate of Flow: Subsonic Conditions. With all of the primary elements described in the preceding paragraphs used under *subsonic* conditions, the same general equation may be used for computing the rate of flow. As developed in Chapter I-5, this equation may be written in several forms, some of which are:

$$m(\text{lb}_m/\text{sec}) = 0.52502 \left(\frac{C\,Y\,d^2 F_a}{\sqrt{1-\beta^4}} \right) \sqrt{\rho_1(p_1 - p_2)}$$

(II-III-15a)

$$= 0.52502\, K\, Y\, d^2 F_a \sqrt{\rho_1\,\Delta p} \quad \text{(II-III-15b)}$$

$$= 0.099702 \left(\frac{C\,Y\,d^2 F_a}{\sqrt{1-\beta^4}} \right) \sqrt{\rho_1\,h_w}$$

(II-III-15c)

$$q_1\ (\text{cfs at } p_1,\ T_1)$$
$$= 0.099702 \left(\frac{C\,Y\,d^2 F_a}{\sqrt{1-\beta^4}} \right) \sqrt{\frac{h_w}{\rho_1}}$$

(II-III-16)

$$m(\text{lb}_m/\text{hr}) = 1865.57\, K\, Y\, d^2 F_a \sqrt{\rho_1\,\Delta p} \quad \text{(II-III-17a)}$$

$$= 358.93 \left(\frac{C\,Y\,d^2 F_a}{\sqrt{1-\beta^4}} \right) \sqrt{\rho_1\,h_w} \quad \text{(II-III-17b)}$$

$$q_1\ (\text{cfh at } p_1,\ T_1) = 358.93 \left(\frac{C\,Y\,d^2 F_a}{\sqrt{1-\beta^4}} \right)$$
$$\sqrt{\frac{h_w}{\rho_1}} \quad \text{(II-III-18)}$$

For a segmental orifice, the d^2 would be replaced with (a = orifice area, in.^{2}), thus giving

$$m(\text{lb}_m/\text{hr}) = 457.0 \left(\frac{C\,Y\,a\,F_a}{\sqrt{1-\beta^4}} \right)$$
$$\sqrt{\rho_1\,h_w} \quad \text{(II-III-19)}$$

II-III-45 In the metering of compressible fluids, the general volume-pressure-temperature relation,

equation (I-3-20) in Chapter I-3, in the form, $q_o = q_1\,[(p_1\,T_o\,Z_o)/(p_o\,T_1\,Z_1)]$, may be applied to equations (II-III-16) and (II-III-18) to give the corresponding volume rates at the reference conditions, p_o, T_o. Using the customary reference conditions for fuel-gas measurements of $p_o = 14.73$ psia, $T_o = 60\,\text{F} = 519.7\,\text{R}$ and dry, equation (II-III-18) becomes

$$q_o(\text{scfh}) = 7708\, K\, Y_1\, d^2 F_a\, Z_o \sqrt{\frac{h_w\,\rho_1}{G\,T_1\,Z_1}} \quad \text{(II-III-20)}$$

As stated in Chapter I-5, Par. I-5-14, this equation gives acceptable values if the metering conditions are not too far from the normal ambient conditions. For more precise measurements of particular gases and gas mixtures, equations similar to (II-III-20) should be developed from equations (I-5-27) and (I-5-33), using the gas constant, R, and the compressibility values for the particular gas.

II-III-46 The equation for the Reynolds number, of which the coefficients are a function, may be written

$$R_d = \frac{d\,V_2\,\rho_2}{12\,\mu} = \frac{48\,m}{\pi\,d\,\mu} \quad \text{(II-III-21)}$$

and

$$R_D = \frac{D\,V_1\,\rho_1}{12\,\mu} = \frac{48\,m}{\pi\,D\,\mu} \quad \text{(II-III-22)}$$

so that

$$R_D = \beta\,R_d \quad \text{(II-III-23)}$$

The symbols and units applying to the above equations are:

a	Area of orifice, nozzle or Venturi throat	in.^{2}
C	Coefficient of discharge	ratio
d	Orifice diameter, also diameter of flow nozzle throat or Venturi throat	in.
F_a	Area thermal-expansion factor	Fig. II-I-3
G	Specific gravity, the ratio of molecular weights for gases	
h_w	Differential pressure	in. water at 68 F
K	Flow coefficient $= C/\sqrt{1-\beta^4}$	ratio

m	Mass rate of flow	lb_m/sec or lb_m/hr
p	Pressure	psia
q	Volume rate of flow	cfs or cfh
R_d	Reynolds number using d	ratio
R_D	Reynolds number using D	ratio
Y	Expansion factor	ratio
Δp	Differential pressure	psi
μ	Viscosity of fluid, absolute	lb_m/ft-sec
ρ	Density of fluid	lb_m/ft^3

II-III-47 For use with metric units,

$$m(\text{kg}_m/\text{sec}) = 0.034783\, K\, Y\, d^2\, F_a\, \sqrt{\rho_1\, \Delta p}$$
(II-III-24)
$$= 0.034752\, K\, Y\, d^2 F_a\, \sqrt{\rho_1\, h_w}$$

$$q_1(m^3/\text{sec at } p_1, T_1) = 0.000\,034572$$
$$K\, Y\, d^2 F_a\, \sqrt{\frac{h_w}{\rho_1}}\quad \text{(II-III-25)}$$

In these two equations,

d	is in cm
h_w	is in cm of water at 20 C
m	is in kg$_m$/sec
p	and Δp are in gm$_f$/cm^2
q	is in m^3/sec
ρ	is in gm$_m$/cm^3

For evaluating the diameter ratio, β, both d and the pipe diameter, D, will be in cm. The Reynolds number with which to determine the coefficient may be evaluated by $R_d = 4000m/(\pi\, d\, \mu)$, in which the viscosity, μ, is in poise. With both p and Δp in the same units, the ratios, $x = \Delta p/p_1$ and x/γ, provide an index for the determination of the expansion factor, Y.

II-III-48 Proprietary Flow Tubes. There are a number of flow tubes that have constructional features that resemble both a flow nozzle and a Venturi tube. By utilizing a boundary layer effect upon the pressure sensed by either the high-pressure or low-pressure tap, these tubes may have a somewhat higher differential pressure than a conventional flow nozzle or Venturi tube of the same

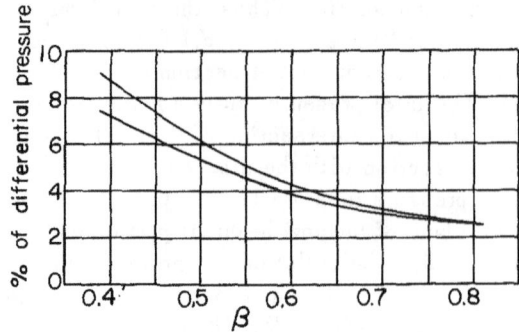

FIG. II-III-29 OVERALL PRESSURE LOSS THROUGH SOME PROPRIETARY FLOW TUBES

diameter ratio. For this reason, the overall pressure loss is rather low, as shown by Fig. II-III-29.

If one of these flow tubes is to be used, it should be calibrated with the piping section in which it is to be used and over the full range of rates of flow to which it will be subjected when in use.

Sonic-Flow Primary Elements

II-III-49 Materials and Form. The materials used for sonic flow nozzles are subject to the same considerations as given for subsonic flow nozzles (Pars. II-III-20 — II-III-22). The form or contour of the inner surface may be any of those shown in Figs. II-III-14 and II-III-22. If a minimal overall pressure loss is important, then the radial inlet Venturi (Fig. II-III-30) should be used. If both subsonic and sonic flows are to be measured with the same nozzle, it is recommended the form be one for which the coefficient is well known, such as the ASME long-radius nozzles (Fig. II-III-14).

II-III-50 Pressure Taps. As discussed in Chapter I-5, the inlet pressure and density only are required for determining the rate of flow. For this purpose, it is recommended that the inlet static pressure be

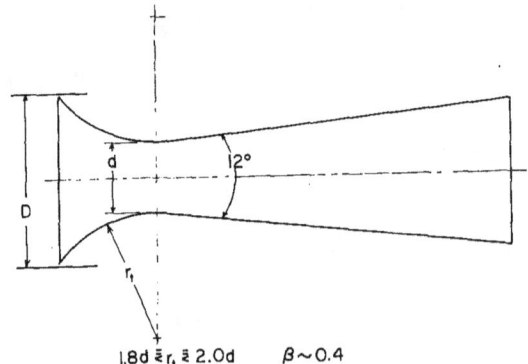

FIG. II-III-30 PROPORTIONS OF INTERIOR SURFACE OF CIRCULAR ARC OR RADIAL INLET VENTURI

measured with a pipe-wall tap located one pipe
diameter preceding the inlet of the nozzle. Alternatively, the total or stagnation inlet pressure
may be measured with an impact or Pitot tube,
mounted at the one pipe diameter distance. The axis
of the open end of the impact tube should coincide
with the nozzle axis if the diameter ratio, β,
exceeds 0.3; but, if β is 0.3 or less, the opening
may be anywhere within the central third of the
cross-sectional area of the inlet channel.

The location of an outlet pressure tap will depend on the purpose for the outlet-pressure
measurement. If the purpose is to insure that the
outlet pressure is low enough so that sonic velocity
is obtained at the nozzle exit, then the tap should
be located in the pipe wall, as shown in Fig.
II-III-15, or at the section of minimum cross-sectional
area of a radial inlet Venturi. On the other hand, if
the purpose is to determine the overall pressure loss,
especially with the radial inlet Venturi, then the
outlet pressure tap should be six to ten pipe
diameters downstream from the throat section.

II-III-51 Installation. It is recommended that
the installation requirements given for flow nozzles
in subsonic service be followed with sonic flow
nozzles in order to give consistent measurement
over the entire flow range. A possible exception
might be when the diameter ratio, β, is less than
0.2, so that the inlet velocity and flow pattern resemble that of a nozzle discharging from a plenum
chamber.

II-III-52 Temperature and Density. For measurement of the fluid temperature, a thermocouple attached to the pipewall or a thermometer, bare or in
a well, desirably a total temperature well, may be
used where a simple instrument can be tolerated in
terms of accuracy and safety. If a total temperature
well is not used, then the stagnation temperature is
to be obtained by applying the appropriate recovery
factor to the thermometer reading. (See Chapter I-3,
Par. I-3-17.)

The preferred location for the thermometer is on
the outlet side of the sonic-flow element. (See Chapter I-5, Par. I-5-43.) If the thermometer is located on
the inlet side of the element, it should be at least
two pipe diameters, and preferably up to 200 stem or
well diameters, upstream of the inlet pressure tap,
so as to have as little effect on the pressure measurement as possible.

If the fluid density is determined with a densitometer or density cell, the instrument should be installed and operated in accordance with the
manufacturer's instructions.

II-III-53 Calibration. As with all types of
primary elements utilizing measurements of a pressure and a differential pressure, a sonic-flow
primary element should be calibrated in the section
of piping in which it will be used. If this is not
possible, the actual piping should be duplicated as
closely as possible, with particular care taken so
that the inlet section is sized so as to maintain
the diameter ratio, β, the *same*, if in service β
is greater than 0.3. If in service β is less than 0.3,
the calibration may be made with a pipe section
giving a lower but not a higher value of β.

The reference method of measuring the fluid
flow will depend on the size and flow capacity of
the element, as well as on the facilities available.
Some of the methods are volumetric gasometers,
weighing containers and the pressure drop due to
the discharge of gas from a container. All of these
involve the measurement of a time interval.

Possibly the simplest procedure is to place the
sonic-flow element in series with another meter
whose flow characteristic is known from a history
of reliable calibration data. This reference meter
may be another sonic-flow nozzle operated under
sonic or subsonic-flow conditions or may be an
orifice. Care must be taken to eliminate the effect
of the discharge-velocity profile of the upstream
element on the inlet-flow profile of the downstream
element. Often screens or flow straighteners are
placed between the two elements to restore a uniform velocity profile.

II-III-54 Coefficients. The discharge coefficient of a sonic-flow element should be obtained
from a calibration as previously outlined. When this
is not possible, an approximate coefficient value of
0.99 may be used. In this case, the uncertainty or
tolerance to the coefficient should be between
± 2 to ± 3 per cent.

II-III-55 Sonic-Flow Computations. If the inlet
static pressure is measured from one or more sidewall pressure taps, then first this value must be
converted to total pressure using the equation

$$\left(\frac{p_1}{p_{1t}}\right)^{2/\Gamma} - \left(\frac{p_1}{p_{1t}}\right)^{(\Gamma+1)/\Gamma} = \beta^4 \frac{\Gamma-1}{2}\left(\frac{2}{\Gamma+1}\right)^{\frac{\Gamma+1}{\Gamma-1}} \quad \text{(II-III-26)}$$

However, if β is less than 0.5, then the following equation may be used

$$p_{1t} = \cfrac{p_1}{\left[1 - \beta^4 \cfrac{\Gamma}{2}\left(\cfrac{2}{\Gamma+1}\right)^{(\Gamma+1)/\Gamma-1}\right]} \quad \text{(II-III-27)}$$

This conversion is not necessary if the total inlet pressure is measured with an impact tube.

With the values of the stagnation pressure and temperature, the mass rate of flow at throat sonic speed may be computed by

$$m\ (\text{lb}_m/\text{sec}) = C\ a\ \phi^*_i\left(\cfrac{\phi^*}{\phi^*_i}\right)\left(\cfrac{p_{1t}}{\sqrt{T_{1t}}}\right) \quad \text{(II-III-28)}$$

where

C = Discharge coefficient at sonic-flow conditions

a = Throat area of sonic-flow primary element, ft^2 or in.^{2}

p_{1t} = Inlet stagnation pressure, psfa or psia

T_{1t} = Inlet stagnation temperature of fluid, °R

ϕ^*_i = Sonic-flow function of an ideal gas having a constant ratio of specific heats, a molecular weight and structure identical to the real gas being metered, as evaluated by equations (I-5-103) and (I-5-105)

$\dfrac{\phi^*}{\phi^*_i}$ = Ratio of the real-gas sonic-flow function to the sonic-flow function of its ideal-gas counterpart, values from Tables II-III-8 through II-III-20

Where the equation of state is represented by tables of gas properties as in the case of steam, it may be more convenient and accurate to use

$$m\ (\text{lb}_m/\text{sec}) = C\ a\ B_F\left(\cfrac{F}{F_i}\right)\sqrt{\cfrac{p_{1t}}{v_{1t}}} \quad \text{(II-III-29)}$$

where

C = Discharge coefficient at sonic-flow conditions

$B_F = \sqrt{g_c}\ F_i$

F_i = Ideal isentropic expansion function, equation (I-5-104)

F = Real-gas isentropic expansion function, equation (I-5-124)

v_{1t} = Specific volume at inlet stagnation conditions

$\dfrac{F}{F_i}$ = Ratios given in Tables II-III-9 through II-III-21

Note: In equation (II-III-29), as given above, a is ft^2 and p is psfa; but, if a is in in.^{2} and p is psia, then the right side must be multiplied by 1/12.

II-III-56 In the special case of the sonic flow of a natural fuel gas, equation (II-III-28) can be modified by replacing $\phi^*_i\ (\phi^*/\phi^*_i)$ with $[(e_c j + b_c)/(e_z j + b_z)]$ $\overline{\sqrt{(gc\ MW)/R}}$ so that

$$m\ (\text{lb}_m/\text{sec}) = C\ a\left(\cfrac{e_c j + b_c}{e_z j + b_z}\right)\sqrt{\cfrac{g_c\ MW}{R}}\cfrac{p_{1t}}{\sqrt{T_{1t}}} \quad \text{(II-III-30)}$$

where

$$j = X_{C_2H_6} + X_{CO_2} - \frac{1}{2}X_{N_2} + 2\,X_{C_3H_8} + 3\,X_{C_4H_{10}} \quad \text{(II-III-31)}$$

X = mole fraction of the subscripted gas and values of e_c, b_c, e_z and b_z are given by Tables II-III-23 through II-III-26 [10].

AIR

Table II-III-8

Values of ϕ^*/ϕ_i, Ratio of the Real-Gas Sonic-Flow Function to the Sonic Flow Function of Its Ideal-Gas Counterpart

$$\phi_i^* = 0.53175$$

T_{1t} (°R)	Inlet Stagnation Pressure, p_{1t} (psia)									
	0	100	200	400	600	800	1000	1200	1400	1500
400	1.0002	1.0073	1.0145	1.0296	1.0458	1.0626	1.0809	1.0994	1.1180	1.1262
450	1.0002	1.0051	1.0100	1.0202	1.0306	1.0413	1.0519	1.0629	1.0733	1.0786
500	1.0002	1.0037	1.0072	1.0142	1.0212	1.0283	1.0353	1.0421	1.0488	1.0521
550	1.0001	1.0025	1.0050	1.0100	1.0150	1.0198	1.0245	1.0291	1.0335	1.0357
600	1.0000	1.0018	1.0035	1.0071	1.0105	1.0139	1.0172	1.0203	1.0232	1.0246
650	0.9998	1.0011	1.0024	1.0049	1.0072	1.0096	1.0119	1.0140	1.0160	1.0169
700	0.9995	1.0004	1.0013	1.0030	1.0049	1.0065	1.0079	1.0093	1.0106	1.0112

Table II-III-9

Values of F/F_i, Ratio of Real Gas to Ideal Gas Isentropic Expansion Function

$$F_i = 0.68473 \qquad B_F = 3.8839$$

T_{1t} (°R)	Inlet Stagnation Pressure, p_{1t} (psia)									
	0	100	200	400	600	800	1000	1200	1400	1500
400	1.0002	1.0019	1.0035	1.0076	1.0138	1.0204	1.0301	1.0418	1.0539	1.0581
450	1.0002	1.0019	1.0036	1.0081	1.0131	1.0195	1.0257	1.0344	1.0422	1.0465
500	1.0002	1.0021	1.0041	1.0081	1.0123	1.0177	1.0230	1.0295	1.0359	1.0390
550	1.0001	1.0016	1.0035	1.0075	1.0115	1.0162	1.0207	1.0262	1.0316	1.0343
600	1.0000	1.0016	1.0032	1.0074	1.0100	1.0157	1.0191	1.0238	1.0283	1.0304
650	0.9998	1.0014	1.0031	1.0066	1.0093	1.0138	1.0175	1.0219	1.0260	1.0279
700	0.9995	1.0012	1.0026	1.0057	1.0091	1.0129	1.0162	1.0200	1.0238	1.0256

STEAM

Table II-III-10

Values of ϕ^*/ϕ_i^*, Ratio of Real-Gas Sonic-Flow Function to Sonic Flow Function of Its Ideal-Gas Counterpart

$$\phi_i^* = 0.40866$$

T_{1t} (°R)	Inlet Stagnation Pressure, p_{1t} (psia)										
	0	1	100	200	400	600	800	1000	1200	1400	1600
600	1.0016	1.0018	1.0089	1.0181	1.0335						
650	1.0006	1.0011	1.0071	1.0139	1.0274	1.0424	1.0589	1.0754			
700	0.9996	1.0003	1.0052	1.0107	1.0218	1.0339	1.0461	1.0603	1.0813	1.1021	1.1230
750	0.9986	0.9995	1.0036	1.0082	1.0174	1.0272	1.0379	1.0487	1.0636	1.0785	1.0934
800	0.9971	0.9986	1.0018	1.0057	1.0136	1.0217	1.0312	1.0397	1.0514	1.0624	1.0738
850	0.9968	0.9978	0.9999	1.0034	1.0103	1.0176	1.0251	1.0326	1.0410	1.0507	1.0597
900	0.9959	0.9969	0.9986	1.0016	1.0074	1.0136	1.0199	1.0263	1.0338	1.0413	1.0487
950	0.9950	0.9960	0.9975	1.0000	1.0052	1.0104	1.0158	1.0212	1.0275	1.0337	1.0401
1000	0.9940	0.9952	0.9964	0.9986	1.0030	1.0076	1.0124	1.0172	1.0224	1.0277	1.0329
1050	0.9930	0.9945	0.9957	0.9976	1.0013	1.0053	1.0096	1.0138	1.0184	1.0229	1.0272
1100	0.9923	0.9935	0.9950	0.9967	1.0001	1.0036	1.0073	1.0109	1.0148	1.0187	1.0226
1150	0.9913	0.9928	0.9940	0.9956	0.9989	1.0019	1.0052	1.0084	1.0118	1.0152	1.0187
1200	0.9903	0.9918	0.9932	0.9946	0.9975	1.0004	1.0033	1.0062	1.0092	1.0123	1.0153
1250	0.9893	0.9908	0.9925	0.9938	0.9964	0.9989	1.0014	1.0038	1.0065	1.0093	1.0120
1300	0.9886	0.9898	0.9915	0.9927	0.9950	0.9973	0.9994	1.0016	1.0040	1.0065	1.0089
1350	0.9879	0.9884	0.9903	0.9914	0.9936	0.9956	0.9975	0.9994	1.0016	1.0038	1.0060
1400	0.9871	0.9879	0.9891	0.9901	0.9920	0.9939	0.9956	0.9974	0.9993	1.0012	1.0031
1450	0.9864	0.9869	0.9879	0.9888	0.9906	0.9922	0.9937	0.9952	0.9969	0.9986	1.0004
1500	0.9857	0.9859	0.9864	0.9872	0.9890	0.9905	0.9918	0.9932	0.9947	0.9962	0.9976
1550	0.9849	0.9847	0.9849	0.9857	0.9873	0.9887	0.9898	0.9910	0.9923	0.9937	0.9950
1600	0.9842	0.9837	0.9832	0.9840	0.9854	0.9869	0.9880	0.9891	0.9902	0.9913	0.9925

STEAM

Table II-III-11

Values of F/F_i, Ratio of Real-Gas to Ideal-Gas Isentropic Expansion Function

$$F_i = 0.66727 \quad B_F = 3.7848$$

T_{1t} (°R)	Inlet Stagnation Pressure, p_{1t} (psia)									
	1	100	200	400	600	800	1000	1200	1400	1600
600	1.0018	1.0005	1.0000	0.9988						
650	1.0010	0.9998	0.9993	0.9984						
700	1.0003	0.9992	0.9987	0.9980	0.9971	0.9961	0.9950			
750	0.9995	0.9984	0.9981	0.9976	0.9969	0.9960	0.9952			
800	0.9987	0.9984	0.9980	0.9974	0.9967	0.9961	0.9954	0.9948	0.9943	0.9937
850	.9979	.9976	.9973	.9968	.9962	.9957	.9952	.9947	.9943	.9938
900	.9970	.9968	.9966	.9962	.9957	.9954	.9950	.9946	.9943	.9940
950	.9962	.9960	.9954	.9956	.9952	.9950	.9946	.9944	.9941	.9939
1000	.9954	.9953	.9952	.9950	.9947	.9945	.9943	.9941	.9939	.9937
1050	.9946	.9945	.9944	.9942	.9941	.9939	.9938	.9937	.9936	.9935
1100	.9938	.9937	.9935	.9934	.9934	.9932	.9932	.9932	.9939	.9937
1150	.9928	.9927	.9926	.9926	.9926	.9926	.9927	.9927	.9936	.9935
1200	.9917	.9918	.9918	.9919	.9920	.9920	.9921	.9922	.9932	.9932
1250	.9908	.9909	.9909	.9911	.9913	.9913	.9914	.9916	.9927	.9927
1300	.9899	.9901	.9901	.9903	.9905	.9906	.9908	.9910	.9911	.9925
1350	.9889	.9890	.9891	.9893	.9896	.9897	.9898	.9900	.9903	.9919
1400	.9879	.9880	.9881	.9884	.9886	.9888	.9890	.9893	.9895	.9913
1450	.9870	.9871	.9872	.9874	.9876	.9877	.9878	.9881	.9883	.9906
1500	.9860	.9861	.9862	.9865	.9865	.9868	.9870	.9871	.9873	.9898

239

METHANE CH$_4$

Table II-III-12

Values of ϕ^*/ϕ_i^*, Ratio of Sonic-Flow Function of Methane to Sonic-Flow Function of Its Ideal-Gas Counterpart

$$\phi_i^* = 0.3885$$

T_{1t} (°R)	Inlet Stagnation Pressure, p_{1t} (psia)						
	0	100	200	400	600	800	1000
450	0.9993	1.0097	1.0209	1.0451	1.0734	1.1057	1.1429
500	0.9975	1.0050	1.0129	1.0297	1.0481	1.0683	1.0902
550	0.9952	1.0008	1.0065	1.0185	1.0315	1.0451	1.0594
600	0.9923	0.9965	1.0010	1.0099	1.0193	1.0289	1.0389
650	0.9891	0.9923	0.9958	1.0025	1.0093	1.0164	1.0237
700	0.9856	0.9882	0.9909	0.9961	1.0011	1.0065	1.0118

Table II-III-13

Values of F/F_i, Ratio of Isentropic Expansion Function to That of Its Ideal-Gas Counterpart

$$B_F = 3.8135 \qquad F_i = 0.6723$$

T_{1t} (°R)	Inlet Stagnation Pressure, p_{1t} (psia)						
	0	100	200	400	600	800	1000
450	0.9993	0.9987	0.9989	0.9989	0.9999	1.0035	1.0100
500	0.9975	0.9974	0.9974	0.9981	1.0000	1.0032	1.0080
550	0.9952	0.9953	0.9956	0.9967	0.9986	1.0016	1.0054
600	0.9923	0.9926	0.9931	0.9944	0.9964	0.9990	1.0025
650	0.9891	0.9895	0.9901	0.9915	0.9934	0.9959	0.9990
700	0.9856	0.9862	0.9868	0.9883	0.9901	0.9925	0.9953

Table II-III-14

Values of ϕ^*/ϕ_i^*, Ratio of Real-Gas Sonic-Flow Function to Sonic-Flow Function of Its Ideal-Gas Counterpart

$$\phi_i^* = 0.52295$$

T_{1t} (°R)	Inlet Stagnation Pressure, p_{1t} (psia)									
	0	100	200	400	600	800	1000	1200	1400	1500
400	1.0000	1.0067	1.0134	1.0275	1.0424	1.0578	1.0736	1.0895	1.1054	1.1131
450	1.0000	1.0046	1.0092	1.0187	1.0282	1.0380	1.0477	1.0572	1.0665	1.0711
500	1.0000	1.0032	1.0064	1.0130	1.0194	1.0259	1.0323	1.0384	1.0442	1.0469
550	1.0000	1.0023	1.0046	1.0090	1.0136	1.0179	1.0222	1.0263	1.0301	1.0320
600	0.9998	1.0015	1.0032	1.0064	1.0094	1.0124	1.0154	1.0180	1.0206	1.0219
650	0.9998	1.0010	1.0021	1.0044	1.0064	1.0084	1.0104	1.0122	1.0138	1.0148
700	0.9997	1.0005	1.0013	1.0028	1.0042	1.0055	1.0067	1.0078	1.0089	1.0093

Table II-III-15

Values of F/F_i, Ratio of Real-Gas to Ideal-Gas Isentropic Expansion Function

$$F_i = 0.68473 \qquad B_F = 3.8839$$

T_{1t} (°R)	Inlet Stagnation Pressure, p_{1t} (psia)									
	0	100	200	400	600	800	1000	1200	1400	1500
400	1.0000	1.0030	1.0038	1.0084	1.0150	1.0229	1.0317	1.0431	1.0556	1.0617
450	1.0000	1.0020	1.0041	1.0091	1.0141	1.0210	1.0276	1.0360	1.0448	1.0491
500	1.0000	1.0019	1.0040	1.0090	1.0132	1.0191	1.0248	1.0316	1.0383	1.0423
550	1.0000	1.0019	1.0040	1.0080	1.0126	1.0176	1.0226	1.0284	1.0339	1.0367
600	0.9998	1.0017	1.0038	1.0079	1.0117	1.0163	1.0209	1.0258	1.0308	1.0333
650	0.9998	1.0015	1.0034	1.0075	1.0109	1.0151	1.0194	1.0238	1.0281	1.0306
700	0.9997	1.0014	1.0032	1.0069	1.0104	1.0143	1.0180	1.0221	1.0263	1.0280

OXYGEN

Table II-III-16

Values of ϕ^*/ϕ^*_i, Ratio of Real-Gas Sonic-Flow Function to Sonic-Flow Function of Its Ideal-Gas Counterpart

$$\phi^*_i = 0.45620$$

T_{1t} (°R)	Inlet Stagnation Pressure, p_{1t} (psia)									
	0	100	200	400	600	800	1000	1200	1400	1500
400	0.9998	1.0082	1.0169	1.0355	1.0560	1.0786	1.1030	1.1294	1.1584	1.1734
450	0.9997	1.0056	1.0117	1.0244	1.0378	1.0520	1.0668	1.0819	1.0974	1.1052
500	0.9995	1.0039	1.0083	1.0172	1.0265	1.0361	1.0459	1.0558	1.0655	1.0705
550	0.9991	1.0022	1.0055	1.0121	1.0188	1.0257	1.0324	1.0393	1.0460	1.0493
600	0.9986	1.0010	1.0034	1.0083	1.0131	1.0180	1.0231	1.0280	1.0327	1.0350
650	0.9979	0.9997	1.0015	1.0051	1.0088	1.0124	1.0161	1.0196	1.0230	1.0247
700	0.9972	0.9985	0.9999	1.0026	1.0053	1.0079	1.0105	1.0131	1.0156	1.0168

Table II-III-17

Values of F/F_i, Ratio of Real-Gas to Ideal-Gas Isentropic Expansion Function

$$F_i = 0.68473 \qquad B_F = 3.8839$$

T_{1t} (°R)	Inlet Stagnation Pressure, p_{1t} (psia)									
	0	100	200	400	600	800	1000	1200	1400	1500
400	0.9998	1.0008	1.0013	1.0031	1.0083	1.0099	1.0157	1.0267	1.0430	1.0524
450	0.9997	1.0008	1.0020	1.0047	1.0088	1.0135	1.0193	1.0267	1.0345	1.0394
500	0.9995	1.0009	1.0022	1.0051	1.0085	1.0124	1.0168	1.0223	1.0286	1.0321
550	0.9991	1.0003	1.0019	1.0050	1.0082	1.0120	1.0158	1.0205	1.0250	1.0275
600	0.9986	1.0000	1.0014	1.0045	1.0073	1.0107	1.0145	1.0185	1.0223	1.0242
650	0.9979	0.9993	1.0006	1.0034	1.0063	1.0096	1.0130	1.0164	1.0198	1.0216
700	0.9972	0.9984	0.9999	1.0027	1.0054	1.0083	1.0112	1.0145	1.0177	1.0193

HYDROGEN

Table II-III-18

Values of ϕ^*/ϕ^*_i, Ratio of Real-Gas Sonic-Flow Function to Sonic-Flow Function of Its Ideal-Gas Counterpart

$$\phi^*_i = 0.14029$$

T_{1t} (°R)	Inlet Stagnation Pressure, p_{1t} (psia)									
	0	100	200	400	600	800	1000	1200	1400	1500
400	1.0096	1.0090	1.0086	1.0075	1.0063	1.0050	1.0039	1.0025	1.0012	1.0005
450	1.0062	1.0055	1.0048	1.0036	1.0022	1.0009	0.9994	0.9980	0.9965	0.9958
500	1.0039	1.0032	1.0025	1.0010	0.9996	0.9981	0.9966	0.9952	0.9937	0.9930
550	1.0023	1.0016	1.0009	0.9994	0.9979	0.9964	0.9949	0.9935	0.9920	0.9913
600	1.0013	1.0006	0.9998	0.9983	0.9969	0.9954	0.9939	0.9924	0.9910	0.9903
650	1.0005	0.9998	0.9991	0.9977	0.9962	0.9949	0.9933	0.9919	0.9904	0.9897
700	1.0001	0.9994	0.9987	0.9973	0.9959	0.9943	0.9930	0.9916	0.9902	0.9895

Table II-III-19

Values of F/F_i, Ratio of Real-Gas to Ideal-Gas Isentropic Expansion Function

$$F_i = 0.68473 \qquad B_F = 3.8839$$

T_{1t} (°R)	Inlet Stagnation Pressure, p_{1t} (psia)									
	0	100	200	400	600	800	1000	1200	1400	1500
400	1.0096	1.0112	1.0132	1.0169	1.0204	1.0243	1.0278	1.0314	1.0342	1.0370
450	1.0062	1.0075	1.0090	1.0125	1.0158	1.0189	1.0219	1.0251	1.0283	1.0299
500	1.0039	1.0052	1.0066	1.0093	1.0122	1.0149	1.0176	1.0208	1.0239	1.0255
550	1.0023	1.0036	1.0048	1.0074	1.0098	1.0122	1.0147	1.0175	1.0200	1.0213
600	1.0013	1.0025	1.0035	1.0057	1.0081	1.0103	1.0126	1.0148	1.0172	1.0184
650	1.0005	1.0016	1.0026	1.0047	1.0068	1.0091	1.0108	1.0129	1.0149	1.0160
700	1.0001	1.0010	1.0019	1.0039	1.0059	1.0075	1.0095	1.0116	1.0133	1.0143

Table II-III-20

Values of ϕ^*/ϕ^*_i, Ratio of Sonic-Flow Function of Natural-Fuel Gas to Sonic-Flow Function of Its Ideal-Gas Counterpart

$$\phi^*_i = 0.3947$$

T_{1t} (°R)	Inlet Stagnation Pressure, p_{1t} (psia)						
	0	100	200	400	600	800	1000
450	0.9980	1.0091	1.0211	1.0476	1.0790	1.1164	1.1611
500	0.9961	1.0041	1.0123	1.0305	1.0507	1.0730	1.0976
550	0.9936	0.9994	1.0055	1.0185	1.0324	1.0472	1.0629
600	0.9906	0.9951	0.9997	1.0093	1.0192	1.0296	1.0404
650	0.9873	0.9906	0.9943	1.0015	1.0088	1.0164	1.0242
700	0.9839	0.9865	0.9893	0.9948	1.0002	1.0059	1.0116

*Composition by mole fraction: Methane (CH_4), 0.960; ethane (C_2H_6), 0.035; carbon dioxide (CO_2), 0.002; and nitrogen (N_2), 0.003.

Table II-III-21

Values of F/F_i Ratio of Isentropic Expansion Function of Natural-Fuel Gas to That of Its Ideal-Gas Counterpart

$$B_F = 3.8135 \quad F_i = 0.6723$$

T_{1t} (°R)	Inlet Stagnation Pressure, p_{1t} (psia)						
	0	100	200	400	600	800	1000
450	0.9980	0.9973	0.9968	0.9964	0.9976	1.0013	1.0090
500	0.9961	0.9959	0.9956	0.9961	0.9977	1.0008	1.0059
550	0.9936	0.9936	0.9937	0.9946	0.9964	0.9992	1.0032
600	0.9906	0.9909	0.9912	0.9924	0.9941	0.9967	1.0001
650	0.9873	0.9876	0.9882	0.9894	0.9912	0.9936	0.9967
700	0.9839	0.9843	0.9849	0.9863	0.9879	0.9903	0.9930

*Composition by mole fraction: Methane (CH_4), 0.960; ethane (C_2H_6), 0.035; carbon dioxide (CO_2), 0.002; and nitrogen (N_2), 0.003.

For use in the calibration of displacement and other volume-indicating meters, equation (II-III-29) can be transformed to give volume rate of flow at the inlet conditions by multiplying all terms by the specific volume, v_{1t}, giving

$$q_1 \text{ (cfs)} = C \, a \, B_F \left(\frac{F}{F_i} \right) \sqrt{\frac{Z_1 R \, T_{1t}}{MW}} \quad \text{(II-III-32)}$$

Let $F = (e_c j + b_c)$ and $\sqrt{Z_1} = (e_z j + b_Z)$.

Then,

$$q_1 \text{ (cfs)} = C \, a \, (e_c j + b_c)(e_z j + b_z) \sqrt{\frac{g_c R \, T_{1t}}{MW}} \quad \text{(II-III-33)}$$

in which a, the nozzle throat area, is in *square feet* and $R = 1545.33$ (see Par. I-3-25). Obviously, the reciprocal of equation (II-III-33) will give the time in seconds for the discharge of 1 cu ft at inlet stagnation conditions.

II-III-57 For gases that are not tabulated here, the sonic flow can be computed with equal accuracy from equation (II-III-34) if Γ, Γ^* and Z are known.

$$m \text{ (lb}_m/\text{sec)} = \frac{C \, a \, F \sqrt{g_c MW}}{\sqrt{Z_1 R}} \left(\frac{p_{1t}}{\sqrt{T_{1t}}} \right) \quad \text{(II-III-34)}$$

$$= 0.1443 \, C \, a \, F \sqrt{\frac{MW}{Z_1}} \left(\frac{p_{1t}}{\sqrt{T_{1t}}} \right)$$

where

$$F = \sqrt{\Gamma^* \left[1 + \frac{\Gamma}{2} \left(\frac{\Gamma - 1}{\Gamma} \right) \right]^{-[(\Gamma+1)/\Gamma - 1)]}} \quad \text{(II-III-35)}$$

The value of Z is fairly easily determined from tables for real gases, but the mean value of Γ between inlet stagnation and throat conditions is more difficult to determine. The value of Γ^* can be determined either from the acoustic velocity data

$$\Gamma^* = \frac{V_s^2}{Z R T g_c} \quad \text{(II-III-36a)}$$

or from

$$\Gamma^* = Z(c_p/c_v) \quad \text{(II-III-36b)}$$

when $p < 450$ psia and also from

$$\Gamma^* = \frac{Z \gamma}{1 - cp^2} \quad \text{(II-III-37)}$$

where c is the coefficient in $Z = 1 + bp + cp^2$ and its value is about $10^{-6}/\text{atm}^2$.

Where the equation of state, or Z, is not well known or where accuracy requirements will permit, the value of Z used in equation (II-III-34) may be determined from "Reduced Coordinates Compressibility Charts" (Fig. II-III-31), which follow the tables. The reduced stagnation pressure is the ratio of the actual pressure to the critical point pressure, and the reduced stagnation temperature is the ratio of the actual temperature to the critical point temperature. Even though the curves in Fig. II-III-31 may not fit a particular gas exactly, they offer a significant improvement over using the ideal-gas equation. The critical properties of some commercial gases are given in Table II-I-5.

Often the deviation between Γ^* and Γ can be neglected. For example, a 7-per cent deviation between Γ^* and Γ would cause less than a 0.20-per cent error in the flow rate computed by equation (II-III-34). For those situations where the uncertainty due to using $\Gamma^* = \Gamma$ is acceptable, values of F_i (Table II-III-22) may be used. The most unfavorable approximation that might have to be accepted would result from computing the sonic flow of a gas using these values of F_i and the value of Z from the charts of Fig. I-III-31.

Table II-III-22 Values of F_i for Selected Values of γ^*

$$F_i = \sqrt{\gamma\left(\frac{1+\gamma}{2}\right)^{-\left(\frac{\gamma+1}{\gamma-1}\right)}}$$

	Increments of γ				
γ	0.00	0.02	0.04	0.06	0.08
1.1	0.6284	0.6325	0.6364	0.6404	0.6448
1.2	0.6483	0.6521	0.6562	0.6599	0.6638
1.3	0.6673	0.6705	0.6741	0.6776	0.6811
1.4	0.6848	0.6880	0.6914	0.6944	0.6978
1.5	0.7010	0.7039	0.7072	0.7128	0.7134
1.6	0.7165	0.7192	0.7222	0.7249	0.7281
1.7	0.7310	0.7337	0.7363	0.7390	0.7420

*These values may be used for real gases where $\Gamma^* = \Gamma = \gamma$.

Table II-III-23 Values of e_c

T_{1t} (°R)	Inlet Pressure, p_{1t} (psia)						
	0	100	200	400	600	800	1000
450	−0.0264	−0.0296	−0.0330	−0.0398	−0.0451	−0.0434	−0.0205
460	−0.0271	−0.0302	−0.0333	−0.0397	−0.0447	−0.0442	−0.0289
470	−0.0278	−0.0307	−0.0337	−0.0397	−0.0444	−0.0447	−0.0341
480	−0.0285	−0.0312	−0.0341	−0.0397	−0.0441	−0.0450	−0.0375
490	−0.0291	−0.0318	−0.0345	−0.0398	−0.0440	−0.0451	−0.0398
500	−0.0297	−0.0323	−0.0349	−0.0399	−0.0439	−0.0452	−0.0414
510	−0.0303	−0.0328	−0.0352	−0.0400	−0.0438	−0.0453	−0.0426
520	−0.0309	−0.0332	−0.0356	−0.0401	−0.0437	−0.0454	−0.0435
530	−0.0314	−0.0337	−0.0360	−0.0403	−0.0437	−0.0455	−0.0441
540	−0.0320	−0.0342	−0.0363	−0.0404	−0.0437	−0.0455	−0.0446
550	−0.0325	−0.0346	−0.0367	−0.0406	−0.0438	−0.0455	−0.0450
560	−0.0329	−0.0350	−0.0370	−0.0408	−0.0438	−0.0456	−0.0454
570	−0.0334	−0.0354	−0.0373	−0.0409	−0.0439	−0.0456	−0.0456
580	−0.0338	−0.0357	−0.0376	−0.0411	−0.0439	−0.0457	−0.0458
590	−0.0342	−0.0361	−0.0379	−0.0412	−0.0439	−0.0457	−0.0460
600	−0.0346	−0.0364	−0.0381	−0.0413	−0.0440	−0.0457	−0.0461
610	−0.0349	−0.0367	−0.0384	−0.0414	−0.0440	−0.0457	−0.0462
620	−0.0353	−0.0369	−0.0385	−0.0415	−0.0440	−0.0456	−0.0462
630	−0.0355	−0.0371	−0.0387	−0.0416	−0.0439	−0.0456	−0.0462
640	−0.0358	−0.0373	−0.0388	−0.0416	−0.0439	−0.0455	−0.0462
650	−0.0360	−0.0375	−0.0389	−0.0416	−0.0438	−0.0454	−0.0461
660	−0.0361	−0.0376	−0.0390	−0.0416	−0.0437	−0.0452	−0.0460
670	−0.0363	−0.0377	−0.0391	−0.0416	−0.0436	−0.0451	−0.0458
680	−0.0364	−0.0377	−0.0391	−0.0415	−0.0435	−0.0449	−0.0457
690	−0.0364	−0.0378	−0.0390	−0.0414	−0.0433	−0.0447	−0.0454
700	−0.0364	−0.0377	−0.0390	−0.0412	−0.0431	−0.0445	−0.0452

Table II-III-24 Values of b_c

T_{1t} (°R)	Inlet Pressure, p_{1t} (psia)						
	0	100	200	400	600	800	1000
450	0.6719	0.6715	0.6713	0.6713	0.6723	0.6747	0.6791
460	0.6717	0.6714	0.6712	0.6713	0.6724	0.6748	0.6789
470	0.6714	0.6712	0.6711	0.6714	0.6725	0.6748	0.6786
480	0.6712	0.6710	0.6710	0.6713	0.6725	0.6747	0.6783
490	0.6709	0.6708	0.6708	0.6712	0.6724	0.6746	0.6780
500	0.6707	0.6706	0.6706	0.6711	0.6723	0.6745	0.6777
510	0.6704	0.6704	0.6704	0.6710	0.6722	0.6743	0.6774
520	0.6701	0.6701	0.6702	0.6708	0.6721	0.6741	0.6771
530	0.6698	0.6698	0.6699	0.6706	0.6719	0.6739	0.6768
540	0.6694	0.6695	0.6697	0.6704	0.6717	0.6736	0.6764
550	0.6691	0.6692	0.6694	0.6701	0.6714	0.6734	0.6760
560	0.6687	0.6689	0.6691	0.6699	0.6712	0.6731	0.6757
570	0.6684	0.6685	0.6687	0.6696	0.6709	0.6727	0.6753
580	0.6680	0.6681	0.6684	0.6692	0.6706	0.6724	0.6748
590	0.6676	0.6678	0.6680	0.6689	0.6702	0.6720	0.6744
600	0.6672	0.6674	0.6677	0.6686	0.6699	0.6717	0.6740
610	0.6668	0.6670	0.6673	0.6682	0.6695	0.6713	0.6735
620	0.6663	0.6666	0.6669	0.6678	0.6691	0.6709	0.6731
630	0.6659	0.6662	0.6665	0.6674	0.6687	0.6705	0.6726
640	0.6655	0.6657	0.6661	0.6670	0.6683	0.6700	0.6721
650	0.6650	0.6653	0.6657	0.6666	0.6679	0.6696	0.6717
660	0.6646	0.6649	0.6652	0.6662	0.6675	0.6691	0.6712
670	0.6641	0.6644	0.6648	0.6658	0.6671	0.6687	0.6707
680	0.6637	0.6640	0.6644	0.6653	0.6666	0.6682	0.6702
690	0.6632	0.6635	0.6639	0.6649	0.6662	0.6678	0.6697
700	0.6627	0.6631	0.6635	0.6645	0.6657	0.6673	0.6692

Table II-III-25 Values of e_z

T_{1t}	Inlet Pressure, p_{1t} (psia)						
(°R)	0	100	200	400	600	800	1000
450	0.	−0.0252	−0.0530	−0.1179	−0.1988	−0.2991	−0.4162
460	0.	−0.0234	−0,0490	−0.1078	−0.1790	−0.2644	−0.3610
470	0.	−0.0218	−0.0454	−0.0989	−0.1622	−0.2360	−0.3175
480	0.	−0.0203	−0.0422	−0.0911	−0.1478	−0.2123	−0.2823
490	0.	−0.0190	−0.0393	−0.0842	−0.1353	−0.1923	−0.2532
500	0.	−0.0178	−0.0366	−0.0780	−0.1243	−0.1752	−0.2288
510	0.	−0.0166	−0.0342	−0.0724	−0.1147	−0.1604	−0.2079
520	0.	−0.0156	−0.0321	−0.0675	−0.1061	−0.1474	−0.1900
530	0.	−0.0147	−0.0301	−0.0630	−0.0984	−0.1360	−0.1743
540	0.	−0.0138	−0.0283	−0.0589	−0.0916	−0.1259	−0.1606
550	0.	−0.0130	−0.0266	−0.0551	−0.0854	−0.1169	−0.1485
560	0.	−0.0123	−0.0251	−0.0518	−0.0798	−0.1088	−0.1377
570	0.	−0.0116	−0.0236	−0.0487	−0.0748	−0.1015	−0.1281
580	0.	−0.0110	−0.0223	−0.0458	−0.0701	−0.0949	−0.1194
590	0.	−0.0104	−0.0211	−0.0432	−0.0659	−0.0889	−0.1116
600	0.	−0.0099	−0.0200	−0.0408	−0.0621	−0,0835	−0.1045
610	0.	−0.0094	−0.0190	−0.0385	−0.0585	−0,0785	−0.0980
620	0.	−0.0089	−0.0180	−0.0365	−0.0522	−0.0739	−0.0921
630	0.	−0.0085	−0.0171	−0.0346	−0.0522	−0.0697	−0.0867
640	0.	−0.0081	−0.0162	−0.0328	−0.0494	−0.0658	−0.0818
650	0.	−0.0077	−0.0154	−0.0311	−0.0468	−0,0623	−0.0772
660	0.	−0.0073	−0.0147	−0.0296	−0.0444	−0.0589	−0.0730
670	0.	−0.0070	−0.0140	−0.0281	−0.0422	−0.0559	−0.0691
680	0.	−0.0067	−0.0134	−0.0268	−0.0401	−0.0530	−0.0654
690	0.	−0.0064	−0.0128	−0.0255	−0.0381	−0.0503	−0.0621
700	0.	−0.0061	−0.0122	−0.0243	−0.0363	−0.0479	−0.0589

Table II-III-26 Values of b_z

T_{1t}	Inlet Pressure, p_{1t} (psia)						
(°R)	0	100	200	400	600	800	1000
450	1.0000	0.9891	0.9780	0.9552	0.9315	0.9075	0.8837
460	1.0000	0.9899	0.9796	0.9585	0.9370	0.9152	0.8938
470	1.0000	0.9906	0.9810	0.9616	0.9419	0.9221	0.9028
480	1.0000	0.9912	0.9824	0.9644	0.9463	0.9283	0.9109
490	1.0000	0.9918	0.9836	0.9669	0.9503	0.9339	0.9181
500	1.0000	0.9924	0.9847	0.9693	0.9539	0.9389	0.9245
510	1.0000	0.9929	0.9857	0.9714	0.9573	0.9435	0.9304
520	1.0000	0.9933	0.9866	0.9734	0.9603	0.9477	0.9357
530	1.0000	0.9937	0.9875	0.9752	0.9631	0.9515	0.9406
540	1.0000	0.9941	0.9883	0.9768	0.9657	0.9550	0.9450
550	1.0000	0.9945	0.9891	0.9784	0.9681	0.9582	0.9490
560	1.0000	0.9949	0.9898	0.9798	0.9702	0.9612	0.9527
570	1.0000	0.9952	0.9904	0.9811	0.9723	0.9639	0.9562
580	1.0000	0.9955	0.9910	0.9824	0.9741	0.9664	0.9593
590	1.0000	0.9958	0.9916	0.9835	0.9759	0.9688	0.9622
600	1.0000	0.9960	0.9921	0.9846	0.9775	0.9709	0.9649
610	1.0000	0.9963	0.9926	0.9856	0.9790	0.9729	0.9674
620	1.0000	0.9965	0.9931	0.9865	0.9804	0.9748	0.9698
630	1.0000	0.9967	0.9935	0.9874	0.9817	0.9766	0.9719
640	1.0000	0.9969	0.9939	0.9882	0.9830	0.9782	0.9740
650	1.0000	0.9971	0.9943	0.9890	0.9841	0.9797	0.9758
660	1.0000	0.9973	0.9947	0.9897	0.9852	0.9812	0.9776
670	1.0000	0.9975	0.9950	0.9904	0.9862	0.9825	0.9793
680	1.0000	0.9976	0.9953	0.9911	0.9872	0.9838	0.9808
690	1.0000	0.9978	0.9956	0.9917	0.9881	0.9849	0.9823
700	1.0000	0.9979	0.9959	0.9922	0.9889	0.9861	0.9836

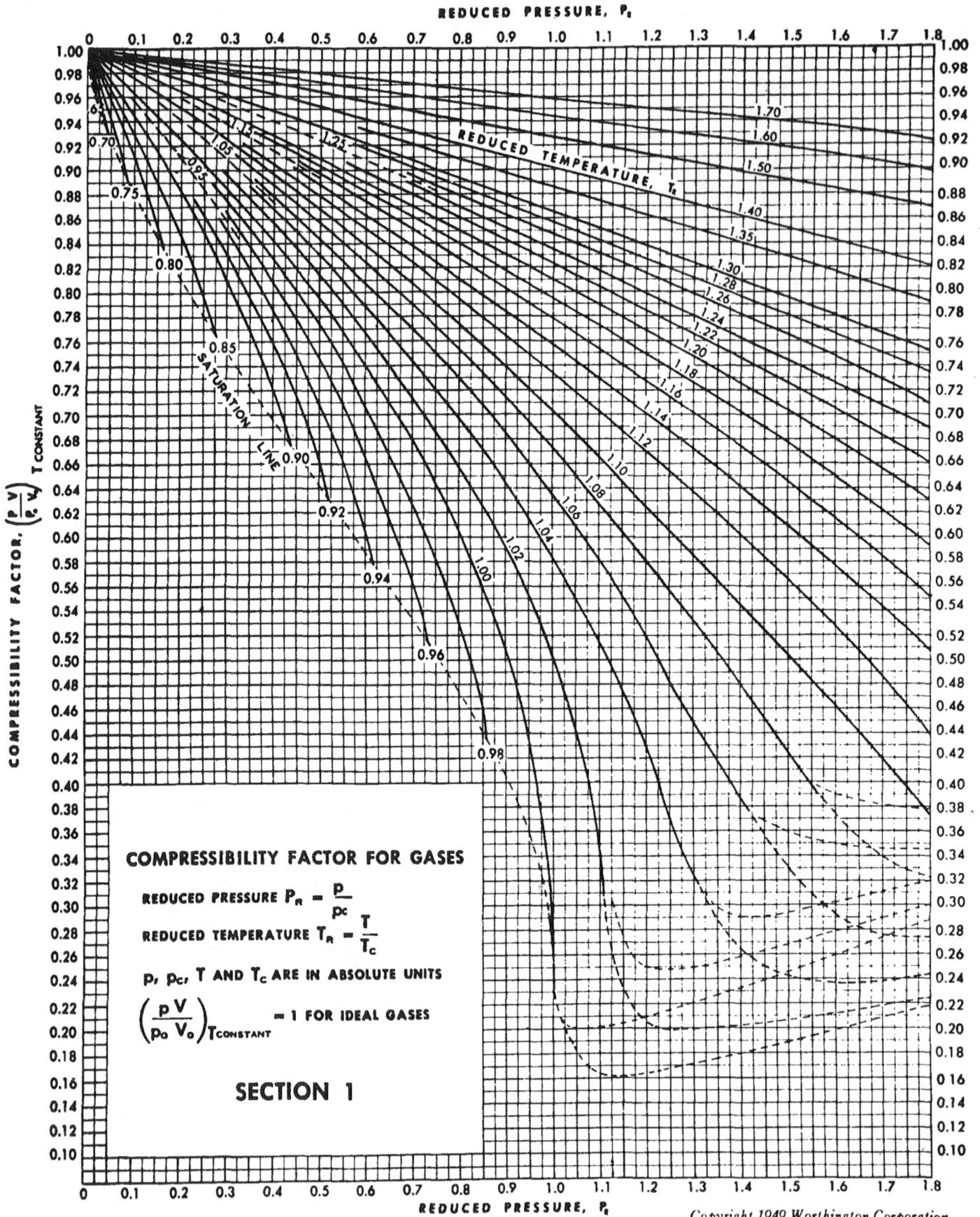

FIG. II-III-31-1 VALUES OF COMPRESSIBILITY FACTOR, Z, AS FUNCTION OF
REDUCED PRESSURE AND REDUCED TEMPERATURE

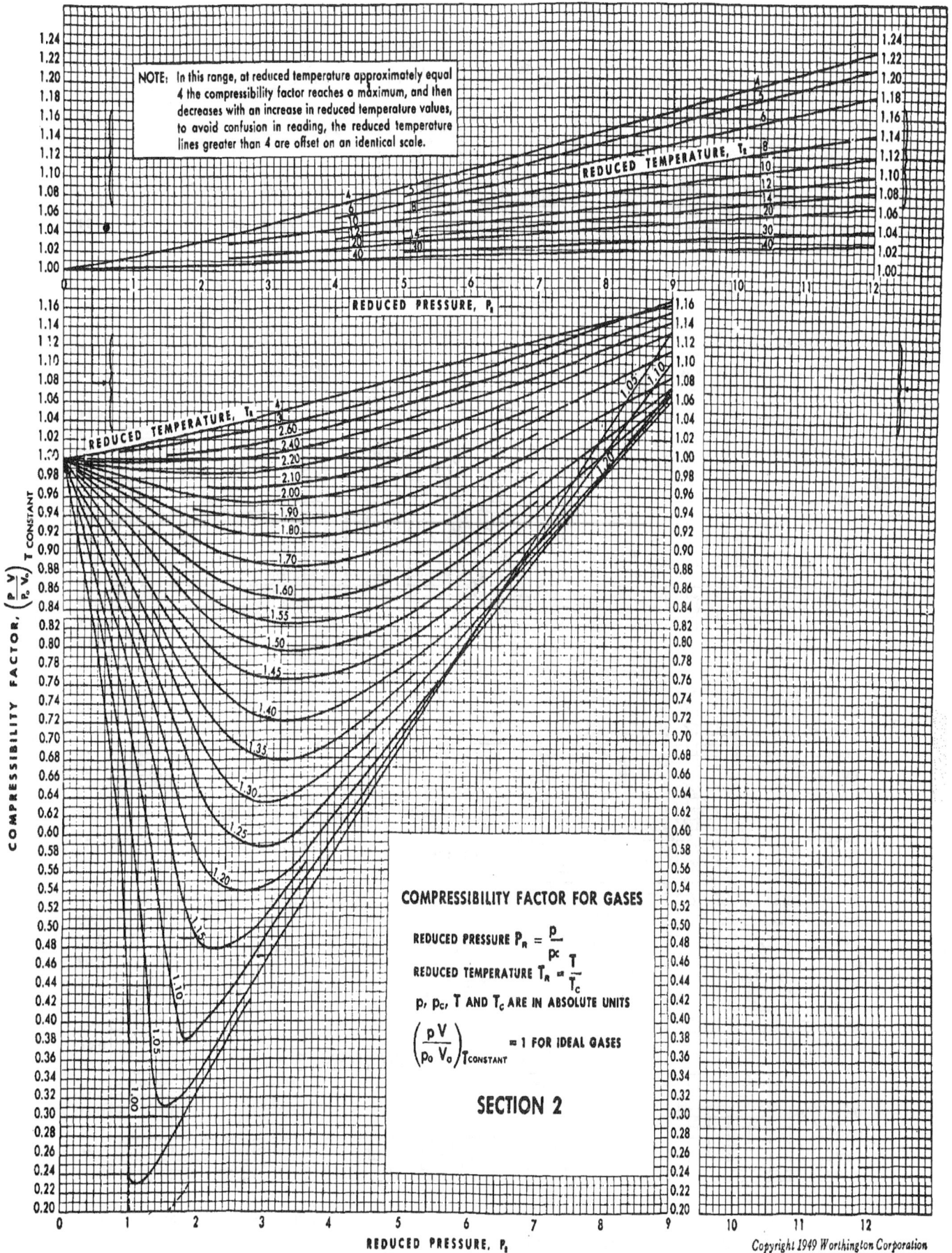

NOTE: In this range, at reduced temperature approximately equal 4 the compressibility factor reaches a maximum, and then decreases with an increase in reduced temperature values, to avoid confusion in reading, the reduced temperature lines greater than 4 are offset on an identical scale.

COMPRESSIBILITY FACTOR FOR GASES

REDUCED PRESSURE $P_R = \dfrac{p}{p_c}$

REDUCED TEMPERATURE $T_R = \dfrac{T}{T_c}$

p, p_c, T AND T_c ARE IN ABSOLUTE UNITS

$\left(\dfrac{pV}{p_o V_o}\right)_{T\,CONSTANT} = 1$ FOR IDEAL GASES

SECTION 2

FIG. II-III-31-2 VALUES OF COMPRESSIBILITY FACTOR, Z, AS FUNCTION OF REDUCED PRESSURE AND REDUCED TEMPERATURE

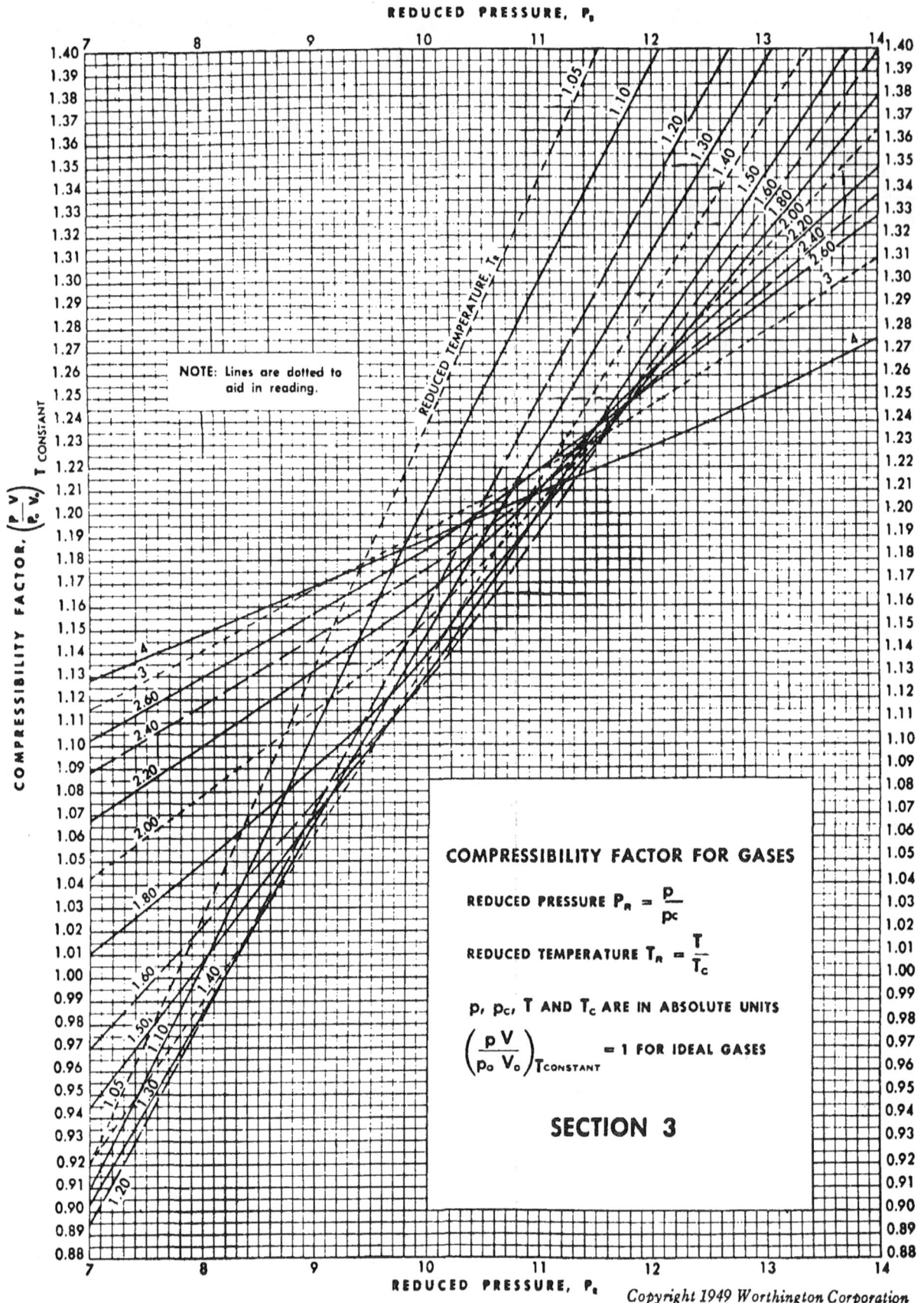

FIG. II-III-31-3   VALUES OF COMPRESSIBILITY FACTOR, Z, AS FUNCTION OF
REDUCED PRESSURE AND REDUCED TEMPERATURE

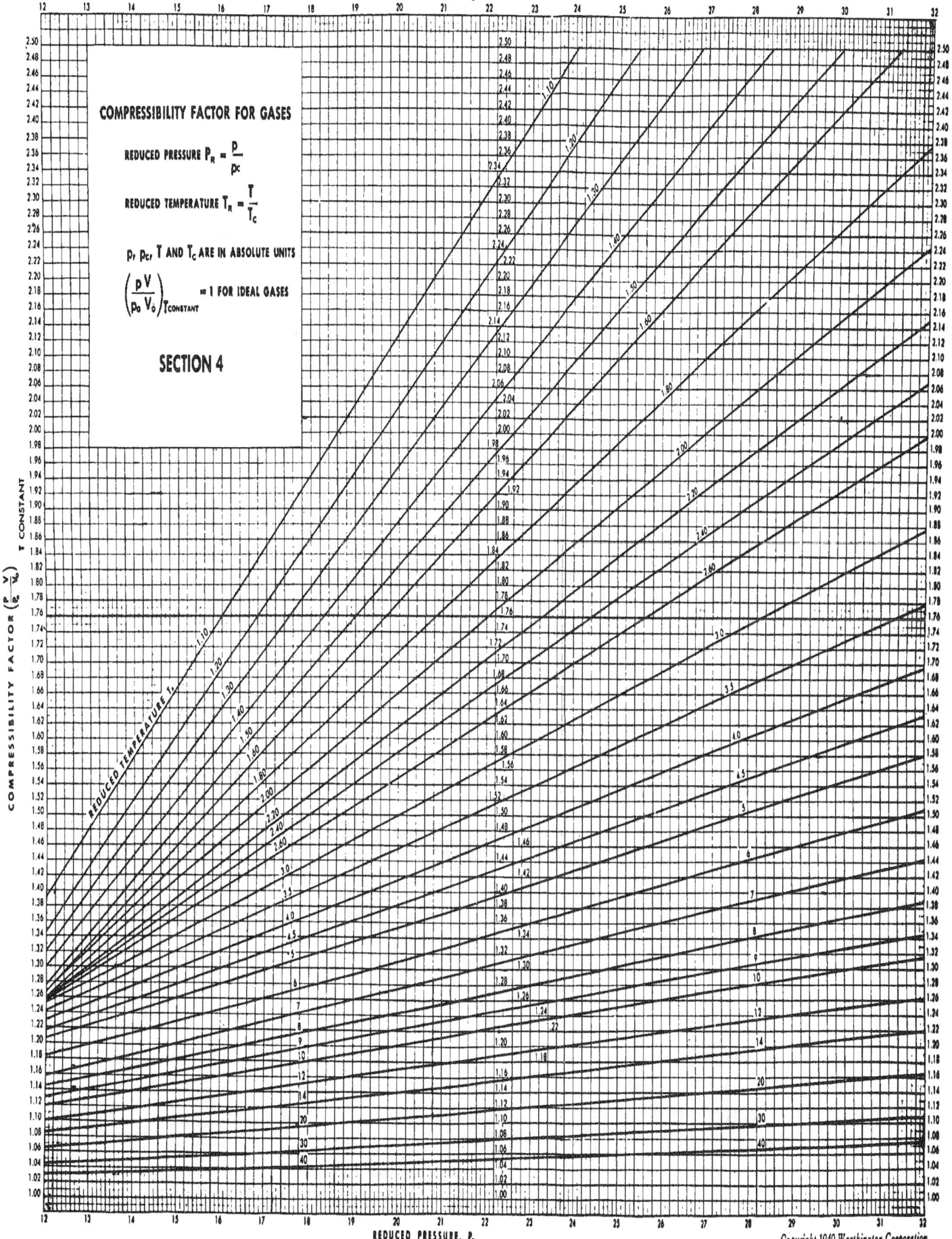

FIG. II-III-31-4 VALUES OF COMPRESSIBILITY FACTOR, Z, AS FUNCTION OF
REDUCED PRESSURE AND REDUCED TEMPERATURE

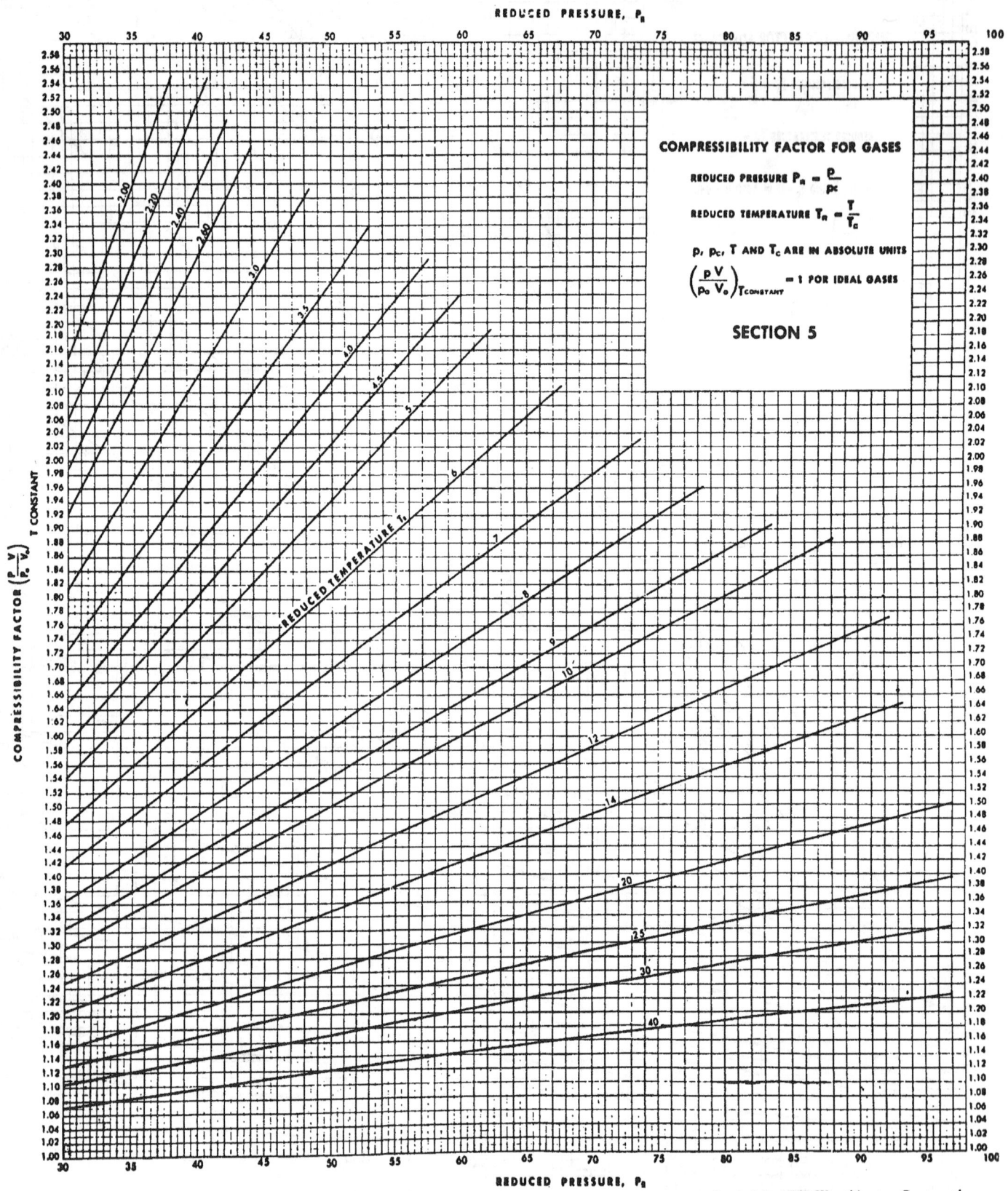

FIG. II-III-31-5 VALUES OF COMPRESSIBILITY FACTOR Z, AS FUNCTION OF
REDUCED PRESSURE AND REDUCED TEMPERATURE

II-III-58 Elbows. Ninety-degree elbows (Fig. II-III-32) may be used for monitoring the steadiness of a fluid flow. If such elbows are adequately calibrated, an accuracy within about ±0.5 per cent may be realized. Because the differential pressure obtained with an elbow is relatively low, their use is usually limited to measuring a liquid flow. If used for a gas flow, even the minimum stream velocity would be high (e.g., 300 to 500 fps).

II-III-59 The recommended locations of the pressure taps are in the outer and inner circumferences of the elbow midplane, 45 deg from the inlet end.

Although elbows may be located in either a horizontal or vertical pipeline, it is desirable that the velocity profile of the fluid stream entering the elbow be fairly uniform and free of swirls. For this reason, the same installation considerations should be followed as given by Fig. II-II-1 for orifices and flow nozzles of 0.80 diameter ratio.

II-III-60 The equation that may be used with elbow meters is

$$q \text{ (cfs)} = \frac{\pi D^2}{4} K \sqrt{\frac{4R}{\pi D} g_c \frac{\Delta p}{\rho}} \quad \text{(II-III-38)}$$

or

$$m \text{ (lb}_m/\text{sec)} = \frac{\pi D^2}{4} K \sqrt{\frac{4R}{\pi D} g_c \rho \Delta p} \text{ (II-III-39)}$$

where

D = Diameter of pipe and elbow, in.

K = Flow coefficient, determined by calibration or by equation (II-III-40)

g_c = Proportionality factor between force and mass = 32.174

Δp = Differential pressure, psi

R = Radius of curvature of elbow center line, in.

ρ = Density of fluid in elbow, lb_m/ft^3

For uncalibrated, 90-deg elbows with pressure taps at the 45-deg section, tap-hole diameters, δ, as given in Table II-II-1, $R/D > 1.25$ (see Fig. II-III-32), and $10^4 < R_D < 10^6$, a value of the flow coefficient may be computed by

$$K = 1 - \frac{6.5}{\sqrt{R_D}} \quad \text{(II-III-40)}$$

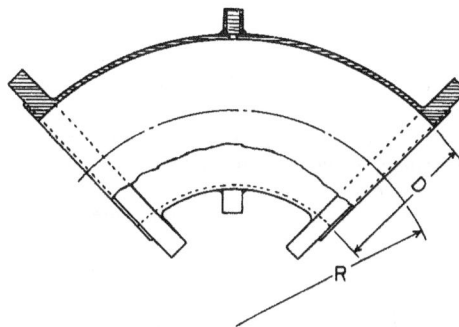

FIG. II-III-32 ELBOW METER

A flow evaluated with an elbow meter and K from equation (II-III-40) will be subject to a tolerance of about ± 4.0 per cent.

II-III-61 Electromagnetic Flowmeters. These meters are suitable for measuring the flow of liquids that have a conductivity greater than 20 micromhos (about 10 ppm of NaCl in water). Variations of the conductivity above this value do not affect the operation of the meter.

II-III-62 No special installation conditions are required inasmuch as the operation of the meter is unaffected by adjacent fittings. A helical flow pattern will have very little if not negligible effect upon the meter indications. Even pulsating flows up to about 10 or 12 cps can be measured with a suitably adjusted receiver. The important requirement is that the flow tube of the meter be completely filled with liquid all the time metering is in progress. Reverse flow can be metered by reversing the leads at the receiver, or a receiver can be adjusted to record flows in both directions.

The pressure loss through these meters is the same as for an equal length of pipe of the same diameter.

An accuracy of ± 1 per cent is to be expected over the normal 10 to 1 range of the unit. However, if these meters are specially adjusted, an accuracy of ± 0.5 per cent and possibly better can be attained over the range.

References

[1] "Testing of Large Diameter Orifice Tubes,"
E. E. Stovall; American Gas Association Transmission
and Storage Conference, 1953, Paper TS-53-8.

[2] "Progress in Large Volume Measurement,"
F. M. Partridge; American Gas Association Convention,
Oct. 1953, Paper OS-53-1.

[3] "Large Diameter Orifice Meter Tube Tests;"
Final Report of Supervising Committee, Research
Project NX-4, American Gas Association, New York,
May 1954.

[4] "Calibration of Eccentric and Segmental Orifices
in 4-in. and 6-in. Pipelines," S. R. Beitler and D. J.
Masson; *Trans. ASME*, vol. 71, Oct. 1949, p. 751.

[5] "Small Diameter Orifice Metering," T. J. Filban
and W. A. Griffin; *Trans. ASME, Journal of Basic
Engineering*, vol. 82, no. 3, Sept. 1960, p. 735.

[6] "ISO Recommendation R541, Measurement of
Fluid Flow by Means of Orifice Plates and Nozzles;"
International Standards Organization, 1st ed., Jan. 1967.
(Apply to American National Standards Institute.)

[7] "Discharge Coefficients of Square-Edged
Orifices for Measuring the Flow of Air;" H. S. Bean,
E. Buckingham and P. S. Murphy; *Journal of Research of
the National Bureau of Standards*, vol. 2, no. 4, Mar.
1929, R.P. 49, p. 561. (Out of print; reference copies
in many libraries.)

[8] "The Impact Tube," S. A. Moss; *Trans. ASME*,
vol. 38, 1916, p. 761.

[9] "ISO Recommendation R781, Measurement of
Fluid Flow by Venturi Tubes;" 1968.

[10] "Calculations of the Flow of Natural Gas
Through Critical Flow Nozzles," R. C. Johnson;
Trans. ASME, Journal of Basic Engineering, vol. 92,
no. 3, Sept. 1970, p. 580.

*The following references should be added to the list
on p. 256.*

[11] "Discharge Coefficients for Long-Radius Flow
Nozzles when Used with Pipe-Wall Pressure Taps," H. S.
Bean, S. R. Beitler, R. E. Sprenkle; *Trans. ASME*, vol. 63,
1941, p. 439.

[12] "A Rational Equation for ASME Coefficients for
Long-Radius Flow Nozzles Employing Wall Taps at 1 D
and ½ D," J. W. Murdock, ASME Paper No. 64-WA/FM-7,
1964 (unpublished).

[13] "ASME Research on High Pressure-High Temperature Steam and Water Flow Measurement," J. W. Murdock;
Trans. ASME, vol. 87, Series D, No. 4, Dec. 1965, p. 1029.

Chapter II-IV

Examples

II-IV-1 As an aid to the use of the equations, tables and figures given in preceding chapters, the following example computations have been prepared. Also, as in the first example, reference and use are made of the computational procedures given in other publications [1, 2]. All of the examples apply to meters of the differential-pressure type.

II-IV-2 A contract for the sale of fuel gas calls for a maximum delivery rate of 14,000,000 scf*d* and a normal flow rate of 9,800,000 scfd. The orifice meter tube is to be a 6-in. schedule 40 pipe with flange pressure taps. The average flowing conditions are expected to be: pressure, measured at the upstream pressure tap, 250 psig; temperature, 81 F; specific gravity, 0.75; and average barometric pressure, 14.70 psia. The secondary element is to be a mercury-type differential recorder, with a range of 100-in. water and using a L-10 charter (i.e., a chart ruled on a square-root scale of 0—10). The reference or base conditions for the measurement are to be 14.70 psia, 60 F and dry.

Wanted: the diameter of the orifice required to provide a direct-reading flowmeter scale.

(a) The solution to this example can be worked readily by using procedures given in the *Flowmeter Computation Handbook* [1]. The equation and table used from that handbook are designated by an asterisk (*). Since only the differential-pressure pen will be read to infer the flow rate, the tacit assumption is that the conditions of measurement will be stable and/or any occasional small variations of pressure, temperature or composition can be neglected.

(b) From Chapter 5*, equation (5-4)*,

$$q_h \text{ (scfh)} = 0.6085 \frac{D^2 \, l}{\sqrt{G}} \frac{T_b}{P_b} \sqrt{\frac{h_w P_f}{T_f}}$$

where

l = principal meter constant
Subscript b refers to the reference condition
Subscript f refers to the flowing condition

(c) $q_h = \dfrac{14,000,000}{24} = 583,333$ scfh

$D = 6.065$ in.

$P_f = 250 + 14.70 = 264.7$ psia

$T_f = 459.7 + 81 = 540.7$ R and $T_b = 519.7$ R

(d) $583,333 = l \left[0.6085 \dfrac{6.065^2}{\sqrt{0.75}} \dfrac{519.7}{14.70} \sqrt{\dfrac{100 \times 264.7}{540.7}} \right]$

$l = 91.2413$

(e) $\beta = 0.62185$ (from Table 1-5*)

(f) $d = 6.065 \times 0.62185 = 3.772$ in. as the (trial) diameter of the orifice

(g) A more exact determination of the required orifice diameter, if desired, may be made by applying a Reynolds number factor, F_r, an expansion factor, Y_1, and an area thermal expansion factor, F_a, as multipliers to equation (5-4)*. (These factors are explained in Appendix C*.) For doing this, the maximum flow rate will be used as above.

(h) $R_D = 0.00424\, w_h/D\, \mu$ (equation A-34*)

(i) $w_h\ (= m_h\ \mathrm{lb}_m/\mathrm{hr}) = 583,333\ (0.75 \times 0.0764) =$
33,430 $\mathrm{lb}_m/\mathrm{hr}$

(j) $\mu = 0.00\,000\,77\ \mathrm{lb}_m/\mathrm{ft\text{-}sec}$ (Fig. II-I-8)

(k) $R_D = (0.00424 \times 33,430)/(6.065 \times 0.00\,000\,77)$
$= 3,035,000$

(l) Using the value of β from (e) above, the Reynolds number adjustment factor, $F_r = 0.997$ (Fig. C-1–4*).

(m) For Y_1, $\dfrac{h_w}{p_f} = \dfrac{100}{264.7} = 0.378$ $Y_1 = 0.9953$
(Fig. C-3-1*)

(n) F_a for 81 F = 1.0002 (Fig. II-I-3 or Fig. C-2-1*)

(o) Adjusted value, $I' = I/\ F_r \times Y_1 \times F_a$
$$= 91.2413/(0.997 \times 0.9953 \times 1.0002)$$
$$= 92.15$$

(p) $\beta' = 0.62405$ (Table 1-5*)

(q) $d' = 6.065 \times 0.62405 = 3.785$ in.

(r) If the normal rate of flow were used instead of the maximum, the value of Y_1 would become 0.9977; the effect on F_r would be too small to read; and no change would occur in F_a, giving an adjusted value of $I'' = 91.70$, for which $\beta'' = 0.62315$ and $d'' = 3.779$ in.

(s) The example may be solved by using the computational procedures given in Gas Measurement Committee *Report No. 3* [2]. The equations, tables and figures from *Report No. 3* will be indicated by a dagger (†). The basic equation is equation (1)† from that report:
$$q_h\ (\mathrm{scfh}) = C'\ \sqrt{h_w\, p_1}$$

(t) $C' = F_b \cdot F_r \cdot Y \cdot F_{pb} \cdot F_{tb} \cdot F_{tf} \cdot F_g \cdot F_{pv}$
$\qquad \cdot F_m \cdot F_a$ (equation (2)†)

(u) F_b = basic orifice factor. It includes the orifice diameter which is sought.

(v) F_r = Reynolds number factor = $1 + b/\sqrt{h_w\, p_1}$ (Table 5†).

The factor b depends on the orifice and pipe diameters; therefore, for this example, it is necessary to *assume* a value of d for the first determination; then, if necessary, a closer

value will be used for a second determination. Assuming $d = 3.875$ in., then for 6-in. pipe $b = 0.0505$ (Table 5†) and
$$F_r = 1 + \frac{0.0505}{\sqrt{100 \times 264.7}} = 1.0003$$

(w) Y_1 = expansion factor = 0.9953 as in (m) above

(x) F_{pb} = pressure base factor = $\dfrac{14.73}{14.70} = 1.0020$ (Table 12†)

(y) F_{tb} = temperature base factor = 1.000 (Table 13†)

(z) F_{tf} = flowing temperature factor = 0.9804 (Table 14†)

(aa) F_g = specific gravity factor = 1.1547 (Table 15†)

(bb) F_{pv} = supercompressibility factor = 1.0175 (Table 16†)

(cc) F_m = manometer factor (mercury type) = 0.9983 (Table 17†)

(dd) F_a = orifice thermal expansion factor = 1.0002 as in (n) above

(ee) $C' = F_b \times 1.0003 \times 0.9953 \times 1.0020 \times 1.000$
$\qquad \times 0.9804 \times 1.1547 \times 1.0175 \times 0.9983 \times 1.0002$
$\qquad = 1.1474\ F_b$

(ff) $583,333 = 1.1474\ F_b\ \sqrt{100 \times 264.7}$

$F_b = 3124.9$ for a 6-in. schedule 40 pipe. (This corresponds to an orifice diameter.) $d = 3.751$ in.

II-IV-3 Compressed air is flowing in a 10-in. pipe under a line pressure of 125 psig, a temperature of 90 F and water vapor saturation; it is metered with a type 316, stainless-steel, concentric, thin-plate, square-edged orifice. The orifice diameter is 6.250 in.; the i.d. of the meter-tube inlet section is 10.02 in., with pressure taps at D and $1/2\,D$. The differential pressure is 30-in. water, and the barometric pressure is 14.7 psia.

Required: the rate of flow in cu ft per hr at the reference conditions of 30-in. Hg abs., 60 F and water vapor saturation.

(a) $\beta = 6.25/10.02 = 0.62375$

(b) $E = 1.0860$ (Table II-I-1 or Fig. II-I-4)

(c) $F_a = 1.0005$ (Fig. II-I-2)

(d) $p_1 = 125 + 14.7 = 139.7$ psia

(e) Since the temperature of the manometer or gage by which the differential pressure is measured is not given, it will be assumed to have been 68 F. Then $x = (30 \times 0.03606)/139.7 = 0.00775$ and $x/\gamma = 0.00775/1.4 = 0.00553$, with which $Y_1 = 0.997$ (Fig. II-I-5).

(f) At 140 psia and 90 F, $Z_1 = 0.998$ (Fig. II-I-11).

(g) The saturation pressure of water vapor at 90 F is 0.698 psia, and the specific humidity is

$$S = \frac{0.622}{1} \times \frac{0.698}{(139.7 - 0.698)} = 0.00312 \text{ (I-3-38)}$$

(h) $\rho_1 = 2.6991 \, (1.0 + 0.00312) \dfrac{(139.7 - 0.7)}{549.7 \times 0.998} \times 1$

$\qquad = 0.6860 \text{ lb}_m/\text{ft}^3$ (I-3-40)

(i) $m = 0.99702 \times 1.086 \times 1.0005 \times 0.997 \times 6.250^2$

$\qquad C\sqrt{30 \times 0.6860} = 19.14 \, C \text{ lb}_m/\text{sec}$
(II-III-16)

(j) At 90 F, the viscosity of air is $\mu = 0.0000127$ (Fig. II-I-8).

(k) For a value of R_d with which to locate the correct tabulated value of C, a trial value of $C = 0.62$ is assumed; then,

$$m = 11.87 \text{ and } R_d = \frac{48 \times 11.87}{\pi \times 6.250 \times 0.0000127}$$

$\qquad = 2,283,000$

(l) $C = 0.6070$ (equations (II-III-3 and -4) or Table II-III-3)

(m) $q_1 = \dfrac{19.14 \times 0.6070}{0.6858} \times 3600 = 60,990$ cfh at the flowing conditions

(n) If the total pressure of this air-water vapor mixture is reduced from 139.7 psia to 30 in. Hg = 14.735 psia, the partial pressure of the water vapor portion will be reduced proportionally, i.e., from 0.698 psia to 0.0736 psia. Now the saturation pressure of water vapor at 60 F is 0.256 psia, so that at 30-in. Hg and 60 F the mixture will not be water-vapor saturated. However, the volume rate of flow of this mixture at 30-in. Hg and 60 F will be

$$q_o = 60,990 \left(\frac{139.7 - 07}{14.735 - 0.0736}\right)\left(\frac{519.7}{549.7}\right)\left(\frac{1.000}{0.998}\right)$$

$\qquad = 547,800$ cfh

(o) On the hypothesis that there would be a condition of water vapor saturation at the reference state given as 30 in. Hg, 60 F and water vapor saturation, the equivalent rate of flow would be

$$q'_o = 60,990 \left(\frac{139.7 - 0.7}{14.735 - 0.256}\right)\left(\frac{519.7}{549.7}\right)\left(\frac{1.000}{0.998}\right)$$

$\qquad = 554,600$ cfh

Note: If the presence of the water vapor is neglected at both the initial and final conditions, the apparent rate of flow would be $q''_o = 547,750$ cfh.

II-IV-4 Fuel oil flowing in a 6-in. pipe is measured with an orifice meter under the following conditions:

1. I.D. of pipe at orifice meter section = 6.065 in.
2. Orifice plate 304 stainless steel, orifice bore = 3.750 in.
3. Differential pressure between vena contracta taps 137-in. water
4. Temperature of the flowing oil = 180 F
5. Oil data, by supplier, viscosity = 150 SSU at 180 F and specific gravity = 0.939 at 180/60, and 0.980 at 60/60

Required: rate of flow in gpm at 60 F, the measurement reference temperature.

(a) $m = 0.099702 \, (C \, E \, Y \, F_a d^2) \sqrt{\rho_1 h_w}$ (equation II-III-15)

(b) $\beta = 3.570/6.065 = 0.5886$

(c) $E = 1.066$ (Table II-I-1 or Fig. II-I-4)

(d) $Y = 1.000$ for a liquid

(e) $F_a = 1.002$ (Fig. II-I-3)

(f) $\rho_1 = 0.939 \times 62.3707 = 58.566 \text{ lb}_m/\text{ft}^3$

(g) $\rho_o = 0.980 \times 62.3707 = 61.123 \text{ lb}_m/\text{ft}^3$

(h) $m = 0.099702 \times 1.166 \times 1.000 \times 3.570^2$

$\qquad \times 1.002 \, C \, \sqrt{58.566 \times 137}$

$\qquad = 132.98 \, C \text{ lb} \cdot_m/\text{sec}$

(i) To compute the Reynolds number with which to determine the value of C, a preliminary value of m may be computed by using an estimated value of C such as $C = 0.61$, thus giving $m \cong 132.98 \times 0.61 \cong 81.12$.

(j) For 150 SSU, $\nu = 0.000345$ ft²/sec (equation (II-I-6) or Fig. II-I-10 and equation (II-I-3)).

(k) $\mu = 0.000345 \times 58.566 = 0.0202$ lb_m/sec-ft

(l) $R_d = (48 \times 81.12)/(\pi \times 3.57 \times 0.0202) = 17,180$

(m) $C = 0.6237$ (equations (II-III-5 and 6) or Table II-III-4)

(n) $m = 132.98 \times 0.6237 = 82.94$ lb_m/sec

(o) gpm $= [m \times 60 \times (7.48052 \text{ gal/ft}^3)]/\rho_o$

$$q_o = (82.94 \times 60 \times 7.48052)/61.123$$

$$= 609.03 \text{ gpm at 60 F}$$

II-IV-5 Calculate the maximum designed capacity, in pounds per hour, of a transmitter of 212-in. water maximum differential pressure range, wet calibrated, used with a 5 per cent chrome-moly steel long-radius, high-ratio flow nozzle of 5.674-in. throat diameter, installed in a pipe of 7.683-in. i.d. and fitted with pipe-wall taps at D and 1/2 D, for measuring steam flow at 900 psig and 900 F.

(a) $m(lb_m/hr) = 1865.6 \; C \; E \; Y_a d^2 F_a \; \sqrt{\rho_1 \; \Delta p}$ (equation II-III-17)

(b) $\beta = 5.674/7.683 = 0.73853$

(c) $E = 1.1931$ (Table II-I-1)

(d) $F_a = 1.0118$ (Fig. II-I-3)

(e) Assume the atmospheric pressure is 14.6 psia; then $p_1 = 914.6$ psia; and at 900 F the specific volume of steam is 0.8362 ft³/lb_m (*1967, ASME Steam Tables*, Table 3), and $\rho_1 = 1/0.8362 = 1.196$ lb_m/ft³.

(f) Assume the transmitter temperature is 70 F; then

$$\Delta p = 212 \times 0.03605 = 7.643 \text{ psi (Fig. II-I-2)}$$

(g) At the maximum range of the transmitter, $x = 7.643/914.6 = 0.00836$, and $Y_a = 0.993$ (Fig. II-III-21).

(h) For the purpose of evaluating the designed capacity, it will suffice to assume that $C = 0.99$. Then,

(i) $m_{max} = 1865.6 \times 0.99 \times 1.1931 \times 0.993$
$$\times 5.674^2 \times 1.0118 \times \sqrt{1.196 \times 7.643}$$
$$= 215,500 \; lb_m/\text{hr, which is approximately 60 } lb_m/\text{sec}$$

(j) Since in use the flow rate may never be at the maximum meter capacity, it is realistic to assume that under-average flow conditions the differential pressure will be about 119-in. water (= 4.29 psi), corresponding to approximately 75 per cent of maximum flow rate. Also, an approximate value of R_d is sufficient for the purpose of establishing the value of C.

(k) Thus, using $m = 45$ lb_m/sec and $\mu = 0.0000192$ (Fig. II-I-7)

$$R_d = (48 \times 45)/(\pi \times 5.674 \times 0.0000192)$$
$$= 6,310,000$$

(l) Although this value of R_d is above the range of values on which equation (II-III-9) and Table II-III-5 were based, an extrapolation by means of equation (II-III-9) may be made along with the application of a larger tolerance. On this basis, $C = 0.993 \pm 2.0$ per cent.

(m) For $\Delta p = 4.29$ psi, $x = 0.00469$ and $Y_a = 0.997$.

(n) $m = 1865.6 \times 0.993 \times 1.1931 \times 0.997$
$$\times 5.674^2 \times 1.0118 \times \sqrt{4.29 \times 1.196}$$
$$= 162,595 \; lb_m/\text{hr assumed normal flow rate}$$

(o) Using $C = 0.993$ in step *(i)* in place of 0.99 gives as the maximum capacity of the transmitter $m_{max} = 216,150$ lb_m/hr.

II-IV-6 Condensate from a steam turbine is metered with a long-radius, low-ratio flow nozzle of type-304 stainless steel. Pressure taps are located at $1 \; D$ upstream and in the nozzle throat. The nozzle had been calibrated, with the calibration going up to a maximum R_d of 4,500,000, at which the value of C was 0.9955. The primary element dimensions and the metering data are:

meter section pipe i.d.	= 12.090 in.
nozzle throat diameter	= 5.000 in.
line pressure at upstream tap	= 250 psig
flowing temperature at nozzle	= 300 F
differential pressure by mercury manometer	= 40.5 in.

temperature of manometer   90 F

barometric pressure, in. mercury
at 32 F   29.40
Required: the rate of flow in pounds per hour.

(a) β = 5.000/12.090 = 0.41356

(b) E = 1.01495 (Table II-I-1)

(c) Y = 1.000 for liquids

(d) F_a = 1.0042 (Fig. II-I-3)

(e) Barometric pressure = 29.40 × 0.4912 = 14.44 psia (Fig. II-I-2), p_1 = 250 + 14.44 = 264.44 psia

(f) v_1 = 0.01744 ft³/lb$_m$, (*1967 ASME Steam Tables*, Table 3), and
ρ_1 = 1/0.01744 = 57.339 lb$_m$/ft³

(g) Δp = 40.5 × 0.45235 = 18.32 psi (Fig. II-I-2)

(h) m = 0.525 C × 1.015 × 5.00² × 1.0042
× $\sqrt{18.32 \times 57.339}$ = 433.59 C lb$_m$/sec

(i) If the value of C by the calibration applies approximately at the actual flowing conditions, then m = 433.59 × 0.9955 = 431.64

(j) μ = 0.000125 (Fig. II-I-5)

(k) R_d = (48 × 431.64)/(π × 5.000 × 0.000125) = 10,550,000

(l) Extrapolation of the calibration curve parallel to the typical curve for a nozzle with throat taps (Fig. II-III-19) shows C = 0.9969 at R_d = 10,550,000.

(m) Pounds per hour = 3600 × 433.59 × 0.9969 ≈ 1,556,100.

II-IV-7 To calculate the rate of flow in pounds per hour of water at 60 F and 95 psig, flowing through a 6.00 × 4.00-in. cast iron Venturi tube which will produce a differential pressure of 100-in. water.

(a) m (lb$_m$/hr) = 358.93 $C E Y d^2 F_a \sqrt{\rho_1 h_w}$
(equation (II-III-17)

(b) β = 4.00/6.0 = 0.6667

(c) C = 0.984 (Par. II-III-34)

(d) E = 1.1163 (Fig. II-I-4)

(e) Y = 1.000 for liquids

(f) F_a = 1.000 (room temperature)

(g) ρ = 62.3707 (Table II-I-4)

(h) m = 358.93 × 0.984 × 1.1163 × 4.00²
× $\sqrt{62.3707 \times 100}$
= 498,190 pounds per hr

(i) μ = 0.00076 (Fig. (II-I-5)

(j) R_D = (48 $\frac{498190}{3600}$)/(π × 6.0 × 0.00076)
= 463,500. Since both R_D and β are within the limits given in Par. II-III-34, the coefficient value of 0.984 is valid.

II-IV-8 The flow of superheated steam through a 3/4-in. bleed line is both controlled and metered with a sonic-flow radial inlet Venturi. The i.d. of the pipe upstream of the Venturi inlet is 0.742 in., and the throat diameter of the Venturi is 0.2569 in. The pressure is measured from a sidewall tap 1 D upstream from the Venturi inlet, and the temperature is measured with a total temperature well and thermometer located downstream. The inlet line pressure is 1200 psig, and the temperature is 950 F. From a calibration the Venturi coefficient is 0.994.
Required: the maximum rate of flow in lb$_m$/sec.

(a) $m = C^* (a/12) B_F (F/F_i) \sqrt{p_{1t}/v_{1t}}$
(equation (II-III-29))

(b) a = 0.05183 in.²; a/12 = 0.004319

(c) β = 0.2569/0.742 = 0.346

(d) Since β is less than 0.5, equation (II-III-27) is applicable for calculation of p_{1t}

$$p_{1t} = \frac{p_1}{\left[1-\beta^4\frac{\Gamma}{2}\left(\frac{2}{\Gamma+1}\right)^{(\Gamma+1)/(\Gamma-1)}\right]}$$

(e) p_1 = 1200 + 15 = 1215 psia, assuming 15 psia as the barometric pressure.

(f) Γ = 1.285 (*1967 ASME Steam Tables*, Fig. 11, p. 298).

(g) $p_{1t} = \dfrac{1215}{1 - (0.346^4)\left(\dfrac{1.285}{2}\right)\left(\dfrac{2}{1+1.285}\right)^{\frac{2.285}{0.285}}}$
$= \dfrac{1215}{1 - 0.01433 \times 0.6425 \times 0.3434}$
= 1220 psia

(h) B_F = 3.7848 (Table (II-III-11)

(i) $F/F_i = 0.9945$ Table (II-III-11)

(j) $v_{1t} = 0.644$ (1967 ASME Steam Tables, Table 3)

(k) $m = 0.994 \times 0.004319 \times 3.7848 \times 0.9945$

$\times \sqrt{1220/0.644}$

$= 0.7033$ lb$_m$/sec or 2531.96 lb$_m$/hr

II-IV-9 Propane (C_3H_8) is measured through a long-radius, low-ratio flow nozzle at sonic-flow conditions. The nozzle is mounted in a pipe of 2.90-in. i.d. and has a throat diameter of 1.1574 in. The total inlet pressure, $p_{1t} = 800$ psia, is measured with an impact tube located 200 stem diameters upstream of the nozzle inlet. The temperature, T_{1t}, measured with a total temperature well located downstream of the nozzle, is 340.3 F (= 800 R). The nozzle has not been calibrated.

Required: rate of discharge in lb$_m$/sec.

(a) Since the factors for propane are not tabulated, the sonic-flow rate must be computed using the "Reduced Coordinates Compressibility Charts" (Fig. II-III-29), an F_i factor from Table II-III-22 in conjunction with equation (II-III-34) and

(b) m (lb$_m$/sec) $= 0.1443\, C\, a\, F_i\, p_{1t} \sqrt{\dfrac{MW}{Z\, T_{1t}}}$

(equation (II-III-34))

(c) For propane, C_3H_8, from Table II-I-5,

molecular weight, $MW = 44.0972$

critical pressure $= 617.4$ psia

critical temperature $= 666$ R

(d) Using $\Gamma^* = \Gamma = \gamma = 1.33$, $F_i = 0.6723$ (Table II-III-22).

(e) Reduced pressure $= 800/617.4 = 1.295$ (use 1.3-)

Reduced temperature $= 800/666 = 1.2$

$Z = 0.748$ (Fig. II-III-29-1)

(f) $a = \dfrac{\pi \times 1.1574^2}{4} = 1.052$ in.^{2}

(g) Nozzle coefficient is assumed to be 0.99 ± 1.0 per cent.

(h) $m = 0.1443 \times 0.99 \times 1.052 \times 0.6723 \times 800$

$\sqrt{44.0972/(0.748 \times 800)}$

$= 21.942$ lb$_m$/sec

II-IV-10 A natural fuel gas is discharged from a sonic-flow nozzle connected to the outlet of a displacement meter. The composition of the gas as given in mole fractions is:

methane	CH_4	0.960
ethane	C_2H_6	0.035
carbon dioxide	CO_2	0.002
nitrogen	N_2	0.003

The conditions under which the nozzle is operated are:

stagnation temperature	= 80.3 F (= 540 R)
stagnation pressure	= 385.6 psig
barometric pressure	= 29.3-in. Hg.

From a calibration with air, the product, Ca, the effective area of the sonic-flow nozzle, was reported to be 0.1930 in.^{2}.

Required: the time in seconds for 1 ft^3 at the displacement meter outlet conditions to be discharged from the displacement meter through the sonic-flow nozzle.

(a) The outlet conditions of the displacement meter are taken to be the same as the inlet conditions to the sonic-flow nozzle.

(b) $q_1 = C\, a\, (e_c j + b_c)(e_z j + b_z) \sqrt{g_c\, T_{1t}\, \dfrac{R}{MW}}$

(equation (II-III-33))

(c) The molecular weight of the gas is (using Table II-III-27)

$(0.960 \times 16.043) + (0.035 \times 30.0701)$

$+ (0.002 \times 44.01) + (0.003 \times 28.013)$

$= 16.6258$

(d) Barometric pressure $= 29.3 \times 0.4912 = 14.4$ psia (Fig. II-III-2)

(e) $p_{1t} = 385.6 + 14.4 = 400.0$ psia

(f) $j = 0.035 + 0.002 - \frac{1}{2}(0.003) = 0.0355$ (equation (II-III-27))

(g) $e_c = -0.0404$; $b_c = 0.6704$; $e_z = -0.0589$;

$b_z = 0.9768$ (Tables II-III-23 through II-III-26)

(h) $(e_c j + b_c) = (-0.0404 \times 0.0355 + 0.6704)$

$= 0.6690$

$(e_z j + b_z) = (-0.0589 \times 0.0355 \times 0.9768)$

$= 0.9747$

(i) $q_1 = \dfrac{0.1930}{144} \times 0.6690 \times 0.9747$

$\sqrt{32.174 \times 540 \times \dfrac{1545.32}{16.6258}} = 1.1106$ cfs

(j) Time to discharge 1 ft³ = 1/1.1106

$= 0.9004$ sec/ft³

II-IV-11 The flow of a fuel oil in a 3-in. pipe is monitored at a long-radius welded elbow, which has been fitted with pressure taps in the 45-deg plane. The pipe is schedule 40. The flowing temperature of the oil is 110 F, and the specific gravity is 0.79 at 110/60. The average differential pressure is 16 in. of water. The elbow was not calibrated.

Required: the approximate rate of flow in barrels per hour at 60 f.

(a) q (cfs) $= \dfrac{\pi D^2}{4} K \sqrt{g_c \dfrac{4R}{\pi D} \dfrac{\Delta p}{\rho}}$

(equation (II-III-38))

(b) $D = 3.068$ in.

$R = 4.5$ in.

$p = 16 \times 0.03605 = 0.577$ psi (Fig. II-I-2)

$\rho = 0.79 \times 62.3707 = 49.27$ lb$_m$/ft³
(Table II-I-4)

(c) Since the elbow was not calibrated, the flow coefficient, K, is to be evaluated by

$K = 1 - \dfrac{6.5}{\sqrt{R_D}}$ (II-III-36)

This requires assuming a value for R_D and, after obtaining a first value of K and completing a computation of q, computing a value of R_D to be compared with the assumed value. Assuming $R_D = 50,000$,

$K = 1 - \dfrac{6.5}{\sqrt{50,000}} = 0.970$

(d) q (cfs) $= \dfrac{\pi \, 3.068^2}{4} \times 0.970$

$\sqrt{32.174 \dfrac{4 \times 3.068}{\pi \times 4.5} \times \dfrac{0.577}{49.27}}$

$= 4.101$ cfs

(e) $R_D = 48m/(\pi D \mu)$. Since no information is given on the viscosity of the oil, a value of 0.009 lb$_m$/ft-sec will be assumed; then

$R_D = \dfrac{48 \,(4.101 \times 49.27)}{\pi \times 3.068 \times 0.009} = 111,800$

(f) $K' = 1 - \dfrac{6.5}{\sqrt{111,800}} = 0.981$

(g) $q' = 4.101 \dfrac{0.981}{0.970} = 4.147$ cfs at 110 F

(h) The observed specific gravity of 0.79 at 110/60 corresponds to a specific gravity of 0.8094 at 60/60 (Table 23 of Ref. [3]).

(i) The volume reduction factor to convert a volume at 110 F to the corresponding volume at 60 F for a specific gravity of 0.8094 is 0.9753 (Table 24 of Ref. [3]).

$4.147 \times 0.9753 = 4.043$ cfs at 60 F

(j) cu ft $\times$ 0.17811 = U.S. barrels (Table 1 of Ref [3]).

$4.043 \times 0.17811 \times 3600 = 2592.4$ bbl/hr at 60 F

(k) Since the value of K is subject to a tolerance (uncertainty) of ± 4.0 per cent, the rate of flow is subject to a tolerance of ± 4.0 per cent or more, and 2592 ± 4.0 per cent bbl/hr would be the reported rate of flow.

References

[1] "Flowmeter Computation Handbook," ASME, New York, 1961.
[2] "Orifice Metering of Natural Gas," Gas Measurement Committee Report No. 3, American Gas Association, New York, 1969.
[3] "ASTM-IP Petroleum Measurement Tables;" ASTM, Philadelphia, Pa., 1952.

Chapter II-V

Tolerances

II-V-1 Tolerances, Their Significance. Except by accident, no two meters, even of the same type, are likely to give *exactly* the same indication when the same quantity of fluid is flowing through each. The degree to which this applies is not the same for all types of meters, applying least to the displacement types and more to the differential-pressure types. For this reason, "tolerances" are assigned to the values of the factors entering into the metering of fluids. (The expressions, "limit of accuracy" or "per cent uncertainty," might well be substituted for "tolerance.") Tolerances have to do with those practically unavoidable differences between ostensibly duplicate primary elements. They do *not* refer to accidental errors of observation, concerning which no general predictions are possible.

In any one measurement, the probability is very small that the departures from 100 per cent accuracy in the individual items will all affect the final result in the same direction; hence, from mathematics, the overall tolerance will be the square root of the sum of the squares of the tolerances on (departures of) the individual factors. In other words, an overall tolerance determined in this way is the most probable amount of departure from the actual quantity, with there being as much chance that the departure will be smaller than larger than this amount.

II-V-2 There have been a number of procedures used for evaluating or assigning tolerances with the result that the per cent uncertainty assigned to an item by one worker has not been exactly comparable to that assigned by another to the same item. In order to provide a uniform basis for assigning numerical values to tolerances, the committee on Fluid Flow Measure-ment of the International Organization for Standard-ization (ISO/TC-30) has adopted the following procedure:

1. The numerical value of a tolerance shall be twice the standard deviation.

2. The standard deviation is to be computed as follows: Sum up the squares of the deviations with respect to *the most probable value;* divide by the number of observations minus one; take the square root of this quotient.

This procedure has been followed in evaluating the tolerances given in this edition of *Fluid Meters.* The *most probable values* of the discharge coefficients of square-edged orifices are, to date, the values computed by equations (II-III-1) through (II-III-6), or read from Tables II-III-2, II-III-3 and II-III-4. Similarly, for flow nozzles used with pipe-wall taps, the most probable values are those computed by equation (II-III-12) or read from Table II-III-5. For low-ratio nozzles with the downstream tap in the throat, the most probable values are those read from the curve of Fig. II-III-19. For Venturi tubes, the most probable values are given in Pars. I-5-35 and II-III-42.

The tolerance values given in Tables II-V-1 and II-V-2 are those recommended as applying to uncalibrated primary elements. When a primary element is calibrated, the tolerance to be used should be computed from the calibration data by the procedure described above.

II-V-3 Prior to the editing of the fifth edition of *Fluid Meters,* tolerance values given by this committee and also by the Gas Measurement Committee of the American Gas Association in their Report

No. 3 were not derived by an evaluation of the standard deviation. Instead, the arithmetic average of the departures of the test values from the correlation curves was computed, and this value, without being doubled, was reported as the tolerance for the particular item. It is of interest that those arithmetic average values are very close to the values of σ obtained in the recent correlation, which is the basis for some of the tolerances given here [1].

II-V-4 The application of the tolerances in the tables and the computation of the overall tolerance to which the measurement of the flow of a fluid may be subject are illustrated by two examples. In doing this the extent or power to which the separate factors affect the total tolerance is taken into account.

Item	Tolerance (per cent)	Effect Factor	Square
Tolerance for Example II-IV-2			
Orifice diameter, d	± 0.08	2	0.0256
Differential pressure, h_w	± 0.25	½	0.0156
Evaluation of density, ρ_1	± 0.50	½	0.0625
Coefficient, C	± 1.1	1	1.21
Expansion factor, Y_1	± 0.5	1	0.25
Area factor, F_a	± 0.02	1	0.0004
			1.5641
Overall tolerance	± 1.25		
Tolerance for Example II-IV-6			
Throat diameter, d	± 0.08	2	0.0256
Differential pressure, h_w	± 0.10	½	0.0025
Value of density, ρ	± 0.10	½	0.0025
Coefficient, C	± 0.70	1	0.49
			0.5206
Overall tolerance	± 0.72		

II-V-5 As may be seen from these examples, the overall tolerance will always be greater than that of the item having the largest tolerance. To say this another way, the final result of a flow-measurement computation cannot be more exact or have a smaller per cent uncertainty than the factor having the greatest uncertainty. Thus, where one factor, usually the coefficient, has a tolerance ranging from ± 0.4 to ± 4.0 per cent, the use of numbers with four to six significant digits does not imply a corresponding high degree of exactness. The use of so many digits improves the agreement between two or more computers and aids in the "rounding off" of the final result.

Reference

[1] "A Statistical Approach to the Prediction of Discharge Coefficients of Concentric Orifice Plates," R. B. Dowdell and Yu-Lin Chen; *Trans. ASME, Journal of Basic Engineering*, vol. 92, no. 3, Sept. 1970.

Table II-V-1 Tolerances for Discharge Coefficients and Flow Coefficients

Primary Element	Coefficient from	Pipe Size, D	R_d or R_D	β	Tolerance (per cent)
Square-Edged Concentric Orifices Flange taps	Equations (II-III-1), (II-III-2) or Table II-III-2				
D & ½ D taps	Equations (II-III-3), (II-III-4) or Table II-III-3	$D \geq 2.0$ in.	$R_d > 5000\, D$	$0.20 \leq \beta \leq 0.70$ $0.11 \leq \beta \leq 0.20$ $0.70 \leq \beta \leq 0.75$	± 1.0 or less — linearly with β ± 2.25 to ± 1.0 — linearly with β ± 1.0 to ± 2.25
Vena contracta taps	Equations (II-III-5), (II-III-6) or Table II-III-4			V.C. taps only $0.70 \leq \beta \leq 0.80$	± 1.0 to ± 2.5 — linearly with β
As above		$1.0 < D < 2.0$ in.		As above	Above tolerances to be multiplied by a factor of 1 to 2 increasing linearly as D decreases
As above			$4000 \leq R_d < 5000\, D$	As above	Above tolerances to be multiplied by a factor of 1 to 2 increasing linearly as R_d decreases
Long-Radius Flow Nozzle (Fig. II-III-14) Pipe-wall taps at D & ½ D	Equation (II-III-12) or Table II-III-5	$2.0 \leq D \leq 16$ in.	$10^4 \leq R_d \leq 2.5 \times 10^6$	$0.2 \leq \beta \leq 0.8$	± 2.0
Long-radius Flow Nozzle (Fig. II-III-14) Taps at 1 D and nozzle throat	Calibration (See Par. II-IV-6)	$2.0 \leq D \leq 16$ in.	$R_d \leq 10^5$	$0.2 \leq \beta \leq 0.5$	As determined (or ± 0.8)
1932 ISA Flow Nozzle (Fig. II-III-22) Corner taps	K by Fig. II-III-23	$2 \leq D \leq 40$ in.	$2 \times 10^4 \leq R_D \leq 10^6$	$0.32 < \beta < 0.8$	± 1.0
Venturi Tube Rough-cast inlet cone	Par. II-III-38	$4 \leq D \leq 32$ in.	$2 \times 10^5 \leq R_D \leq 10^6$	$0.3 \leq \beta \leq 0.75$	± 0.75
Venturi Tube Machined inlet cone	Par. II-III-38	$2 \leq D \leq 10$ in.	$10^5 \leq R_D \leq 10^6$	$0.4 \leq \beta \leq 0.75$	± 1.0
Venturi Tube Welded sheet metal inlet cone	Par. II-III-38	$8 \leq D \leq 48$ in.	$2 \times 10^5 \leq R_D \leq 2 \times 10^6$	$0.4 \leq \beta \leq 0.7$	± 1.5

Eccentric Orifice
  Flange taps
  Vena contracta taps
	Fig. II-III-9	$4 \leq D \leq 14$ in.	$10^4 \leq R_D \leq 10^6$	$0.3 \leq \beta \leq 0.8$	$D = 4$ in. ± 1.9
					$D > 4$ in. ± 1.4

Segmental Orifice
  Flange taps
  Vena contracta taps
	Fig. II-III-10	$4 \leq D \leq 14$ in.	$10^4 \leq R_D \leq 10^6$	$0.35 \leq \beta \leq 0.85$	± 2

Table II-V-2 Tolerances for Expansion Factors

Primary Element	Factor from	D	β	x or $h_{w/p}$	Fluids	Tolerance* (per cent)
Square-Edged Concentric Orifices Flange taps	Equations (II-III-4), (II-III-5)	$D \lesssim 1.0$	$0.11 \lesssim \beta \lesssim 0.75$	$x < 0.4$, $h_{w/p} < 11.1$	Gases for which γ (or Γ) is known	0.0 to ± 0.75
D & ½D taps, Vena contracta taps	Figs. II-III-5, II-III-6			$h_{w/p_2} < 18.5$	Gases for which γ (or Γ) is uncertain	0.0 to ± 1.5
Flow Nozzles Venturi Tubes	Equation (II-III-10) Tables II-III-6, II-III-7	$D \lesssim 1\frac{1}{2}$ in.	$0.1 \lesssim \beta \lesssim 0.8$	$x < 0.4$, $h_{w/p_1} < 11.1$	Gases for which γ is (or Γ) known	0.0 to ± 0.4
	Figs. II-III-20, II-III-21				Gases for which γ is (or Γ) uncertain	0.2 to ± 1.2
Eccentric Orifice	Fig. II-III-11	$4 \lesssim D \lesssim 16$	$0.3 \lesssim \beta \lesssim 0.8$	$x < 0.3$	Gases for which γ is known	0.0 to ± 1.0
Segmental Orifice	Fig. II-III-12			$h_{w/p_1} < 8.3$	Gases for which γ is uncertain	± 1.0 to ± 4.0

*Tolerances increase linearly as x increases.

Note: Numbers followed by (T) indicate Table.

Acceleration, due to gravity, 153
Actual rates of flow, equations for, 53, 55, 233—245
    foot-pound-second-Fahrenheit units, 53, 233
    kilogram-meter-second-Celsius (S.I.) units, 55, 234
    sonic flow, 235—245
    subsonic flow, 233
Acoustic ratio, 33
Anemometers, 91
  alignment of, 95
  cup type, 91
  density and viscosity effects, 96
  hot-wire, 105
  propeller type, 92
  range, 94
  vane type, 92
Area expansion factor, 153, 156
Area meters, 6, 81
  coefficients, 87
  operating adjustment factors, 88
Atmosphere, standard, 11

Baumé, degrees, 25
Bellows meters, 5, 39
Boundaries, effects of, 96

Calibration, differential pressure meters, 196
  displacement meter with sonic flow element, 254
  flow nozzle, subsonic, 227
  pipeline flowmeters, 100
  quantity meters, 44
Capacities, quantity meters, 44
Capillary seal, 37
Capillary tubes, 77
Celsius, temperature scale, 24
Center-line average, CLA, 60
Centrifugal meters, 75
Characteristics of fluid flow, 58
Classification of fluid meters, 3
 table, 4
Coefficients, discharge, D and $1/2D$ taps, 204-205(T)
  definition of, 53
  determination of, 56
  correlation of, 61
  eccentric orifices, 210, 212
  flange taps, 202-203(T)
  flow nozzles, 64, 222-223(T)
  ISA 1932 nozzle, 227
  segmental orifices, 210, 212
  small bore orifice meter, equation for, 214
  sonic flow elements, 235
  vena contracta taps, 206-207(T)
  Venturi tubes, 64, 232

Coefficient, flow, 53
Compressibility of gases, 25
  factor Z, air, 165
    carbon dioxide, 167
    gases, reduced coordinates, 250-254
    hydrogen, 166
    methane, 168
Conical-edge orifice, 73
Connecting nipples, valves and reservoirs, horizontal
    pipes, 187, 189
    vertical pipes, 188, 190
Construction, effects of, 60
Conversion factors, density, gm_m/cc to lb_m/ft^3, 153
  inch of mercury to psi, 153, 155
  inch of mercury under water to psi, 153, 155
  inch of mercury to inch of water, 153
  inch of water to psi, 153, 155
  metric (S.I.) units to ft-lb units, 153
  psi local gravity to psi standard gravity, 153
  viscosity, absolute, poise to lb_m/ft-sec, 153
    kinematic, stoke to ft^2/sec, 153
Corner taps, 60
Corresponding states, theorem, 27
Critical density, 27
Critical pressure, 27
Critical temperature, 27
Critical volume, 27
Current meters, 93
  alignment of, 95
  boundary effects, 96
  cup types, 93
  density and viscosity effects, 96
  propeller type, 94
  range of, 94
  temperature effects, 96
Cylinder and piston meter, 82

D and $1/2D$ taps, 60, 199
Density, 24, 154
  air, dry, 171(T)
  gas, dry, equation for, 26, 27
    moist, equation for, 30
  mercury, 170(T)
  water, 172(T)
Differential pressure, 21
  measurement of, 21
  meters, 47
Dimensional analysis, 32
Dimensional formulas, 33(T)
Displacement meters, positive, 38
Distribution of velocities, 96
Dynamic response, velocity meters, 97

Eccentric orifice, 73

Effects, of boundaries, 96
  of construction, 60
  of installation, 60
Elbow meter, 75
Elements, primary, xi
  secondary, xi
Equations, actual rates of flow, 53
    area meters, 87
    elbow meter, 77
    foot-pound$_m$ units, 53
    head-area meters, 115, 116, 117
    linear resistance meters, 78
    maximum sonic flow, 70, 235—245
    metric (S.I.) units, 55
  area meters, 84
  compressible fluids, theoretical, 51, 52, 85
  energy, 50
  hydraulic, theoretical, 51, 58
  of state, 25
  sonic flow elements, 65—72, 235—245
Examples, elbow meter, 263
  flow nozzle, 260, 261
  orifice meter, 257—260
  sonic flow element, 261, 262
Expansion factor, 61
  flow nozzle, 52, 220, 224—226
  orifice meter, square-edged, 62, 208, 209
  Venturi tube, 52, 220, 224—226

Flange taps, 60, 199
Force meters, 6, 128
Forces, types of, 56
Floats, 145
Flow, characteristics, 58
  pulsating, 34
  steady, 34
Flow nozzle, 48, 216
Flowmeters, pipeline, 97
  axial flow, 97
  propeller type, 97
  radial flow, 97
  turbine, 98
Flow rate, sonic, maximum, real gas, 68-72, 235—245
  maximum theoretical, ideal gas, 66—68
Flow rate, subsonic, liquids and real gases, 53, 54, 233
  theoretical ideal gas, 51—53
Fluid temperature, 23
Flumes, 120—123
  Palmer-Bowlus, 122
  parabolic discharge, 122
  Parshall, 120—122
Fundamental units, 9

Gallon, U.S. standard, 11
Gases, physical data, 29, 176
  reference pressure, 11
  reference temperature, 11
Gaskets, 185
Gear meter, 43

Gibson, pressure-time method, 137—144
Gravity, standard, 9

Head-area meters, 6, 113
Hodgson number, 35
Hot-wire anemometer, 105
  calibration, 107
  circuits, 106-107
  probe design, 105
Humidity, 28
Hydrometric pendulum, 128

Impeller meter, 43
Installation, effects of, 60
  orifice plates and flow nozzles, high pressure-high
    temperature, 184
    low pressure-low temperature, 183, 185
  pipeline flowmeters, 100
  recommended minimum arrangements, 180
  quantity meters, 45
Instrumentation, velocity meters, 97
ISA nozzle, 1932, 227

Jet contraction, 58

Kilogram-meter-second-Celsius (S.I.) units, 55
Kinematic similarity, fluid flows, 56

Letter symbols, 12—16
Linear resistance meter, 77
Liquid-sealed drum meter, 39

Mach number, 33, 58
Magnetic and electromagnetic flowmeters, 125—128
  fluids, 127
  range, capacity, 126
  theory of operation, 126
  tolerances, 128
Mass flowmeters, modified orifice meter, 64
  transverse momentum, 130
Materials, flow nozzles, 216
  orifice plates, 198, 211
  small bore meter tube, 211
  sonic flow primary element, 234
  Venturi tube, 230
Maximum flow rate, ideal gas, 66—68
  real gases, 68—72, 235—245
Mean molecular heat, 30
Metering tank, 39
Meters, anemometers, 91
  area, 6, 81
  bellows, 5, 39
  centrifugal, 75
  current, 93
  differential pressure, 5, 47
  elbow, 75
  electromagnetic flowmeter, 7, 125
  force, 6, 128
  gear, 43
  head-area, 6, 113

270

impeller, lobed, 43
linear resistance, 77
liquid sealed, 5, 39
magnetic flowmeter, 125
nutating disk, 5, 41
orifice, 6, 47, 151, 198
reciprocating piston, 3, 40
ring or rotary piston, 5, 41
rotating vane, 43
sliding vane, 43
tanks, 3, 39
thermal, 6, 131
vane, 129
velocity, 6, 91
volumetric, 3, 38
weighing, 3, 37
Methods, determining rates of flow, 125
Mixtures, method of, 136
Modifications of Venturi tubes, 72

Natural fuel gas, sonic flow computation, 236
Neutral position of pressure, 21
Nozzle-Venturi, 72, 232
Nutating disk meter, 5, 41

Orifices, eccentric, 73
  conical-edge, 73
  quadrant-edge, 73
  segmental, 73
  square-edged, 47, 48, 198
Orifice and plug meter, 83

Palmer-Bowlus flume, 122
Parabolic discharge flume, 122
Parshall flume, 120—122
Petroleum products, liquid, 11
Physical data, gases, 29, 176
Pipe diameter, internal, 181
Pipe section, 78
Pipe surface, 181
Piston meter, reciprocating, 3, 40
  ring, 5, 41
Pitot tube, 101
  directional, 102, 103
Pitot-static tube, 101
  computation, velocities, 104
  reversed type, 103
Pitot-Venturi, 103
Porous plug, 77
Positive seal, 37
Pound force, 9
Pound mass, 9
Poundal, 10
Pressure, critical, 27
  definition, 9
  differential, 21
  negative, suction, 21
  neutral position, 21
  stagnation, 20
  static, 19

total, 20
  units, 33
  velocity, 20
Pressure connecting pipes, 191, 192
Pressure connections, 185, 186
Pressure instruments, 192
Pressure loss, differential pressure elements, 74
  flow nozzles, 220-221
  orifices, 200-201
  proprietary tubes, 234
  Venturi tubes, 232
Pressure, variation of, 20
Pressure taps, construction, 219
  diameter of hole, 185
  locations of, 60
    concentric orifice, 200
    D and $1/2D$ taps, 60, 199
    eccentric orifice, 210
    flange taps, 60, 199
    flow nozzle, 216
    proprietary flow tubes, 234
    segmental orifice, 210
    small bore meter tube, 214
    sonic flow primary element, 234
    vena contracta taps, 60, 199
    Venturi tube, 230, 231
Pressure-temperature-volume relation, gases, 25
Pressure-time method, 7, 137—144
Primary above secondary, 192
Primary below secondary, 193
Primary element, xi
Propeller type meter, 94
Proportionality constant, g_c, 9-10
Provers, accuracy of, 45
  displacement meters, 44
  gas meters, 44
  mechanical displacement, 45
Proving liquid displacement meters, 44-45
Psychrometer, wet and dry bulb, 30
Psychrometry, 29
Pulsating flow, 34-35
  effects of, 35, 100

Quantity meters, 3, 37
  installation, 45
  operation, 45
  tolerance, 44

R, gas constant, 49
Radioactive tracers, 135
Range, pipeline flowmeters, 99
Rate of flow, elbow meter, 255
  sonic flow, equations for, 68—72, 235—245
  subsonic, equations for, 53, 233
Reciprocating piston meters, 3, 40
Recovery factor, 24
Rectangular notch weir, 115
Reduced coordinates, pressure, temperature, volume, 27
Reference conditions, 11
Reservoirs, 191

Reynolds number, 32, 57
Rotary or ring piston meters, 5, 41

Salt velocity method, 132
Saybolt viscosimeters, 156, 158, 164
Screens, moving, 145
Seal, capillary, 37
  positive, 37
Secondary element, xi, 6
Settling chambers, 194, 195
Shutoff valves, 181
Similitude, application, 58
Slip, 37
Slotted cylinder and pistons, 82
Slug, 10
Sonic flow primary elements, 234
  calibration, 235
  coefficient, 235
  computations, 235—245
  density determinations for, 235
  forms, 234
  materials, 234
  temperature measurements, 235
Sound and light velocity method, 7, 144
Specific gravity, 25
Specific heats, ratio of, 32
Specific weight, 25
Stagnation pressure, 21
Stagnation temperature, 24
Static pressure, 19
  changes close to orifice, 59
  measurement of, 19, 21
  variations of, 20
Standard conditions, 11
Standard U.S. gallon, 11
Steady flow, 34
Strouhal number, 35
Suction, 21
Symbols, letter, 12, 16

Tables, classification of fluid meters, 4
  coefficients, flow nozzles, 222-223
    orifices, 202—207
  computation of $(\pi - x_{Ymax})/$ 143
  density, dry air, 171
    mercury, 170
    water, 172—175
  dimensional formulas, 33
  dimensions, Parshall flumes, 121
  expansion factory, Y_a, 224
  letter symbols, 13—17
  mixture method substances, 137
  physical data, gases, 29, 176
  sonic flow functions, 237—244, 246—249
  station locations for averaging, 109
  tolerances, differential pressure meters, 267, 268
  tracer materials, 136
  values of r_c, 68
  velocity of approach factors, 169

Tanks, 3
  tilting tank, 38
Tapered tube and float, 81
Temperature, 10
  absolute, 24
  Celsius, 24
  metering, 39
  of a fluid, 23
Theoretical equation, compressible fluids, 51
  liquids, 51
  sonic flow, 65—70
Theory, flow of fluids, 47
Thermal meters, 6, 131
Tolerances, differential pressure primaries, 267, 268
  displacement meters, 45
  weirs, 119
Total pressure, 21
Tracer method, 7, 132
Tracers, carbon dioxide, nitrous oxide, 135
  dyes, 136
  hot and cold injections, 135
  radioactive materials, 135
  salt, 132
Transverse momentum flowmeters, 130
Triangular notch weir, 116
Turbine flowmeters, 97—99
  axial flow, 98
  tangential flow, 99

Units, absolute system, 9
  fundamental, 9
  gravitational, 9
  measurement of fluids, 10

Vane meters, 129
Velocities, averaging, 107—110
  distribution around a cylinder, 102
  locations of instruments, 110
  spatial distribution, 96
Velocity meters, 6, 91
Velocity pressure, 20
  measurement of, 20, 21
Vena contracta, location, 59
Vena contracta taps, 60, 199
Venturi tube, 48, 230
  circular arc Venturi, 72
  coefficient, 64, 232
  expansion factor, 52, 220, 224—226
  fabrication, 230
  grouping, I.S.O., 230
  pressure loss, 232
  pressure taps, 230, 231
  proportions, 230, 231
  throat, 230
Viscosity, absolute, 31, 153
  kinematic, 31, 153
Vortex flowmeter, 99

Weighing meters, 3, 37

Weirs, 113
  contracted weir, 114
  equation of flow, 115, 117
  installation, 118
  measurement of head, 115
  round-crested, 118
  special notches, 118
  special terms, 114
  suppressed rectangular weir, 114
  thin-plate, 114
  tolerances, 119
  trapezoidal, Cippoletti, 118

www.ingramcontent.com/pod-product-compliance
Lightning Source LLC
Chambersburg PA
CBHW061348210326
41598CB00035B/5921